How Monkeys See the World

HOW MONKEYS SEE THE WORLD

Inside the Mind of Another Species

Dorothy L. Cheney and
Robert M. Seyfarth

University of Chicago Press
Chicago and London

The University of Chicago Press, Chicago 60637
The University of Chicago Press, Ltd., London

 Published 1990
Paperback edition 1992
Printed in the United States of America
02 01 00 99 98 97 96 95 94 93 3 4 5 6 7 8

ISBN: 0-226-10245-9 (cloth); 0-226-10246-7 (paper)

Library of Congress Cataloging in Publication Data

Cheney, Dorothy L.
How monkeys see the world / Dorothy L. Cheney and Robert M. Seyfarth.
p. cm.
Includes bibliographical references.
1. Cercopithecus aethiops—Kenya—Amboseli National Park—Psychology. 2. Cercopithecus aethiops—Kenya—Amboseli National Park—Behavior. 3. Primates—Psychology. 4. Cognition in animals. 5. Primates—Behavior. I. Seyfarth, Robert M. II. Title.
QL737.P93C44 1990
599.8′2—dc20 90-30295
CIP

♾ The paper used in this publication meets the minimum requirements of the American National Standard for Information Sciences—Permanence of Paper for Printed Library Materials, ANSI Z39.48-1984.

CONTENTS

Acknowledgments

The best part of writing a book is the opportunity it provides to thank the many people who have contributed, however inadvertently, to it. This is particularly true when one's research is conducted in somewhat remote places and when friends and colleagues have over the years been asked to transport crucial spare parts, hand deliver mail, loan a Land Rover without charge for 6 weeks, or carry a 12-foot stuffed python through two police checks.

We should begin by expressing our sincere thanks to the Office of the President and the Ministry of Tourism and Wildlife of the Republic of Kenya for permission to conduct research in Amboseli National Park from 1977 until 1990. While in Kenya, our work was conducted under the auspices of the Institute of Primate Research of the National Museums of Kenya. We are deeply indebted to Jim Else, Mohammed Isahakia, Richard Leakey, Maria Buteyo, Margaret Omoto, and Mary Sefu for their support of our work and for their tireless efforts in helping us obtain research clearance. Jim and Margaret Else also provided hospitality and moral support at all stages of our research.

In Nairobi, we thank Andrew Hill and Claire Infield for their hospitality throughout our work, Cynthia Moss for allowing us to share her house during 1980 and 1983, and Debbie Snelson for help and generosity of many different sorts, including suggestions for vervet names. Among many other things, Andrew helped procure a stuffed leopard and stuffed python for our experiments, introduced us to head waiters at many of Nairobi's finest restaurants, and took us to a party where a woman danced with a terrier. Claire tolerated our sudden arrivals, our delayed departures, and our (perfectly behaved) children; Cynthia and Debbie have accompanied us through more of life's bizarre and disjointed events than even they would like to remember. In Naivasha, Debbie and Charles Nightingale provided the perfect place for a break from fieldwork. They also helped us to tape record the vocalization of cows, without asking any questions.

Many people in Amboseli made our research both more efficient and far more enjoyable. We are grateful to all of Amboseli's wardens and rangers during the period of our research; special thanks are due to Joe Kioko and

Bob Oguya for their assistance and kindness. Throughout the years, the managers and staff of Serena Lodge generously allowed us to store food in their freezer and to use their water supply. We also thank Mercy Sikawa and Zippy Sindiyo, who helped us to obtain housing in Ol Tukai.

We thank Cynthia Moss for sharing her camp with us, and Cynthia, Cyn Jensen, Phyllis Lee, Keith Lindsay, and Joyce Poole for their friendship during the times that we lived in the camp with them. Susan Alberts, Jeanne and Stuart Altmann, Chris Gakahu, Moses Gumpumulu, Glenn Hausfater, David and Gillie Jackson, Christine Kioko, David Maitumo, Michael Milgroom, Norah Njiraini, Margaret Pertet, Raphael Mututua, Francis Saigilo, Dorcas Saita, Amy Samuels, Joan Silk, and Bev and Jeff Walters all provided friendship and crucial logistical assistance at various stages of our work. Patrick Mutua helped to keep our vehicle and Ol Tukai's generator in running order, sometimes with rubber bands and small pebbles, and always despite his assistants' most imaginative efforts to thwart him. We are especially grateful to Soila and Serah Sayialel, initially for their excellent care of our children and subsequently for their friendship. Together with Christine and Norah, Soila and Serah opened up to us a new perspective on life in Amboseli's village, Ol Tukai, and verified the anecdote about the "third chappati."

We thank Masaku Sila, Cynthia's and our major domo for many years, for his superb cooking and cleaning and for sharing with us his unique view of the world—both today's world and the world "before there were watches."

The long-term study of vervet monkeys in Amboseli has been a joint effort and would not have been possible without the assistance and cooperation of those who studied the vervets during periods when we were in the United States. We thank Sandy Andelman, Marc Hauser, Lynne Isbell, Phyllis Lee, and Shari Milgroom for keeping up the long-term demographic records and contributing to our research in ways too numerous to mention. It is always a privilege to have collegues who are excellent scientists; it is more unusual to have colleagues who prove to be good friends. We also owe a great deal to our research assistant, Bernard Musyoka Nzuma, who began working on the project in 1981. Bernard always did his level best in habituating and finding monkeys, collecting behavioral data, and carrying out playback experiments. He spotted lions and buffaloes before we did, found errant male vervets before we could, and shared with us the agony of field experiments on uncooperative monkeys.

Special thanks are due also to David Klein, who worked on Amboseli vervets from 1974 to 1976 and who provided crucial advice and demographic data in the early months of our study. Tom Struhsaker and Lysa Leland also played an important role in our work, visiting Amboseli frequently between 1977 and 1986. Tom's research on Amboseli vervet monkeys provided a starting point for our own, and he has consistently provided help and good counsel over the past 11 years. We also thank Richard Wrangham, who worked on the vervets in 1978, for his friendship over the years and in particular for the loan of his Land Rover during the summer of 1988.

For the past 6 years we have supplemented research on the Amboseli vervets with studies of captive primates. We are grateful to Michael McGuire and Michael Raleigh for permission to conduct research on the captive vervets housed at the Sepulveda Veterans Administration Hospital in the San Fernando Valley, California. We also express our thanks to C. R. Cornelius, Roy Henrickson, Andrew Hendrickx, and Bill Mason for permission to work on captive rhesus and Japanese macaques at the California Primate Research Center (CPRC). At the CPRC we thank Jacqueline Dieter and Michael Owren for support and Jerry Adams, Mark Alves, Art Cabrera, John Steele, and Paul Telfer for able technical assistance.

We began research on vervet monkeys as post doctoral fellows at Rockefeller University, where we received support from a National Science Foundation (NSF) grant to Peter Marler, a grant from the National Geographic Society to Peter Marler, an NSF fellowship to Dorothy Cheney, and a National Institutes of Health (NIH) fellowship to Robert Seyfarth. When these fellowships ended we received crucial funds to continue our work from the Harry Frank Guggenheim Foundation and, 1 year later, the NSF. From 1980 to the present our work has been supported by grants from the NSF; from 1982 to the present we have also received support from the NIH and from the University of Pennsylvania. We are especially grateful to Fred Stollnitz, director of the animal behavior program at NSF, for his support and encouragement and for understanding that research doesn't always proceed, and money isn't always spent, exactly as originally planned.

It is always an imposition to ask anyone to read a manuscript of this bulk, and we are lucky to have friends patient enough to have made extensive and constructive comments on previous drafts. The book was immeasurably improved by the criticism and comments of Verena Dasser, Lynn Fairbanks, Lila Gleitman, Sandy Harcourt, Robert Hinde, Peter Marler, Barbara Smuts, Kelly Stewart, and an anonymous reviewer. Selected chapters profited from the comments of Jeff Cynx, Randy Gallistel, Marc Hauser, Joe Macedonia, Paul Rozin, John Smith, Charles Snowdon, and Andrew Whiten. We also benefited from fruitful discussions with Brian Bornstein, Robert Boyd, Tim Caro, Gretchen Chapman, Cindy Fisher, Robert Harding, Bill Mason, Michael Owren, Susan Rakowitz, Steve Robbins, Paul Rozin, and Joan Silk. We owe special thanks to Daniel Dennett, not only for many ideas and discussions over the last 8 years and for reviewing the book in its entirety, but also for his offer to help us clean out, or at least rearrange, the detritus in our philosophical Augean stables. Chapter 1 owes much of its historical review to his editorial assistance, almost to the point that we can claim, for once, that any mistakes are his, and not ours.

Robert Hinde and Peter Marler have done far more for us than just read and comment on our papers. As the director of our graduate work, Robert taught us how to carry out research in animal behavior. He also guided and influenced many of our ideas about the relation between individual behavior and social structure in primates. Peter Marler introduced us to the study of vocal communication, helped carry out our initial field experiments, and has for the past 13 years been a constant source of ideas, encouragement, and helpful suggestions. Of the many sorts of luck that have aided our research, perhaps the greatest good fortune has come from our associations with Robert and with Peter.

Finally, we would like to thank the subjects of our research, the Amboseli vervet monkeys. Although it cannot be said that they welcomed us into their lives, they tolerated us with apathetic aplomb. Far more of successful research is owned to luck than we sometimes like to think, and we have benefited tremendously from being able to study such fascinating animals. We only wish that they could have led more pleasant lives. While studying vervets we have also been immensely lucky to have enjoyed a way of life that is fast disappearing. We feel privileged to have experienced the beauty and solitude of Amboseli as few in the future will be able to do.

CHAPTER ONE What Is It Like to Be a Monkey?

On the 5th of September, 1379, as two herds of swine, one belonging to the commune and the other to the priory of Saint-Marcel-le-Jeussey, were feeding together near that town, three sows of the communal herd, excited and enraged by the squealing of one of the porklings, rushed upon Perrinot Muet, the son of the swinekeeper, and before his father could come to his rescue, threw him to the ground and so severely injured him that he died soon afterwards. The three sows, after due process of law, were condemned to death; and as both the herds had hastened to the scene of the murder and by their cries and aggressive actions showed that they approved of the assault, and were ready and even eager to become *participes criminis,* they were arrested as accomplices and sentenced by the court to suffer the same penalty (Evans 1906 / 1987: 144).

The preceding account is by no means unusual. Throughout Europe during the Middle Ages animals ranging from insects to horses were commonly brought to trial, provided with defense lawyers, and charged with crimes such as the willful destruction of crops, murder, and sodomy. People in the Middle Ages were not careless anthropomorphizers; they clearly recognized that animals were not people. Nevertheless, the behavior of their animals often led people to believe that animals could be aware of what they did and held accountable for their acts. Why are we sometimes willing to attribute intent, beliefs, and consciousness to animals? What aspects of their behavior suggest the operation of complex mental processes?

We borrowed the title for this chapter from an article written in 1974 by the philosopher Thomas Nagel, who asks: "What is it like to be a bat?" Not a particularly auspicious beginning, since Nagel's conclusion was, essentially, "We can never know." Bats use sonar echoes to navigate through their environment and regularly attend to high-pitched sounds that fall outside the range of human hearing. Because their sensory world is so different from our own, Nagel argues, we can never really understand what it is like to be a bat, no matter how much we may learn about the biological processes that underlie a bat's existence. Likewise, Ludwig Wittgenstein (1958) felt that the close intermingling of customs, concepts, and meaning in any society would always create barriers between different cultures or between different species. "If a lion could talk," he said, "we would not understand him."

From a quite different perspective, other philosophers have argued that, even if there are things we cannot learn about the mind of another animal (or another person), the very inaccessibility of these presumably deeper facts renders them scientifically negligible—or even nonexistent. In his book *Word and Object,* W. V. O. Quine (1960) imagines that a linguist has entered an unknown land where none of the sounds people make are familiar. The linguist's goal is to construct a dictionary of the local language, to learn what each word means, and in this way to begin to understand how the people think. But if a native shouts "Gavagai!" when a rabbit appears, can the linguist conclude that *gavagai* means the same as *rabbit* in English? Quine believes that with no words in common and only gestures like pointing to go on, the linguist can never know. He might be able to specify the conditions under which gavagai is uttered, but his assessment of precisely what this sound means to the native will always be an approximation. Of course this residual uncertainty, Quine adds in mitigation, is no worse than the uncertainty we have about what our own words mean. We can never be sure that the others who (apparently) speak our language match objects and words in exactly the same way we do. So Quine concludes that the barriers to understanding another species are virtually insurmountable, but in principle no greater than the barriers to understanding our friends and neighbors.

We are, obviously, different from other animals, and these differences inevitably complicate any attempt to understand how animals communicate and how they see the world. Nevertheless, we hope to convince you that Nagel and Wittgenstein have been too pessimistic and have declared impossible what is merely difficult—and fascinating. Their reservations need not preclude research; on the contrary, we hope to show that we can learn a great deal about the mind of another species. We do this by describing the work of a number of ethologists, including ourselves, who—like Quine's imaginary linguist—have gone into the field intent on deciphering the gestures and sounds made by different creatures, particularly monkeys and apes.

Humans have always been curious about the behavior of nonhuman primates. During the past 30 years, however, there has been renewed interest in primate behavior and communication and in what such research might reveal about how animals think. The motivation to probe into the minds of monkeys and apes has come from at least four sources.

First, the richly detailed, descriptive accounts of primate behavior produced during the past 20 years (e.g., Goodall 1983, 1986; de Waal 1982, 1989; Fossey 1983; and Smuts 1985) raise important—and largely neglected—questions about how animals communicate and how they might deceive, plot strategies, or attribute motives to others. Much of the material in these books is anecdotal, but there is a growing awareness among ethologists that such anecdotes should be taken seriously and can, at the very least, serve as working hypotheses for further observation and experimentation (Byrne and Whiten 1988b, 1988c; Whiten and Byrne 1988c).

Second, the well-known ape language projects (e.g., Gardner and Gardner 1969; Premack 1976; Rumbaugh 1977; Patterson 1978; Terrace et al. 1979; Miles 1983; Savage-Rumbaugh 1986) have demonstrated clearly that apes in the laboratory can learn elements of human language. In some cases, research on captive apes has also revealed cognitive abilities not yet demonstrated for the same species in their natural habitat. Some people involved in these projects explain the disparity between performance in the laboratory and the field simply as a function of training: experience in the use of signs provides animals with skills they would, under natural circumstances, neither need nor possess (e.g. Premack 1983b; Savage-Rumbaugh et al. 1983). This conclusion, however, may be premature, because no one has yet gone into the wild and systematically presented apes with the same sort of problems they face in the laboratory.

To evolutionarily oriented ethologists, by contrast, the disparities between field and laboratory are puzzling and provide an intriguing challenge. Evolutionary theory holds that general skills, such as the ability to use signs as representations of objects or to make judgments based on analogical reasoning, do not evolve unless they serve some adaptive function (e.g., Humphrey 1976; Dawkins 1986). The evolutionary view assumes as its working hypothesis that the skills uncovered by research on captive apes *are* part of the animals' natural behavior and are used regularly under natural conditions. The lack of systematic data on such abilities simply means that primatologists have not looked closely enough at the problems their subjects face in the wild and the skills they use to solve them. At the very least, this evolutionary view provides a framework within which field and laboratory studies can be compared and interesting questions can be addressed.

Third, spurred on by developments in evolutionary theory, ethologists have demonstrated that many animals, from insects to apes, have an extraordinarily sophisticated knowledge of certain aspects of their environment and their social companions (reviewed in, for example, Griffin 1984). In their natural habitats animals act as if they can recognize kin, assess the suitability of mates or the size and aggressiveness of opponents, remember past social interactions, and solve complex problems that allow them to locate and harvest food efficiently (see, e.g., Holmes and Sherman 1983; Cheney, Seyfarth, and Smuts 1986; Kamil, Krebs, and Pulliam 1987; Waldman, Frumhoff, and Sherman 1988 for reviews). The picture of animal intelligence that emerges from fieldwork is richer and more complex than earlier studies, conducted mainly on laboratory animals, would have led us to expect. This very richness, however, makes it all the more challenging to specify how animals and humans differ. Toads, ground squirrels, and monkeys may respond differently to full- and half-siblings, but their understanding of kinship is unlikely to be the same as our own. How should we characterize the difference?

Fourth, there is increasing awareness that studies of animal social behavior can contribute to the growing area of research that has been broadly labeled *cognitive science.* Here the ultimate object of interest—for philosophers, linguists, computer scientists, anthropologists, and psychologists—is the human mind. How does the mind represent and process information? Can we build a machine, find another species, or write a computer program that duplicates the mind's performance? What would it mean if we could? One research strategy, pursued in a variety of forms, has been to investigate "almost minds," such as the minds of children or the "minds" of computer programs, to see what makes them different from our own and what would be needed to elevate them to adult human status. As we hope to demonstrate, the social behavior of nonhuman primates offers a glimpse of almost minds at work—a glimpse that may ultimately tell us how, in the course of our own evolution, some minds gained an advantage over others.

How We Approach The Problem

Many scientists implicitly assume that, among all animals, the behavior and intelligence of nonhuman primates are most like our own. Nonhuman primates have relatively larger brains and proportionately more neocortex than other species (e.g., Passingham 1982; Martin 1983; Jerison 1985), and it now seems likely that humans, chimpanzees, and gorillas shared a common ancestor as recently as 5 to 7 million years ago (Weiss 1987). This assumption about the unique status of primate intelligence is, however, just that: an assumption. The relation between intelligence and measures of brain size is poorly understood, and evolutionary affinity does not always ensure behavioral similarity. Moreover, the view that nonhuman primates are the animals most like ourselves coexists uneasily in our minds with the equally pervasive view that primates differ fundamentally from us because they lack language; lacking language, they also lack many of the capacities necessary for reasoning and abstract thought (Premack 1983b).

To make our own position clear at the outset, we accept the broad evolutionary view which argues that, just as one can learn about the origin of the human hand or brain by studying the anatomy of nonhuman primates, so research on nonhuman primate behavior can shed light on the origins of human language, cognition, and self-awareness. This does not mean that modern monkeys and apes present us with a complete and accurate picture of the ancestral human condition, nor that similar behavior in human and nonhuman species is necessarily caused by similar underlying mechanisms. At a purely descriptive level, for example, there are many similarities between the interactions that occur within monkey matrilines and the interactions that occur within human families. Such data do not, however, prove that human and nonhuman primates recognize kin or compare social relationships in exactly the same way (e.g., Hinde 1987). An evolutionary perspective draws our attention to the many apparent similarities between human and nonhuman primate behavior. Given these observations, the goal of our (as well as many other scientists') research is to probe more deeply and establish where such similarities break down.

Perhaps more important, taking an evolutionary perspective means that we approach the study of primate intelligence from a practical, functional perspective. What are the problems monkeys face in their daily lives? What do they *need* to know, and how might one method of obtaining and storing knowledge give certain individuals a reproductive advantage over others? Done with care, such analysis can suggest not only how human intelligence evolved but *why*.

Regardless of the theoretical view one adopts, the problem of investigating intelligence empirically remains. The difficulty arises in part because field and laboratory studies of intelligence have traditionally used different methods to address fundamentally different questions.

In the laboratory, learning and intelligence have usually been measured by discrete and quite specific experiments (see Skinner 1974; Rescorla 1985; Herrnstein 1970; Essock-Vitale and Seyfarth 1987; Kamil 1987; and Roitblat 1987 for reviews). Whether they measure performance in the formation of associations, "learning sets," oddity discrimination, analogical reasoning, or language learning, such experiments are attractive for a number of reasons. Their precision and control, for example, are unlikely to be matched by any study conducted in the field. Different experiments can focus precisely on different cognitive skills and allow one to state explicitly what would constitute evidence that an individual possesses a particular ability. Laboratory tests provide one means of comparing intelligence across species (but see Warren 1973) and, since similar tests can be conducted on humans, also allow comparison between human performance and that of closely related species.

At the same time, the very controlled, objective nature of laboratory experiments creates problems. For example, the goal of many studies of captive primates has been to formulate one measure or set of measures that permits comparisons of intelligence in different species (e.g., Harlow 1949; Schrier 1984). Although they are tightly controlled and systematic, such experiments are typically divorced from questions of evolutionary function, and their relevance to the animals' natural social behavior is often unclear. Moreover, cross-species comparisons generally ignore the fact that different species have evolved in different social and ecological environments. Given any one task or set of tasks, some species will invariably perform better than others, and it will always be possible to argue that one species is predisposed to learn certain tasks more easily than others (Seligman and Hager 1972; Hinde and Stevenson-Hinde 1973), that one species is simply more familiar with the stimuli in question, or that one species has more experience with the type of question being posed (e.g., Bitterman 1965; see also Kamil 1987). Of course, this kind of "species relativism" easily becomes tautological; to compare species fairly, one must control all possible contextual features that might result in a performance difference, which in turn makes it almost impossible to determine if a species difference exists. As Niko Tinbergen (1951 : 12) remarked, "One should *not* use identical experimental techniques to compare two species, because they would almost certainly not be the same to *them*."

In their search for an objective measure of performance, laboratory psychologists have traditionally designed tests using arbitrary stimuli like lights, tones, or different-shaped blocks that animals would never encounter in their natural habitat. This method allows tests on different species to be compared directly, but it also increases the likelihood that results will underestimate or fail to reveal a subject's true ability. An animal may not understand the problem or it may simply lack the motivation to perform in what it regards as an unfamiliar or even hostile environment (e.g., Terrace et al. 1979).

Finally, despite overwhelming evidence that primates are social creatures, laboratory tests have generally ignored the social dimension of primate intelligence. With some notable exceptions (which we discuss in subsequent chapters), few tests have been designed to mimic the social problems faced by group-living animals and few tests have measured an individual's performance using social stimuli, like the pictures or voices of familiar conspecifics.

The research we describe here takes a different approach to the study of how animals think. Drawing on our own work with East African vervet monkeys (*Cercopithecus aethiops*), and the research of scientists studying other species, we document what monkeys and apes do in their natural habitats and consider what sorts of underlying mental operations might possibly account for this behavior. Although the virtue of this approach is that it considers behavior and communication within an evolutionary framework, its drawback is its heavy reliance on observations, experiments, and anecdotes that may suggest, but can never definitively prove, what an animal's mental abilities are.

To put our approach in perspective, we begin by outlining a philosophical debate that we are not really qualified to discuss and will not pursue at great length. For more detailed reviews of the history of ethology and comparative psychology, together with the debates that have punctuated their development, we refer the reader to Hinde (1982), Dewsbury (1984), Boakes (1984), and Roitblat (1987).

For more than 200 years, scientists concerned with the nature of mind have divided roughly into two camps, although the ground rules of the dispute have shifted markedly over the years. Originally, debate focused on the different interpretations made by rationalists, such as Descartes and Leibniz, and empiricists, such as Locke, Berkeley, and Hume. The rationalists followed Descartes' lead in starting from personal introspection, which revealed a wealth of mental items—thoughts, sensations, dreams, desires, "concepts"—being manipulated and appreciated by an active mind or self. Many of these mental items, the rationalists claimed, were innate and not derived from experience. Empiricists were extremely dubious about some of the powers claimed by the rationalists for the mind, but they could not deny the apparent revelations of introspection. Instead, they attempted to secure the scientific standing of their categories of mental items by posing strict "empirical" conditions for their occurrence. What was in the mind, empiricists argued, was only what had entered it through sensory experience, and what happened in the mind was due solely to *association*—mechanical principles of attachment caused by proximity and repetition in experience. The mind was not an active thinking thing, but rather a passive arena in which sensory inputs combined according to simple laws to yield behavior. This hard-headed, skeptical minimalism of the empiricists provoked a counterrevolution by Kant, who articulated the doctrine that the mind actively organizes experiences and formulates behavior on the basis of preexisting schemes (Gardner 1987).

In the twentieth century, the descendants of the radical empiricists, the behaviorists, have taken a still more extreme stance, arguing in some cases that mental processes like thought and consciousness do not exist at all but have been mistakenly inferred from behavior (e.g., Skinner 1957, 1974). Others have not denied the existence of putative mental items but have insisted that they can be "operationally defined" only in terms of measurable stimulation and behavior; instead of attempting to conduct science in terms of these otherwise unmeasurable and illusory phenomena, we should concentrate on predicting and explaining behavior (Fancher 1979; see also Ryle 1949 for a different sort of behaviorism).

Although behaviorism dominated American comparative psychology, some European and American psychologists eschewed the behaviorists' paradigm and continued in the mentalists' tradition. An extreme proponent of mentalism in animals, George Romanes (1882), saw evidence of consciousness in the behavior of virtually all animal species. In a more restrained form, the mentalistic approach to animal behavior adopted to varying degrees by psychologists like Margaret Washburn (1908), Robert Yerkes (1916, 1925), and Wolfgang Kohler (1925) anticipated modern comparative cognition by emphasizing the ways that mental events shape and interpret behavior. Arguing against the view that most learning can be explained by relatively simple stimulus-response associations, Kohler's pioneering study of captive chimpanzees stressed learning through insight, purpose, and recognized goals.

The abstemious antimentalism of the behaviorists owed much of its appeal to the (justified) distrust of Descartes' metaphysical dualism, the doctrine that minds, unlike bodies, were made of a different, nonphysical sort of stuff. Among many philosophers this antimentalism held sway until, with the advent of computers, a new materialistic, indeed mechanistic, brand of mentalism was formulated. Today this approach is best exemplified by Jerry Fodor (e.g., 1975), who has argued not only that the mind organizes thought but also that the mind has an innate "language of thought" that both represents the world and affects behavior (see also Churchland 1984; Beer 1990). Thus, while an investigator in the empiricist-behaviorist tradition seeks to explain behavior solely in terms of experience (how prior, public, external events affect subsequent activity) and avoids any "irreducible" reference to the mind, rationalists and their descendants view mental operations as having true causal power, both in structuring what organisms perceive and in determining what they do. Indeed, much of modern comparative cognition concerns itself with questions of how animals represent their world and how they perceive relations between stimuli (Roitblat 1987).

By far the most important challenge to behaviorism, however, came from the European ethologists. Studying animals in their natural habitats, scientists such as Konrad Lorenz, Niko Tinbergen, and Karl von Frisch viewed behavior as the product of natural selection operating on specific traits in specific environments, and they stressed the close interaction between a given species' perceptual abilities and its adaptation to a particular habitat. In a sense, the evolutionary perspective was more mentalist than behaviorist, since it focused on species-specific adaptations and traits that organized each animal's experiences. Indeed, as one of our reviewers has pointed out, the title of our book recalls an imaginative essay written in 1934 by the ethologist von Uexkull entitled *A Stroll Through the Worlds of*

Animals and Men, as well as Tinbergen's classic book *The Herring Gull's World* (1953). Both stress the unique perspective of each species' subjective world, or *Umwelt* (von Uexkull 1934).

Research by Lorenz, Tinbergen, and others did not, however, presage a return to the uncritical mentalism of Romanes. Guided by predictions derived from evolutionary theory, their work was firmly grounded in the experimental approach of the empiricists and avoided questions of conscious thought. Indeed, when a given behavior was considered as an adaptation to a particular set of environmental factors, the question of whether a species did or did not have humanlike consciousness became irrelevant (Roitblat 1987).

As the early European ethologists clearly recognized, assumptions about the mental processes of others are controversial enough when we discuss our own species; when the minds of animals are at issue the problem becomes even more intractable. Given these difficulties, our solution has been adopted as much from necessity as from conviction: we borrow the methods of the empiricists but place them tentatively within the framework of a more mentalistic approach. Since we cannot interview our subjects and ask them for introspective reports of their current or recent mentation, we must, in conducting our observations and experiments, adopt a nonmentalistic position and study communication and behavior operationally, in terms of the responses they evoke in others. This method, of course, carries with it all of the limitations pointed out by Quine. Can we ever really know, from a study of responses alone, what gavagai means?

On the other hand, though armed with behaviorist methods, we are less agnostic about the mind as a causal agent than behaviorists might like us to be, and we cautiously adopt such words as *attribution, representation, consciousness,* and *strategy*. In using this terminology our approach is, philosophically speaking, mentalist rather than behaviorist, since we assume that mental states are characterized not just by external environmental influences but also by reference to other mental states (Churchland 1984). This approach need not imply that psychological terms like *believes* and *fears* introduce contents with intrinsic or invariant meaning. Indeed, if there are "semantic" properties to mental representations, they probably cannot be interpreted except "syntactically," in terms of their relation to each other (see Dennett 1987 for more discussion of this issue).

Our openness to the use of mental terms also means that, in spite of our efforts to avoid it, the word *cognition* occasionally appears and therefore requires definition. For present purposes, we adopt Markl's (1985) definition of cognition as the ability to relate different unconnected pieces of information in new ways and to apply the results in an adaptive manner. This definition is useful to those who study animals because it examines cognition in terms of what individuals *do* without specifying or being limited to any particular mental mechanisms that might underlie behavior.

Intellectually, then, our approach to animal intelligence is a hybrid, combining the methods of behaviorists (who don't believe in mentality) with many interpretations of mentalists (who feel that the behaviorist approach is inadequate). The uneasy alliance of two historically different viewpoints recalls the comment of Saki's character, Reginald, who, when asked to name his religion said, "The fashion just now is a Roman Catholic frame of mind with an Agnostic conscience: you get the medieval picturesqueness of the one with the modern conveniences of the other."

Despite its hybrid status, our method is neither original nor unique. It echoes a growing conviction in ethology that many aspects of animal behavior cannot be explained without ascribing some types of complex mental processes to animals (e.g., Griffin 1982, 1984). At this point we agree with Kamil (1987) that it is probably premature and unproductive to debate whether associative learning paradigms can or cannot adequately explain all behavior or to try to specify every way in which animals might form mental representations of their world. In fact, the dichotomy between behaviorist and cognitive approaches is no longer as clear as it once was. Modern theories of association have a strong cognitive component (Rescorla 1985; Kamil 1987), and mental representations may well play an important role in the formation of associations between two stimuli (Dickinson 1980; Mason 1986; Rescorla 1988). Rather than committing ourselves to one or more specific processes underlying the acquisition of knowledge, we have more modest goals: to document what monkeys seem to know, what they can do with this knowledge, and why they *need* this knowledge under natural conditions.

This last point is crucial. As we have mentioned, we believe that it will prove difficult if not impossible to understand the mental processes of animals unless we adopt a functional stance that focuses on the context in which these abilities evolved (Seligman 1970; Seligman and Hager 1972; Hinde and Stevenson-Hinde 1973). Once again, this approach is not at all novel. Ethology (or behavioral ecology, as the functional analysis of animal behavior is commonly called) assumes as its working hypothesis that most patterns of behavior, like most morphological structures, have evolved and serve some adaptive function. Such a Panglossian paradigm (Dennett 1983, 1987) is not as naive as it might appear. Adaptationism is an assumption, not a conclusion, whose primary purpose is to suggest testable predictions that can ultimately reveal both the advantages conferred by a given trait and the constraints within which evolution has operated. The naturalistic approach can reveal inconsistencies, suggest new experiments, and shed considerable light on research that might otherwise be conducted in an evolutionary vacuum.

For example, consider tests that examine whether primates can solve problems of transitive inference (If A is bigger than B and B is bigger than C, is A bigger than C?). Although squirrel monkeys and chimpanzees do learn to solve these problems in captivity, why they should be able to do so remains unclear (McGonigle and Chalmers 1977; Gillan 1981; D'Amato and Colombo 1988). Is there any possible function to solving problems through transitive inference, or is this ability simply devoid of function—an artifact dependent upon human training? Even a cursory examination of primate social behavior under natural conditions, however, reveals that monkeys and apes regularly confront problems of transitive inference in their interactions with each other. Field observations have suggested (although not yet definitively proved) that monkeys infer dominance hierarchies among the members of their group based on their observation of interactions among pairs of individuals (chapter 3; see also Cheney 1978; Seyfarth 1981; Datta 1983a; Gouzoules, Gouzoules, and Marler 1984). There is some evidence that even birds may be aware of each others' relative dominance ranks (Popp 1987) and hence that the ability to solve problems of transitive inference is widespread. Such results suggest that the tasks faced by animals in the laboratory and in the field may be formally similar. We need both approaches if we are fully to understand why an ability has evolved and how it benefits an animal in its daily life.

Further evidence of the need for ecological validity in laboratory experiments comes from tests of spatial memory in rats and birds. Originally, experiments tested whether rats could remember which arms of an 8- or 12-arm radial maze they had already visited. Results suggested that these very opportunistic foragers have extremely accurate spatial memories (Olton and Samuelson 1976). In contrast, even after extensive training, pigeons seemed unable to retain information about which arms they had visited, a result that prompted some researchers to conclude that spatial memory might be poorly developed among birds in general. However, when Balda and Kamil (1988) conducted a similar test using Clark's nutcrackers, a bird that uses spatial memory to revisit places where it has previously cached seeds, the birds performed as well as rats (chapter 9). Because of results like these, the need for an evolutionary, ecological perspective is becoming increasingly recognized in studies of animal learning (e.g., Johnston 1981).

While some students of animal learning have been criticized for ignoring function, there has also been a tendency among evolutionarily oriented behavioral ecologists to minimize the importance of the proximate mechanisms governing behavior (e.g., Wilson 1975). This view seems shortsighted, because mechanisms (such as how an animal thinks) may either constrain evolution or open up new opportunities for adaptation. A bird's foraging efficiency, for example, depends on its ability to remember which food patch it has already visited, monitor the rate at which it is depleting food in its present patch, compare this rate with rates of return in other patches, and so on (see, e.g., Krebs and McCleery 1984; Kamil and Roitblat 1985; Kamil, Krebs, and Pulliam 1987; and Gallistel 1989a and 1989b for reviews). Limits to these abilities will determine how readily the bird can colonize or adapt to new environments.

As a second example of the close link between evolutionary theory and the mechanisms governing behavior, consider the evolution of cooperative interactions, which in most animal species occur primarily among close genetic relatives. For such discriminative altruism to evolve, there must be some mechanism by which individuals can distinguish their genetic relatives from others. And indeed, kin recognition has now been demonstrated in an extraordinary variety of animals, including social insects, amphibians, reptiles, birds, and mammals (e.g., Gouzoules 1984; Sherman and Holmes 1985; Waldman, Frumhoff and Sherman 1988). In most species, the mechanism underlying kin recognition is close association during development. If nonkin are reared together, they treat each other as relatives; if kin are reared apart, they behave as if they were unrelated (Sherman and Holmes 1985). In other species, however, a kind of *phenotypic matching* seems to occur, so that siblings associate preferentially even if they have not been reared together. The mechanisms involved in phenotypic matching are not well understood but may depend at least in part on visual or olfactory discriminations (Bateson 1980; Sherman and Holmes 1985).

Although many animals can distinguish between their own kin and other individuals, the mechanisms governing kin recognition in nonhuman primates apparently allow them to go several steps further. Monkeys not only distinguish their own kin but, like humans, also seem to recognize the kin associations that exist among *other* individuals (chapter 3). Compared with simpler forms of kin recognition, this more complex recognition clearly confers a greater ability to monitor patterns of social interactions within groups and to predict, for example, which group members are likely to support each other during aggressive interactions.

Kin recognition seems to have evolved independently in many insects, birds, and mammals because natural selection confers an advantage on individuals that behave differentially toward close genetic relatives (Hamilton 1964). Taking a broad evolutionary view, variation in the underlying mechanisms is unimportant, because the functions, or evolutionary consequences, of kin recognition are similar across species. At the same time, natural selection has clearly favored different degrees of behavioral flexibility—and hence different levels of social knowledge—in different species. Distinguishing different mechanisms is important because a given mechanism may set the stage for increasingly complex social interactions, which in turn may create selective pressures for more complex and elaborate systems of recognition and classification.

How This Book is Organized

This book examines the communicative and cognitive abilities of nonhuman primates living under natural conditions. Its emphasis is functional, for our interest is in what monkeys know and what they need to know in their social interactions. From descriptions of social behavior and interpretations of its consequences, we go on to speculate about the possible factors that might have favored the evolution of primate intelligence.

There are several things this book does not do. First, we have not attempted a complete review of the literature on animal learning. We do discuss some laboratory studies, but our treatment focuses selectively on those that relate directly to primate behavior under more or less natural social conditions. Second, we make no attempt to evaluate the various ape language projects, except when their results are directly relevant to the natural communication of primates. Finally, we avoid whenever possible any comparison of nonhuman primate intelligence with that of other animals (including, for the most part, humans), since this issue has the twin defects of being both highly controversial and, at least for the moment, insoluble. Whenever we *do* drift into a comparison among species, our ideas are mere speculation and not opinions derived through strong conviction, much less from hard data.

We begin this book as we began our research, by immersing ourselves in the social organization and behavior of East African vervet monkeys. Chapter 2 reviews what we learned about the behavior and ecology of vervet monkeys over the eleven-year period during which we and our students continuously observed a number of vervet groups in Amboseli National Park, Kenya. Although the primary focus is on vervets, we also draw on other research, emphasizing in particular studies of baboons and macaques. One goal of chapter 2 is to describe the richness of vervet monkey social organization and the problems animals face in attempting to survive, reproduce, cooperate, and compete with one another.

Chapter 2, however, is not just descriptive; it also seeks to explain. Indeed, it is almost impossible to describe what vervet monkeys (and other primates) do without implying that specific purposes and strategies underlie the animals' behavior. Hence a second goal of chapter 2 is to infer some of the rules that vervet monkeys seem to be applying in their social interactions. Our working hypothesis is that natural selection has favored some strategies over others and that the behavior we observe today is the result of this selection. Having inferred a set of rules, we discuss why one set of rules may have evolved rather than another.

But all this talk of rules, strategies, and social knowledge raises important questions—in fact, some of the central questions of our research. When we use these words we describe primate social organization in human terms, from a human perspective. How does it look from the monkey's point of view? Do we really mean to imply that monkeys know where they stand in their group, have long-term goals, and believe that other animals have similar intentions? Are they in any sense aware of their own behavior? Having suggested links between evolution and social behavior in chapter 2, we develop a critical examination of the links between behavior and cognition in chapters 3 through 9.

In chapter 3 we examine the monkeys' knowledge of group membership, dominance, kinship, reciprocity, and social alliances. Given the complexity of primate social behavior, we take a critical look at the underlying mechanisms and ask what monkeys actually know, not only about their own social relations but also about the social relations of others in their group. We consider whether monkeys have social concepts and distinguish among different *types* of relationships independent of the particular individuals involved. In examining the monkeys' knowledge of their social companions, we join others in emphasizing the distinction between what Jackendoff (1987) has called "intelligent sensitivity" and "conscious awareness." Monkeys may well exhibit flexible, innovative behavior and recognize subtle distinctions between others without necessarily realizing what they are doing (see also Dennett 1983; Premack 1988). Along with asking how much monkeys know about what they do, therefore, we must also ask how much they know about what they know.

In chapter 4 we examine in greater detail the vervet monkeys' system of vocal communication. Part of our interest in vocalizations derives from a concern with vocal communication itself. Nonhuman primates are, after all, our closest living relatives. Hence, if there exists anywhere in the animal kingdom a system of communication like our language—or a system of communication that can give us hints about how our language evolved—it is likely to be found among the monkeys and apes. As reasonable as this view seems, for years it was not generally accepted. Despite the many parallels between human and nonhuman primates in morphology, physiology, and behavior, fundamental differences in vocal communication were believed to exist. We disagree, and in chapter 4 we review evidence indicating that vervet monkeys and other primates use calls in a manner that has many important parallels with language use in humans.

Regardless of whether monkey vocalizations qualify as language (and we believe they do not), once it can be shown that some of the information contained in the calls of monkeys denote objects or events in the external world, vocalizations become an important tool for understanding how monkeys think. Just as language has been called a "window on the mind" of humans, so the study of communication among monkeys and apes may help us to understand how these animals see the world and what they know about it (Griffin 1984). Chapter 4 therefore examines vocal communication not only for its own inherent interest, but also because vocalizations provide us with a picture of the world from the monkeys' point of view.

The evidence presented in chapter 4 introduces, once again, the problem of studying monkeys from a human perspective and describing their behavior using our own, value-laden terms. What, precisely, do we mean when we say that a vervet monkey gave a particular vocalization to warn others that there was an eagle nearby? Can we honestly claim that the monkey knew that others hadn't seen the eagle and intended to inform them? Do monkeys that do this sort of thing actually understand the relation between sounds and the things for which they stand?

In chapter 5 we consider in greater detail what monkeys mean when they vocalize to one another. We do so by examining the question of meaning in three complementary ways. First, we consider what a caller "intends" to communicate when he vocalizes by examining how call production is affected by the presence or behavior of others. An important distinction is made between communication that is affected by other animals' behavior and communication that is affected by other animals' *minds*. Second, we examine what monkeys know about the relation between their calls and the things for which they stand. Here we draw upon experiments in which vervets were asked to compare two vocalizations that differ in meaning or acoustic properties. Results suggest that monkeys, like humans, classify calls not just as physical entities but as sounds that *represent* things. Third, we compare the use of vocalizations by monkeys with the very earliest goos, gurgles, and wordlike noises made by human infants.

To this point, we have made two claims about the knowledge that underlies social behavior and communication in vervets and other nonhuman primates. First, we have suggested that in their social interactions monkeys do not simply associate some individuals with others but instead classify relationships into types. *Mother-offspring bonds* or *bonds between the members of family X* are abstractions that allow different relationships to be compared with one another. Second, we have argued that monkeys classify sounds according to the objects and events they denote. As in their assessment of social relationships, the monkeys seem to represent a sound's meaning in their mind and compare different sounds on the basis of these representations.

The notion of *mental representation*—in children, adults, animals, and even machines—is central to much of modern cognitive science (e.g., Gardner 1987; Johnson-Laird 1988). Representations can be thought of as images, rules, propositions, or symbols that guide an organism's (or a machine's) behavior. The problem is to determine what these representations actually are—what information they contain, and how the information is coded, stored, transformed, or compared. Surprisingly, while cognitive scientists have studied many aspects of behavior, they have paid relatively little attention to one ubiquitous part of our lives: social interactions and the representation of social relationships. Similarly, while the representations that underlie human language have received considerable attention, there has been no comparable work on the simpler communicative systems of nonhuman species, perhaps because of the misguided view that animals never signal to each other *about* things.

With these omissions in mind, in chapter 6 we examine the evidence for mental representations in monkeys as they apply both to social relationships and to the meaning of vocalizations. We consider what sort of information such representations might contain, how it might be stored, and how the monkeys' use of concepts might be limited when compared with our own. From the speculation that monkeys can represent features of other animals' *behavior,* we then turn, in chapters 7 and 8, to what monkeys may know about other animals' *minds.*

Communication in our own species is used not only to inform others but also to manipulate, argue, persuade, and even deceive. Are such uses unique to humans? For many years ethologists assumed that animal signals provided reliable information in the sense that communicative displays allowed accurate assessment of subsequent behavior. This view has recently been challenged, however, on empirical and theoretical grounds. Empirically, a growing list of examples indicates that animal signals frequently provide inaccurate or even false information (reviewed in Markl 1985; Byrne and Whiten 1988b, 1988c; Whiten and Byrne 1988c; Cheney and Seyfarth 1990). From a theoretical perspective, models drawn from game theory strongly suggest that in any contest between truthful and deceptive signalers, the deceivers are likely to win (Dawkins and Krebs 1978; Andersson 1980; Maynard Smith 1982; Krebs and Dawkins 1984). In chapter 7, we review the evidence for deceptive signaling in animals. We describe the forms that apparently deceptive signals take, how modifiable and flexible such signals are, and how animals assess their meaning.

When humans try to deceive each other, we try among other things to alter what another individual thinks. Conversely, to detect deception we must be able to read another's mind: to distinguish, for example, between a person who seeks help and genuinely intends to reciprocate and one who seeks help for more selfish reasons. The supplicant's behavior in each case may be the same—it is the difference between states of minds that must be detected. In other words, deceit and its detection assume (at least in the case of humans) that individuals can attribute mental states to others. Can the same be said of animals? Is it valid to assume, in David Premack's words, that animals have "a theory of mind"? In chapter 8 we consider the question of attribution and ask whether monkeys and apes recognize the existence of mental states in others. There is evidence that nonhuman primates recognize that other individuals have social relationships. Do they also recognize that other individuals have emotions, schemes, beliefs, or intentions? Are they aware of similar states of mind in themselves? How would we know if they were? If we are compelled to infer mental processes solely through the responses that they evoke in others, such questions may remain largely intractable, simply because it may always be possible to explain behavior in terms that do not depend on abstract mental representations.

Although the question of attribution in animals has not yet received much systematic treatment and still relies too much on anecdotes, we do not believe that these anecdotes should be derided as inevitably leading to inflated anthropomorphism (e.g., Thompson 1986, 1988). To the contrary, as chapter 8 shows, anecdotes often reveal puzzling gaps and lapses in monkeys' and apes' theories of mind. Nonhuman primates seem to be experts at analyzing each others' *behavior;* they seem much worse at analyzing each others' *thoughts.* It remains for future research to determine where these gaps and lapses lie.

One of the striking features of adult human intelligence is our ability to apply what we have learned in one context to other, quite different, circumstances. We take it for granted, for instance, that the basic rules of physics apply whether we are playing billiards, hanging a picture, or building a house. In short, the knowledge we possess is in many cases accessible to us: we are aware of it, or know that we know it. As a result, one hallmark of human intelligence is our ability to generalize from one narrowly defined problem to logically similar ones in novel circumstances (Rozin 1976).

In contrast, animals often seem to have a kind of "laser beam" intelligence—extraordinarily powerful when focused in a single domain but much less well developed outside that narrow sphere. Birds like homing pigeons can navigate enormous distances using the sun, the stars, geographical markers, or a magnetic sense to guide them (reviewed in Gould 1982). Honey bees dance to inform one another about the distance, size, and location of a food source, compensating as the sun changes its angle throughout the day so that the flight from hive to food is always indicated as a straight line (von Frisch 1967; Gould and Gould 1988). And yet we rarely think of animals like homing pigeons or bees as intelligent in the human sense, primarily because their sophisticated performance seems limited to specific, highly circumscribed spheres. In Paul Rozin's terms, the homing pigeon's navigational and the bee's communicative skills are "inaccessible": the animals don't know what they know and cannot apply their knowledge to problems in other domains.

Most monkeys and apes are social creatures, and their expertise in the domain of social interactions is striking. But how accessible is their intelligence? Do they understand how much they know? Our first attempt to grapple with the question of metaknowledge comes in chapters 3 and 8, where we explore the extent to which monkeys not only engage in flexible, adaptive behavior but also *recognize* that they and others do so. Chapter 9 approaches the issue from a different angle, by comparing the performance of nonhuman primates in their social interactions with their behavior outside the social domain. By comparing social and nonsocial performance we hope to specify more precisely both the richness of primate intelligence and its limitations.

In their social interactions, monkeys and apes face problems that are formally similar to the problems with which they have been tested in the laboratory. Understanding a dominance hierarchy, for instance, mirrors problems in transitive inference, while understanding relations within kin groups mirrors problems in concept formation. Yet we have no idea whether the skills used by primates in social interactions can ever be extended to other, nonsocial tasks. If we can demonstrate that a baboon recognizes the relative ranks of other group members, does it necessarily follow that she will also be able to rank containers of water? To put it another way, we already know that monkeys make excellent primatologists; does their knowledge extend outside this domain to other spheres of activity? Do they make equally good naturalists?

It has been hypothesized (e.g., Jolly 1966; Humphrey 1976) that primate intelligence (including our own) originally evolved to solve social problems and was only later extended to problems outside the social domain. If this hypothesis has any validity, at least some of the abilities that vervets and other primates demonstrate in their social interactions ought to be relatively inaccessible and not easily generalized to nonsocial problems. In chapter 9, we attempt to compare vervets' knowledge of other vervets with their knowledge of different species like leopards, hippopotami, starlings, and humans. We find that vervets are sometimes puzzlingly ignorant about the behavior of other species, even when they have ample opportunity to observe them and when learning about them would have great survival value. Though wizards in their social interactions with each other, the monkeys seem largely unaware of their skills and hence unable to apply them outside this narrow domain.

Our goal in the succeeding chapters, then, is not just to describe the marvelous things that monkeys do but to probe more deeply into the mechanisms that underlie their behavior and to clarify not only what monkeys know but also what they do not know. Armed with this information we may gain a better idea of how a monkey's view of the world differs from our own.

CHAPTER TWO | SOCIAL BEHAVIOR

The main subjects of this book are East African vervet monkeys (*Cercopithecus aethiops*), members of the largest family of Old World monkeys, the Cercopithecidae. Vervets are close—but by no means the closest—nonhuman primate relatives of modern humans. The ancestors of modern Old World monkeys diverged from the common ancestor of Old World monkeys, apes, and humans roughly 20 million years ago (Fleagle 1988). By contrast, humans and the great apes are far more closely related; it is thought that humans, chimpanzees, and gorillas shared a common ancestor as recently as 5 to 7 million years ago (Weiss 1987). For nonbiologist readers, we should note that apes are *not* monkeys. The term *monkey* is reserved for Old and New World monkeys, whereas *ape* refers to chimpanzees (*Pan troglodytes*), bonobos (*Pan paniscus*), gorillas (*Gorilla gorilla*), orangutans (*Pongo pygmaeus*), and gibbons (*Hylobates* spp.). We use the term *primate* or *nonhuman primate* to refer collectively to monkeys and apes.

Adult vervet monkeys (fig. 2.1) are approximately the size of a domestic cat—smaller and lighter than other semiterrestrial Old World monkeys. Adult males weight about 4 to 5 kg and females 3 to 4 kg; infants weigh under 1 kg. Unlike other members of the genus *Cercopithecus*, which are forest dwelling and arboreal, vervets spend roughly equal amounts of time during daylight hours on the ground and in trees. They are one of the most common primate species in Africa. Although their primary habitat is savanna woodland, vervets are a highly adaptable species and are also found in forests and semideserts throughout much of sub-Saharan Africa. Two subspecies of vervets are recognized, *Cercopithecus aethiops sabaeus* in West Africa and *Cercopithecus aethiops johnstoni* in the rest of the continent. There is also a population of West African vervet monkeys living on the Caribbean islands of Saint Kitts, Nevis, and Barbados, their ancestors having traveled across the Atlantic, probably on slave boats, during the seventeenth and eighteenth centuries (Denham 1987).

In contrast to many forest-dwelling primates, vervet monkeys have been known to Africans and Europeans for centuries. Most African tribes that inhabit regions where vervets are found have a specific name for the vervet monkey: among the Maasai, the predominant tribe in the area where our research was conducted, vervets are called *enderei*. Vervetlike monkeys, presumably brought as pets from Africa, are represented in frescoes of the late Minoan age (about 1500 B.C.) in Knossos, Crete (Hill 1966). In western Europe, vervets have turned up in taxonomical classifications and drawings since at least the seventeenth century.

Despite their ubiquity, however, vervet monkeys seem to have been almost invisible creatures. Vervets appear in almost no African myths or legends, and Europeans seem almost to have gone out of their way to ignore them. The nineteenth- and early twentieth-century European explorers and colonialists in East Africa were ardent naturalists, and they compiled exhaustive lists of the birds they collected and the large game animals they shot. Notably absent from their descriptions, however, were primates, in particular vervet monkeys. To the extent that vervets were mentioned at all, they were characterized as mischievous, crop-raiding pests whose skins were of little interest or value.

It was not until the early 1960s that the first long-term studies of vervet monkeys were initiated by K. R. L. Hall and Stephen Gartlan (1965) on Lolui Island in Lake Victoria and by Tom Struhsaker (1967b, 1967d) in Amboseli, Kenya. Since then vervets have become one of the most intensively studied primate species; long-term research on vervets has been conducted at a number of sites in East, West, and southern Africa, as well as on several Caribbean islands and in captive colonies in the United States.

The Monkeys and Their Habitat

We conducted our research in Amboseli National Park, which lies at the foot of Kilimanjaro in southern Kenya (fig. 2.2). Between 1977 and 1989 we monitored demographic changes in eleven social groups of vervet monkeys living at the western end of the park. We studied three of these

Figure 2.1. Adult female Teapot Dome. Adult vervets are about the size of a domestic cat.

Figure 2.2. Amboseli National Park lies in southern Kenya, at the base of Kilimanjaro. A herd of impalas grazes in the foreground.

groups intensively and continuously for the entire 11-year period and an additional three groups for a 5-year period between 1983 and 1988. The behavioral data discussed in this book are drawn from research we conducted from 1977 to 1978 (16 months), 1980 (9 months), 1983 (9 months), 1985 to 1986 (9 months) and 1988 (3 months). When we were not in the field, our colleagues Sandy Andelman, Marc Hauser, Lynne Isbell, Phyllis Lee, Shari Milgroom, and Richard Wrangham and our research assistant Bernard Musyoka Nzuma continuously observed the primary study groups. Our debt to each of these scientists is enormous, since without this teamwork and cooperation, collection of the long-term data on vervet behavior would not have been possible. Special acknowledgment is due also to Tom Struhsaker, who first studied vervets in Amboseli, and to David Klein, who initiated a study of group A in 1975 and generously provided us with information on their demography when we began our research in 1977.

Vervet monkey groups in Amboseli occupy territories ranging from 11 to 100 ha. Each territory is vigorously defended against incursions by neighboring groups (for a more detailed overview of vervet demography in Amboseli, see Cheney and Seyfarth 1983, 1987; Cheney, Lee, and Seyfarth 1981; and Cheney et al. 1988). On average, a vervet group is comprised of between one and seven adult males, between two and ten adult females, and their offspring. Females become sexually mature between 4 and 5 years of age, whereas males reach full adult size at around 6 years. Although mortality rates are high, two females are known to have lived until they were at least 17 years old.

The social structure of vervet monkeys is similar to that found in other Old World monkeys, particularly baboons (*Paplo cynocephalus*) and the various species of macaques (genus *Macaca*). Adult females typically remain throughout their lives in the groups in which they were born, maintaining close bonds with ther maternal kin even as adults. As a result, the stable core of any vervet group consists of several families of closely related adult females (mothers, sisters, and their adult daughters) and their dependent offspring.

Unlike females, vervet males migrate from their natal groups at around sexual maturity and transfer to a neighboring group. Migration in Amboseli is risky, and during adolescence the mortality rate among males is higher than that among females. Some males simply disappear, others are known to have been taken by predators, and still others are the targets of considerable aggression from adult males and females in their new groups. Perhaps to minimize the cost of transfer, many males, particularly those transferring for the first time, transfer to neighboring groups in the company of age mates or maternal brothers. Often the group they join is one that has received migrants from the males' natal group in years past. Transfer to a neighboring group minimizes the distance traveled and increases the probability that a male will have allies in his new group.

Males transfer two, three, or more times during their lives. Fully adult males are more likely to transfer alone and to travel far, crossing a number of vervet territories before joining a new group (Cheney and Seyfarth 1983). As far as we know, no male ever returns to his natal group. These characteristics of male transfer are not unique to vervets but have been documented in other Old World monkeys with similar social organizations (reviewed in Pusey and Packer 1987).

Among baboons, the female's sex skin inflates to maximum size around the time of ovulation. As a result, male baboons can identify those females who are about to ovulate, and dominant males can monopolize access to sexually receptive females (e.g., Hausfater 1975). In contrast, female vervets undergo no visible changes in physiology or behavior around the time of ovulation (Andelman 1985). Perhaps as a result, high-ranking male vervets are unable to maintain exclusive access to females around the time of ovulation. High-ranking vervet males do have an advantage over others in obtaining matings, but the benefits of high rank are not absolute: over a 5-year period more than half of all males mated with at least 80% of the females in their group (Andelman 1985).

Although Amboseli appears at first to contain a uniform habitat of open savanna broken by *Acacia* woodlands, there are in fact striking local differences in vegetation, caused mainly by the scattered presence of permanent swamps. Swamps occur where underground streams, fed by rainfall on nearby Kilimanjaro, bubble to the surface (Western 1983). Some of the swamps in Amboseli are small, no larger than 10 m^2 of slushy wet ground, whereas others are the size of lakes and maintain permanent populations of 10 or more hippopotami.

Some vervet groups in our study population lived alongside permanent swamps and had access to water throughout the year. Other groups, however, had no permanent source of water within their territories and could only drink from temporary pools that formed during the rainy seasons of November to December and April to May. Differences in the availability of food and water between "wet" and "dry" groups produced significant intergroup differences in survival, growth, reproduction, and mortality (Cheney et al. 1988). In wet groups, most females had their first offspring at 4 years of age and generally produced one offspring per year during the birth season (October to February). By contrast, although females in dry groups adhered to the same birth season, they did not usually give birth until they were 5 or 6 years of age and typically produced an offspring only once every other year. Vervets in wet groups encountered predators more often and were more likely to die of predation than were vervets in dry groups. By contrast, vervets in dry groups, while not so susceptible to predation, were more likely than vervets in wet groups to die of disease. Deaths from disease were most common in the long dry season (June to November), when the monkeys were without water for the longest period of time.

Regardless of whether they live in wet or dry groups, life for vervets in Amboseli is tough. During our study period, over 60% of all infants died during their first year. Of the remaining animals, only 27% lived to reach sexual maturity. Predation by leopards, eagles, pythons, and baboons accounted for at least 70% of all mortality (Cheney et al. 1988; Isbell 1990).

Predation

Since the mid-1960s, when Tom Struhsaker first initiated his study in Amboseli, there has been a steady decline in vervet numbers (Struhsaker 1967c, 1973, 1976). This decline is due to a marked deterioration in Amboseli's acacia woodlands, caused by the death of *Acacia xanthophloea* and *A. tortilis* trees. Acacia tree mortality can in turn be related to ecological and social factors. A rising water table (due, presumably, to increased water runoff from Kilimanjaro, the cause of which is unknown) has pushed mineral salts to the surface, killing many fever trees (*A. xanthophloea*), the vervets' primary source of food. In addition, an increasing human population outside the park now prevents Amboseli's 700 elephants from ranging beyond park boundaries as they once did. The elephants' more intensive use of the park has largely destroyed the young, regenerating trees not killed by the rising water table. In our study area, habitat deterioration reduced vervet numbers from 215 individuals in 11 groups in early 1978 to 35 individuals in 4 groups in 1988.

Wow!

Methods of Observation and Analysis

It is no accident that the most detailed behavioral studies of nonhuman primates have been conducted on ground-dwelling species like baboons, macaques, vervets, chimpanzees, and gorillas. In marked contrast to the unfortunate investigator whose job it is to follow an arboreal species that ranges high in the canopy of a dense rain forest, those of us who study semiterrestrial species of monkeys and apes have been spoiled by our ability to observe subjects at close range on the ground. After a 2- to 3-month period of habituation, all of our vervet groups became accustomed to observers on foot, and we were able to follow the animals at distances of as close as 2 to 3 m. We did not mark or capture the animals in any way but instead identified individuals by fur color, nicks and breaks in tails and ears and, most important, their facial features. The names and genealogies of all adult females and juveniles in our three main study groups are listed in appendix A.

To collect systematic data on the social behavior of vervet monkeys we used a method of observational sampling originally described by Jeanne Altmann (1974; see also Hinde 1973; Dunbar 1976; and Cheney et al. 1987 for a more complete description of sampling procedures). On a typical day we would select a group for study and drive to its sleeping trees, arriving between 7:00 and 7:30 in the morning. There, from a previously prepared list of the individuals in that group, each of us would select one animal and follow it for 10 minutes, recording all social interactions, the identities of the individuals involved and the duration of grooming bouts. These *focal animal samples* were supplemented by ad libitum data gathered on behaviors of particular interest, such as alliances (when two animals join together and direct aggression against a third) or encounters between neighboring groups. In addition, we tape recorded as many vocalizations as possible, noting the time of the call, the animals involved, and the behavior immediately preceding and following each vocalization. Observations continued until all individuals in the group had been sampled at least once. We then moved on to another group.

The advantages of focal animal sampling are that it preserves information about sequences of interactions and permits direct calculation of rates of behavior. Moreover, by keeping the length of our focal samples relatively short (10 minutes as opposed to 30 minutes or an hour, for example) we decreased the probability that a subject would go out of sight during the sample period and minimized the length of time between samples. During a typical month we sampled each individual in every group twice every 3 days, providing a fairly uninterrupted record of behavioral changes over time. Observation sessions with each group were scheduled so that by the end of each month an equal amount of data had been collected on all individuals and the data from each group had been drawn from a similar distribution of time periods. Ideally, such a sampling regime yielded, for every group, a representative cross section of vervet behavior throughout the day.

Our analysis owes much to the work of Robert Hinde (1976a, 1976b, 1983a, 1983b, 1987), who not only pointed out the limits of purely descriptive work but also introduced a theoretical framework for studying what he called the "deep structure" of nonhuman primate groups. The goals and methods of this approach are now well documented, widely used, and form the basis for much of this chapter.

Characteristic Features of Vervet Monkey Social Behavior

Kinship

Studies of the social behavior of Japanese macaques (*Macaca fuscata*) (Kawai 1958; Yamada 1963; Koyama 1967, 1970) were the first to document the close social bonds maintained among adult female kin. Later research by Donald Sade (1965, 1967) on the rhesus macaques (*Macaca mulatta*) of Cayo Santiago, Puerto Rico, provided further evidence that bonds among female kin can lead to the formation of large, closely bonded matrilines of grandmothers, mothers, daughters, and their offspring. Behavior within such matrilines is measurably different from behavior shown by the same animals toward unrelated individuals.

In vervet monkeys, as in macaques, baboons, and many other primates, bonds among kin have their genesis in the close relationship between a mother and her offspring (Kummer 1971; fig. 2.3). Vervet monkey mothers

Figure 2.3. Members of adult female Carlyle's matriline gather around Carlyle's newborn infant. Photograph by Marc Hauser.

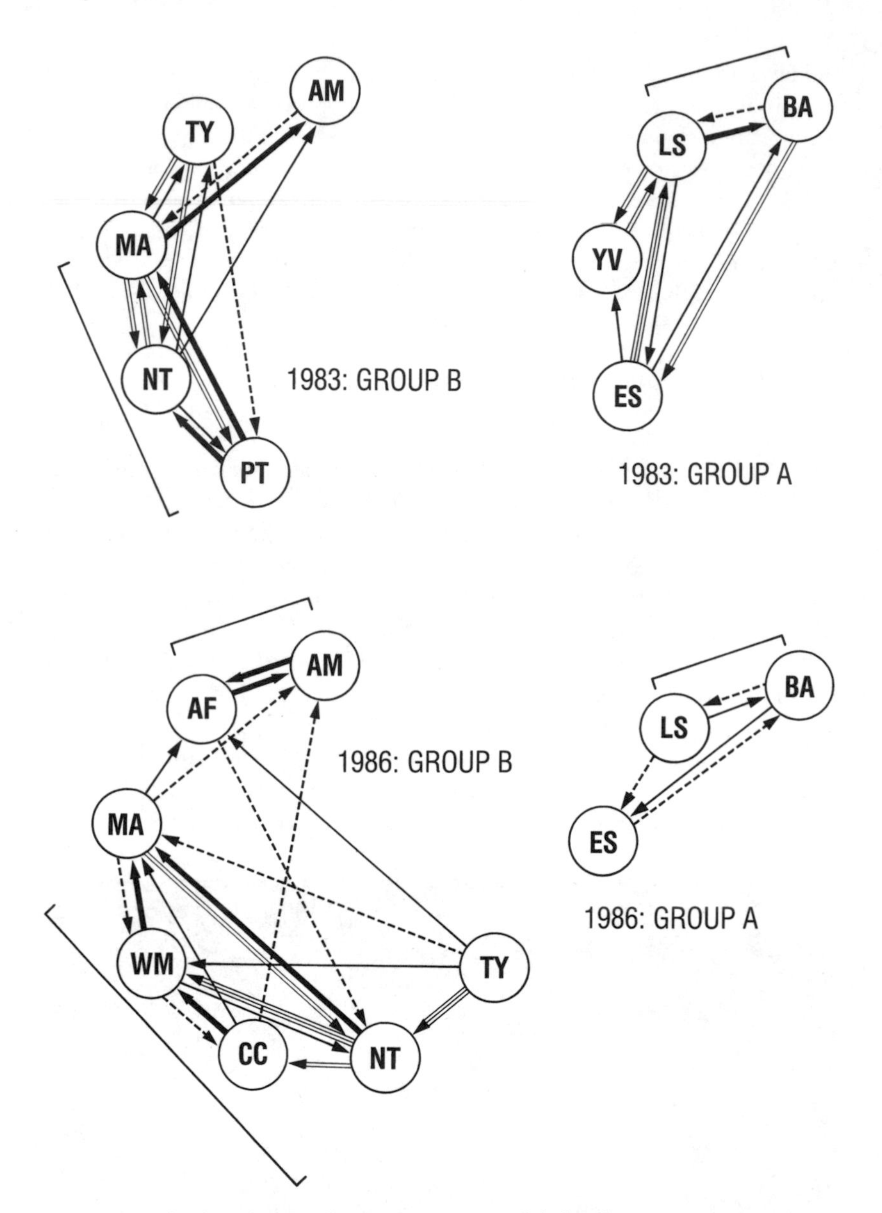

Figure 2.4. The distribution of grooming among adult females in two vervet groups during 9 months in 1983 and 9 months from 1985 to 1986. Females are designated by circles, which give each individual's initials. Females are arranged in descending rank order, reading counter clockwise from the top. Close kin (either mother and daughter or sisters) are linked by brackets. Thickness of lines indicates the amount of grooming given to each individual as follows: broken line, 5 to 10 minutes; single line, 10 to 15 minutes; double line, 15 to 20 minutes; triple line, 20 to 25 minutes; thick solid line, >25 minutes.

spend a great deal of time near their offspring, groom them at high rates (Struhsaker 1971; Lee 1983a; Fairbanks and McGuire 1985; Hauser and Fairbanks 1988) and support them by coming to their defense when the infants show signs of distress (Cheney and Seyfarth 1980). High rates of cooperative social interactions persist as the infants grow older, particularly if the infant is female. Juvenile and subadult daughters reciprocate their mother's grooming, join her in the formation of alliances, and serve as temporary caretakers of their mother's subsequent offspring (Lee 1983a; Fairbanks and McGuire 1985). As a result, bonds are formed not only between mother and offspring but also among maternal siblings. In most social groups, the majority of grooming among adult females is exchanged between close genetic relatives (fig. 2.4).

The same is true of alliances. An alliance occurs whenever two individuals are involved in an aggressive interaction and a third, previously uninvolved animal intervenes to aid one of them in attack or defense (fig. 2.5). If we calculate the rate of alliance formation between all pairs of individuals (that is, the probability that B will come to the aid of A given that A is involved in a fight with a third party), rates of alliance formation among kin are consistently higher than those among unrelated animals (fig. 2.6).

In provisioned populations where mortality rates are low, mutual cooperation between a mother and her daughters persists even beyond the mother's reproductive years. In one captive vervet monkey colony studied by Lynn Fairbanks (1988), grandmothers provided considerable assistance in the care of their grandchildren. Females with a living mother were less

Figure 2.5. Two females, Disney and Leslie, form an alliance against an adult male.

protective of their infants, and infants with grandmothers became socially independent at an earlier age. The effects of such "grandmothering" can have long-term consequences. Fairbanks and McGuire (1988) have shown that infants with less restrictive mothers are more interested in their environment and less afraid of novel situations as many as 3 years later.

Although individuals related through the maternal line interact at high rates and maintain close social bonds, there is little evidence that vervets recognize their paternal kin. Adult males interact only rarely with infants and show no special preference for those infants that are likely to be their offspring (but see Hauser 1986 for an exception). And, although there is often a high probability that two infants born in the same year to different mothers will be half-siblings (Altmann 1979), we have no evidence that paternal half-siblings are more cooperative than other age mates. As in many other species of nonhuman primates, the primary mechanism for kin recognition in vervet monkeys seems to be close association with a common mother during development (Gouzoules 1984; Gouzoules and Gouzoules 1987). The monkeys behave as if they have no way of recognizing who their fathers are and hence no way of distinguishing kinship through the paternal line. Therefore, when we speak of kin throughout this book we are referring only to individuals related through the maternal line.

Data from vervets and many other primates demonstrate that cooperative behavior among maternal kin can affect reproductive success. For example, if an infant is orphaned, its siblings are the animals most likely to care for it, and such care can make the difference between survival and death (see Lee 1983c for vervets; Hamilton, Busse, and Smith 1982 for baboons; Berman 1983 for rhesus macaques; Goodall 1983 and Pusey 1983

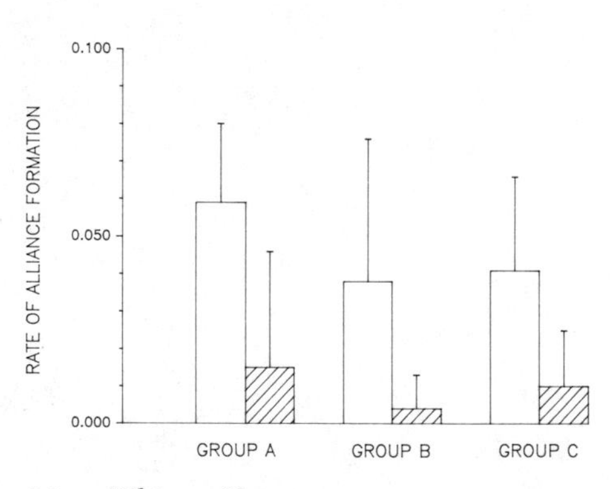

Figure 2.6. Rates at which adult females in three vervet groups formed alliances with each of their kin (open histograms) and nonkin (shaded histograms), 1985 to 1986. Data from groups A, B, and C are based on 3, 6, and 7 females, respectively. Histograms show means and standard deviations. For groups B and C, the difference between rates of alliances formed with kin and nonkin was statistically significant (two-tailed Mann-Whitney U test, P <0.05). For group A, the difference approached significance (P = 0.067).

for chimpanzees). Second, close bonds among maternal kin can help high-ranking matrilines maintain their status or allow low-ranking matrilines, as a unit, to rise in rank over others (see pp. 32–33). Since reproductive success is often correlated with dominance rank, a clear link may exist between strong matrilineal kin relations and fitness. Third, the care provided to infants by their older siblings and/or grandmothers may not only promote independence but also ease the trauma of weaning (Fairbanks 1988) and improve the mother's foraging efficiency at a time when she is under considerable nutritional stress (Whitten 1982). Finally, cooperative behavior among kin may have more subtle, indirect effects on individual reproductive success. One bout of grooming or a single alliance, while not immediately augmenting the fitness of the individuals involved, may nevertheless contribute incrementally over longer periods of time to a social relationship that itself has adaptive consequences for the participants (Hinde 1976a, 1976b; Seyfarth, Cheney, and Hinde 1978).

Close bonds among related females are also crucial to the successful defense of a group's resources against incursions by neighboring groups. A vervet group can be thought of as an alliance of several matrilines against other groups of allied matrilines (Wrangham 1980). In the absence of strong alliances within and between families, a group may be unable to defend its resources, to the detriment not only of the females' reproductive success but also of their survival. Female kin are more likely to aid each other than to help unrelated group members during aggressive intergroup encounters, and groups that include a number of large matrilines seem better than smaller groups at defending and even expanding their ranges (Cheney and Seyfarth 1987).

Dominance

One of the most common social interactions among vervets occurs when one animal approaches another and the other moves away. Such approach-retreat interactions (Rowell 1966) are a subtle form of competition and occur in a variety of contexts. In some cases, the dominant animal may take over the subordinate's food, resting place, mate, or grooming partner. In other cases, no obvious resource is being contested. In vervet monkeys, the direction of approach-retreat interactions between any two individuals is relatively stable over time. If one individual is dominant to another on a given day, the chances are good that she will still be dominant days, weeks, months, or even years later. In our study we define dominance in terms of the direction of approach-retreat interactions between two individuals and use the term dominance to describe a relationship in which this direction is stable and predictable over time.

Dominance in vervet monkeys is a useful concept because, once defined in terms of the direction of approach-retreat interactions, it has considerable predictive value in many other contexts. If one vervet, for example, is dominant to another when competing for food, the same animal will also be dominant when competing for grooming partners, mates, or resting sites (Seyfarth 1980; Andelman 1985). If one animal is dominant to another in approach-retreat interactions, she will also be dominant in fights. Finally, in most vervet relationships the subordinate both grooms the dominant and forms alliances with her at higher rates than vice versa (see pp. 35–39). These predictable features of dominance relations, however, are not found in all species. Among male baboons, for instance, dominance relations are quite unpredictable from one context to the next (Strum 1982; Smuts 1985). Among vervets, dominance relations between males and females are similarly context dependent: males are generally dominant to females in one on one, or dyadic, interactions, but two allied females can easily drive away an adult male.

Dominance relations among vervets are not only consistent across contexts but also transitive: if A is dominant to B and B is dominant to C, A is invariably dominant to C. As a result, males and females in each group can be arranged in a linear dominance hierarchy based on the direction of approach-retreat interactions.

Among female vervets, as among female baboons and macaques, there is a clear link between dominance rank and kinship; the dominance hierarchy among females and juveniles is in fact a hierarchy of families (reviewed by Walters and Seyfarth 1987). At the top of the hierarchy is the oldest adult female in the highest-ranking family. Her offspring rank immediately below her, with daughters ranked in inverse relation to their ages (unpublished data). Next in rank comes the oldest adult female in the second-ranking family, her offspring, and so on. The same relation between kinship and dominance rank is also common in other populations of vervets (Horrocks and Hunte 1983; Fairbanks and McGuire 1984). The phenomenon of rank by family has been most thoroughly documented among the rhesus macaques of Cayo Santiago, where infant survival is high and adult life span is long (Sade 1965, 1967, 1972a; Datta 1983a). Under these conditions, family genealogies can be quite extensive and include grandmothers, mothers, grandoffspring, and even great-grandoffspring (Berman 1982).

Acquisition of rank. Among infants of both sexes, rank is acquired at least partly through alliances with family members (Cheney 1977, 1983a; Horrocks and Hunte 1983; Lee 1983b). In over 20% of the aggressive interactions involving female or juvenile vervets, a third individual intervenes on the behalf of the aggressor or the recipient of the aggressive act and helps her alliance partner drive away the opponent. The majority of alliances (65%) are formed among close genetic relatives, either a mother and her offspring or two siblings. In a typical interaction involving two playing infants, one or both of the infants will scream when play becomes rough, and both mothers will come running. The dominant mother will then threaten or supplant the subordinate mother and her infant, and the subordinate pair will retreat. In these circumstances the infant quite literally attains the rank its mother acquires for it.

There is, however, more to rank acquisition than a high-ranking mother simply bullying subordinates into accepting her offspring as dominant (Walters and Seyfarth 1987). For example, from a very early age group members behave differently toward the infants of high- and low-ranking mothers. High-ranking infants are often more sought after as play and grooming partners, and in many other ways interactions with them are carried on in a more careful manner than are interactions with infants of lower rank (Lee 1983a, Whitten 1982; see also Nicolson 1987). Perhaps as a result, even juveniles whose mothers have died often acquire their mothers' rank (Lee 1983c; Berman 1983). In rhesus macaques (Datta 1983b), juveniles consistently challenge adults who rank below their mothers but rarely challenge adults who rank above their mothers. This suggests that a juvenile monkey learns about her "expected" dominance relations with others at a very early age. She seems to do so both through her own experiences and by observing interactions between her mother and other group members (Datta 1983b; see also Gouzoules 1975; Altmann 1980; Berman 1980; Horrocks and Hunte 1983).

Maintenance of rank. Although rank acquisition in infancy is similar for males and females, the maintenance of dominance rank among older animals is strikingly different in the two sexes. For females, the behaviors that maintain dominance rank are subtle and difficult to describe precisely. High-ranking females are usually not larger, heavier, or, as far as we can tell, in better physical condition than others, nor are high-ranking females necessarily more aggressive (Seyfarth 1980). Although females apparently do not maintain their status by size or aggressiveness, however, the threat of alliances among them does seem important (Chapais 1988a, 1988b). In this respect, high-ranking females have two advantages. First, in many populations (though not among the Amboseli vervets), high-ranking females have larger families (e.g., Fairbanks and McGuire 1984; Drickamer 1974; Sade et al. 1976). Since most alliances are formed among family members, the "implied threat" of receiving aggression from many individuals may be enough to produce deference in lower-ranking animals. Indeed, in a study of rhesus macaques, Datta (1983c) found that high-ranking aggressors were especially likely to receive their relatives' support against lower-ranking opponents when they were young and small. Though these interventions did not affect the immediate outcome of the dispute (since the younger animal was already winning the fight), they may have allowed high-ranking females to aid their younger relatives in maintaining their rank against potential challenges.

Second, even when high-ranking families are not larger, high-ranking females are better than low-ranking females at recruiting allies from among *unrelated* individuals (see pp. 35–36). Here again, the threat of retaliation from a large number of animals may discourage low-ranking animals from challenging the established hierarchy.

Compared with male rank, female dominance rank is relatively stable over time. Over a 4-year period, for example, adult females in three vervet groups changed ranks at the rate of one rank change per female every 10 years. The comparable figure for males was seven times higher (Cheney 1983a). The stability of the female dominance hierarchy is also illustrated by the fact that, as a rule, alliances cannot override existing dominance ranks: two females acting together can only drive away a third if the third animal ranks lower than at least one of them (Cheney 1983a; for comparable data on rhesus macaques, see Datta 1983c; Walters and Seyfarth 1987). A logical question, then, is why do females form alliances at all, since in most cases alliances do not help an individual gain anything that she could not already achieve on her own?

Despite their relative stability, female ranks are not immutable. Moreover, the circumstances leading to female rank reversals provide some explanation of why coalitions among females are so important. Among baboons, macaques, and vervets, most rank reversals occur when a female with few or no close kin loses her status as the result of persistent, aggressive challenges by a larger, lower-ranking matriline (Koyama 1979; Chance, Emory, and Payne 1977; Gouzoules 1980; Smuts 1980; Ehart and Bernstein 1986; Samuels, Silk, and Altmann 1987). Occasionally, events due largely to chance—not uncommon in small groups—increase the likelihood that a lower-ranking matriline can successfully mount such a challenge. For example, in our group B, adult female Marcos, who originally ranked sixth out of seven females, gave birth to three surviving daughters between 1977 and 1981. During the same period, adult female Duvalier, who ranked fifth out of seven (higher than Marcos), gave birth to two surviving daughters. In 1981, however, Duvalier was killed and eaten by a python. For two and a half years Duvalier's two young daughters retained their ranks above the Marcos matriline, but during this time Marcos gave birth to two more daughters, and her oldest daughter also had two daughters. Then, in 1984, one of Duvalier's daughters died. The seven females in the Marcos matriline immediately began to mount aggressive challenges against the sole surviving female in the Duvalier matriline. Within a few months the entire Marcos matriline had risen in rank above her.

This example illustrates that, while successful challenges to an established female hierarchy are undoubtedly rare, they do occur. And challenges to the established hierarchy are very likely if high- or middle-ranking females are unable to recruit kin as allies (Chapais 1988b). The maintenance of dominance rank among females may seem subtle to us because aggression occurs at low rates and actual rank reversals are infrequent. The circumstances surrounding known rank changes, however, indicate that the process of "challenge and defend," though subtle, is going on among females all the time.

Although a young male's dominance rank depends on his female relatives' assistance, as he grows older (and particularly after he leaves his natal group) his rank increasingly depends upon age, size, strength, and other determinants of fighting ability. Males typically are highest ranking during the years when they have reached full adult size and have gained social experience (roughly ages 6 to 10 in vervets), with rank declining thereafter (Henzi and Lucas 1980; for data on baboons see Packer 1979; Rasmussen 1980; Busse and Hamilton 1981; Strum 1982; Bercovitch 1988). As noted earlier, dominance ranks among males are less stable than among females, and a male may change rank many times during his life. Moreover, in some species, coalitions can have a marked effect on a low-ranking male's ability to gain access to resources from which he might otherwise be excluded. Among baboons, for example, two allied low-ranking males can drive away a male who normally ranks higher than either of them (Hall and DeVore 1965; Packer 1977; Noe 1986; Bercovitch 1988). Male-male alliances among baboons therefore differ strikingly from female-female alliances, which seldom result in even temporary reversals of the existing dominance hierarchy. Given the prevalence of male-male alliances in many baboon populations, as well as the clear advantages that low-ranking males potentially derive from them, it is surprising that these alliances are extremely rare among vervets and macaques. During our entire study only one pair of males ever formed alliances with any regularity, even though many of the groups included brothers or males who had originally transferred from the same natal group.

The function of rank. Why should male and female monkeys strive to maintain high dominance rank? In vervets the advantages of high rank, as measured by its immediate behavioral consequences, are clear. High-ranking individuals have preferred access to resting places, food, mates, social partners, and other desirable resources that are in limited supply.

Motive

The long-term, evolutionary consequences of high rank are more difficult to assess. In our study, a female's life span and the survival of her offspring were the major factors influencing reproductive success (Cheney et al. 1988). The major causes of mortality were predation (which accounted for at least 70% of all deaths) and disease or starvation. Predation accounted for the majority of deaths in the top three rank quartiles, whereas animals in the lowest-ranking quartile were equally likely to die of predation or illness. In this respect it is possible that high-ranking females had an advantage: the probability that an individual would be taken by a predator was unrelated to her rank, whereas the probability that a female would die of disease or starvation was greatest for individuals in the lowest rank quartile. In Whitten's (1983) study of vervets in the Samburu Reserve in Kenya, dominant females had greater access to clumped food resources and gave birth to more infants. Nevertheless, Whitten found no relation between a female's dominance rank and the survival of her infants, whereas in our study neither infant survival nor life span was greater for high-ranking females. Similarly, among males there was no significant correlation between dominance rank and frequency of copulation.

Reviews that assess the relation between dominance rank and reproductive success in nonprimates (Dewsbury 1982; Clutton-Brock 1988) or in primate species other than vervets (Fedigan 1983; Robinson 1982; Silk 1987) provide more consistent evidence that high status is correlated with high reproductive success. Among both males and females a positive association often exists between rank and one or more measures of reproductive activity. Across many studies and species, positive correlations outnumber the negative ones. From this we may conclude that, on average, high-ranking individuals of both sexes obtain a reproductive advantage by virtue of their rank (Silk 1987).

At the same time, in some populations, including the Amboseli vervets, dominance ranks of males and females are unrelated or even negatively correlated with reproductive success (McGinnis 1979; Taub 1980; Strum 1982; Smuts 1985). In the case of females, we need to learn more about the relationships among different measures of reproductive activity. Reproductively successful females are generally assumed to be those who begin to reproduce at an early age, continue to reproduce for many years, give birth at shorter intervals, produce more infants, or produce healthier infants. Some of these measures, however, may be negatively correlated with each other; for example, in cases when interbirth intervals are longer after surviving infants than after those that die (Cheney et al. 1988; see also Altmann, Altmann, and Hausfater 1978; Altmann, Altmann, and Hausfater, 1988 for baboons).

It is often difficult to measure male reproductive success because of the problem of establishing paternity. Genetic studies have in some cases indicated that there is not always agreement between indirect behavioral measures of reproductive activity, such as who copulates most often, and the actual father of a given infant (Estep, Johnson, and Gordon 1981; Stern and Smith 1984; see also Duvall, Bernstein, and Gordon 1976). Furthermore, simple relations between rank and reproductive success are complicated by the effects of a male's age, length of tenure in the group, female preferences, and male parental care, all of which have been shown to affect a male's access to mates. Finally, as Hausfater (1975) points out, because males change rank frequently during their life times it is necessary to obtain data on a male's *life time* reproductive activity before all the factors affecting reproductive success can be fully assessed. For the moment, it seems safest to conclude that, among both males and females, dominance rank explains some but by no means all of the variation in reproductive success (Silk 1987).

Status striving. It is depressingly common to find popular articles or television shows on animal behavior in which a commentator explains knowledgeably, "Among these animals life would be a constant struggle for survival were it not for the dominance hierarchy, which sets the rules about who may challenge whom and thus brings peace to the group." In the commentator's mind, it is as if the animals have gotten together and agreed that one should rank first, another second, another third, and so on, leading to a kind of Panglossian paradise in which no one has to fight because everyone accepts the status quo. Ethologists cringe when they hear this sort of thing because nothing could be further from the truth.

Vervet monkeys, for example, not only challenge higher-ranking individuals in attempts to improve their status but also employ a variety of subtle strategies apparently designed to win some of the benefits associated with high rank even if their actual status remains low. Evidence for status striving comes from two common types of social interactions: alliances and grooming.

Alliances. Since aggressive interactions occur often, individual vervets are given many opportunities to "choose" whether to intervene on behalf of another. In an analysis of data from 1977, 1978, and 1980, we found that females received support in aggressive disputes in direct relation to their rank (Cheney 1983a). In other words, in addition to showing a preference for aiding kin, female vervets also preferred to form alliances with high-ranking individuals. In figure 2.7, we present data from 1985 and 1986, when all kin relations were known. Here again, in addition to forming alliances at high rates with their kin, adult females and juveniles formed alliances with higher-ranking individuals at a higher rate per opportunity than they formed them with lower-ranking individuals.

We have argued that preferences for high-ranking alliance partners represent attempts to form long-term bonds with high-ranking individuals (Cheney 1977; Seyfarth 1977). Though such bonds may not actually increase an individual's rank, they may nevertheless allow lower-ranking animals to acquire some of the benefits enjoyed by higher-ranking group members. This argument rests on three points.

First, in species such as baboons, macaques, and vervets, high-ranking females win the majority of the disputes they enter and have preferential access to scarce resources. Second, even among unrelated animals there is often a strong positive correlation between alliances and other cooperative behavior such as grooming or tolerance at feeding sites (e.g., Packer 1977; Seyfarth 1977; Colvin 1983b). Animals that form alliances at high rates are also those that groom, feed, or play together most often (Seyfarth 1980; Cheney 1983a; Lee 1983d; see also Walters and Seyfarth 1987, who review data on other species). Third, we argue that these correlations reflect *causal* relations: if a low-ranking animal forms an alliance with a higher-ranking one, the higher-ranking animal will subsequently be more likely to support her partner in an alliance or allow her partner access to a resource she otherwise could not obtain. Alternatively, the higher-ranking animal may simply be less aggressive in the future toward her partner or her partner's offspring (Silk 1982). Experiments with vervets have shown that when one individual grooms an unrelated animal, the recipient is more willing to respond to the groomer's subsequent solicitations for support (Seyfarth and Cheney 1984a; see also chapter 3). Of course, completely reciprocal relations between females of different ranks are not inevitable or even highly predictable, at least partially because of the asymmetrical benefits that each partner has to offer the other: a low-ranking female has much more to gain from a high-ranking female than vice versa. Even if high-ranking animals reciprocate their low-ranking partners only rarely, however, it may still be advantageous for low-ranking females to strive to establish bonds with those who can potentially return the greatest benefit (Cheney 1983a; Seyfarth and Cheney 1988b).

Grooming. Grooming (fig. 2.8) is the most common affinitive behavior among primates. Although one function of grooming is obviously the re-

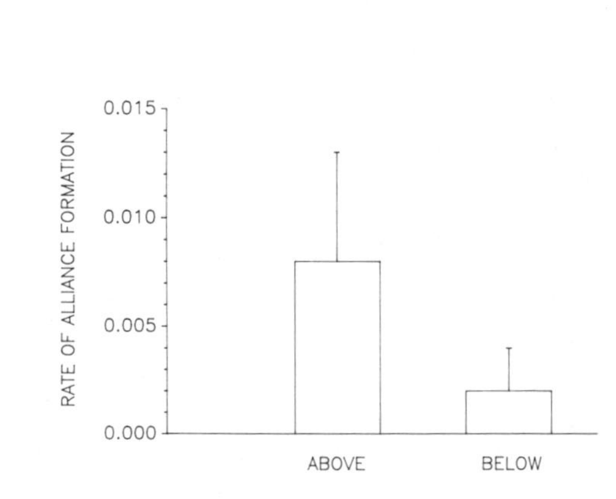

Figure 2.7. Rates at which adult females formed alliances with each unrelated adult female who ranked above and below median rank from 1985 to 1986. Data are drawn from all adult females ($N = 7$) *who had an opportunity to interact with high- and low-ranking adult female partners. All seven individuals formed alliances at a higher rate with females above median rank (two-tailed Wilcoxon test,* $P < 0.05$).

Figure 2.8. Members of the Borgia matriline groom each other.

moval of ectoparasites (e.g., Freeland 1976; Hutchins and Barash 1976; McKenna 1978), a variety of evidence supports Carpenter's original (1942) view that the primary function of grooming is to aid in establishing and maintaining close social bonds (see also Marler 1965). Across many primate species, for example, there is no correlation between time spent grooming and body size, whereas there is a significant correlation between time spent grooming and group size (Dunbar in press). Among vervets, animals who receive the least grooming from others do not groom themselves most often (Seyfarth 1980), and in all species individual monkeys show strong preferences for particular grooming partners.

If grooming functions primarily to establish and maintain social bonds, one would predict that a large portion of a female's grooming would be given to her kin. This is in fact the case (see fig. 2.4). Following the argument presented for alliance formation, however, we might also predict that grooming should provide evidence of status striving. Females should groom those of high rank, since grooming with high-ranking individuals potentially brings the greatest benefit.

While the formation of alliances, however, is unconstrained by competition (any number of animals can join an alliance—no one is prevented form doing so), in vervets and other Old World monkeys, grooming is primarily done in pairs. As a result, there can be competition for grooming partners. We might therefore predict that the interaction between competition and a preference for high-ranking partners would lead both to frequent grooming of high-ranking animals and to frequent grooming between animals of adjacent rank. This is because high-ranking females, who are unconstrained by competition, should distribute the majority of their grooming of nonkin to others of high rank. Middle-ranking females, in turn, should meet competition for access to those of high rank and compromise by grooming nonkin of middle rank. Finally, low-ranking individuals should meet competition for access to all individuals and should consequently groom those of low rank (Seyfarth 1977, 1980, 1983).

Because adjacently ranked females are often kin, and because kin relations are generally unknown in the early stages of any long-term study (ours included), it is often difficult to tell whether high rates of grooming between any two individuals result from kin-based preferences or from the compromises associated with competition for high-ranking animals. Bearing this in mind, we can make three predictions about the distribution of grooming and related behaviors in vervets and other monkeys. First, there should be high rates of grooming among kin. Second, when females groom nonkin, they should prefer high-ranking partners. Finally, even when adjacently ranked females are unrelated, they should groom each other at high rates.

To a great extent, these predictions are supported by data from studies of vervets and other monkeys. Among vervet females, for example, the majority of grooming occurs among kin. Nevertheless, these same females compete less over access to their own kin than for the opportunity to groom high-ranking individuals.

In our study, we defined a competitive interaction over access to a grooming partner as occurring whenever a female approached two grooming partners, supplanted one of them, and then groomed the other. In figure 2.9, we illustrate the distribution of such competitive interactions among females in three social groups during 1977, 1978, 1980, 1983, and 1986. Females of all dominance ranks supplanted each other over access to grooming partners, and the most attractive partners were high-ranking individuals.

High- and middle-ranking females also distributed their grooming as predicted. High-ranking females groomed unrelated animals in direct relation to their ranks, grooming high-ranking females the most and low-ranking females the least (fig. 2.10). Middle-ranking females, in turn, groomed those ranking lower than themselves in direct relation to their ranks and those ranking higher than themselves in inverse relation to their ranks (fig. 2.10). We had predicted that low-ranking females would also groom higher-ranking females in inverse relation to their ranks, so that they would groom the highest-ranking female less than they groomed middle-ranking females. However, low-ranking females distributed their grooming roughly equally among those higher ranking than themselves (fig. 2.10). Apparently, competition did not inhibit the grooming of low-ranking females as much as predicted, and they were able to gain some access to the highest-ranking members of their group.

Patterns of grooming that reflect an interaction among kinship, competition, and the attractiveness of high-ranking animals are not restricted to vervet monkeys. As already noted, frequent grooming among kin is widespread in primates. And, although some studies have not found a correlation between rank and attractiveness as a grooming or alliance partner (for data on baboons, see Rowell 1966 and Walters 1986; for data on gelada baboons [*Theropithecus gelada*] see Dunbar 1983a), this correlation has been documented in many others (Fairbanks 1980 for vervet monkeys; Seyfarth 1976 and Pereira 1988 for baboons; Kummer 1968 and Stammbach 1978 for hamadryas baboons [*Papio hamadryas*]; Kummer 1975 for gelada baboons; Sade 1972b, Chapais 1983, de Waal and Luttrell 1986 for rhesus macaques; Oki and Maeda 1973 and Mehlman and Chapais 1988 for Japanese macaques; Silk, Samuels, and Rodman 1981 for bonnet macaques [*Macaca radiata*]).

Status striving, moreover, is not limited to adult females, nor is it manifested only in grooming and the formation of alliances. Frans de Waal (1982, 1987), who first used the term, describes a variety of interactions among male and female chimpanzees over a 2-year period that cannot be

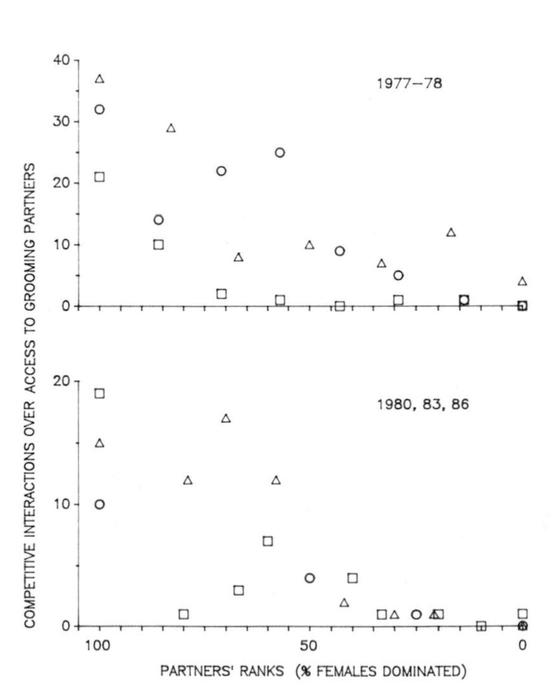

Figure 2.9. The relation between a female's dominance rank and the number of competitive interactions that occurred over access to her as a grooming partner. Female rank is measured as the proportion of other adult females dominated (100% = highest ranking). Data in the top graph are drawn from 15 months in 1977 to 1978, and taken from Seyfarth 1980; data in the bottom graph are pooled from 9 months' observation in each of the years 1980, 1983, and 1985 to 1986. Females in group A are depicted by circles, in group B by triangles, and in group C by squares.

explained except by assuming that the animals were selectively forming bonds (by means of grooming, alliances, tolerance, reconciliation, and reassurance) with those from whom they could potentially derive the most benefit (see also Nishida 1983).

More recently, Eduard Stammbach (1988a, 1988b) trained low-ranking females in a captive group of longtailed macaques (*Macaca fascicularis*) to press a series of levers to obtain a highly desirable food (popcorn) from a machine. Each low-ranking "specialist" was trained individually, and an experimenter controlled the setting of the machine so that on any given day only one of the specialists and no other group member was able to obtain popcorn. In the first experimental test period, when the food-dispensing machine was still active, high-ranking females began following specialists whenever they approached the machine. They also refrained from chasing

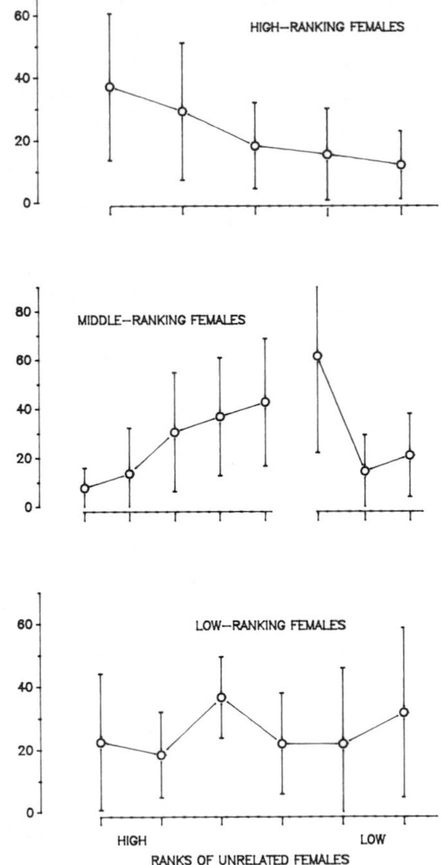

Figure 2.10. The distribution of grooming given by high-, middle-, and low-ranking females to unrelated female partners. Top graph shows that high-ranking females groomed others in direct relation to their ranks; middle graph shows that middle-ranking females groomed those lower ranking than themselves in direct relation to their ranks and those higher ranking than themselves in inverse relation to their ranks; bottom graph shows that for low-ranking females there was no relation between a partner's rank and the amount of grooming given. Graphs show means and standard deviations for all females in 1983 and 1985 to 1986.

specialists from the machine. In many dyads grooming and spatial proximity increased, with the greatest increase occurring in those dyads where the high-ranking female obtained the greatest amount of food as a result of the specialist's behavior. More interestingly, even after the food machine had been turned off and there was no prospect of an immediate reward, high-ranking females continued to groom and remain near specialists. Stammbach concludes, "At least some of the group members became aware of the skills of the specialists and adapted their behavior accordingly as if to maximize benefits" (1988b; 265).

Complementing this experimental evidence, we offer two case histories from vervet monkeys that illustrate how an apparent awareness of the real or potential "power" of certain individuals can affect social interactions. In both accounts certain females consistently violated the relation between high rank and attractiveness, receiving either more or less grooming than might be predicted given their ranks. In each case, it is as if other females recognized these individuals' future potential to recruit allies and rise in rank.

Our first account focuses, once again, on Marcos, a female who through guile and the production of many daughters and granddaughters gradually rose to prominence in group B. In 1977, when we began our research, group B had an odd demographic composition; six of the seven adult females were nulliparous, and there were no juveniles in the group. In descending rank order the adult females were Bokassa, Somoza, Amin, Franco, Duvalier, Marcos, and Pinochet. Although kinship among these females was not known, we strongly suspected, from association patterns and physical resemblance (Walters 1981), that the group included two matrilines of close relatives, probably sisters. The first matriline contained the three highest-ranking females—Bokassa, Somoza, and Amin; the second contained the two lowest-ranking females, Marcos and Pinochet.

Over the next 10 years, Marcos not only produced more surviving offspring than any other female in the group but also produced more surviving daughters. As a result, during each of the subsequent periods when we collected grooming data (1980, 1983, and 1986), Marcos always had more living female relatives than any other female in the group. Other females seemed to recognize that Marcos, because of her larger family, was destined to recruit more allies, and they groomed her at consistently higher rates than might have been expected given her low status. As we noted earlier, in 1984 Marcos did rise in rank, becoming dominant over the lone descendant of Duvalier. By 1986, Marcos was the second-ranking female in the group, subordinate only to Amin and her adult daughter, Aphro. In 1987 both Amin and Aphro's daughter died, leaving Aphro as the sole survivor of the highest-ranking matriline. Almost immediately, Marcos' matriline rose over Aphro to become the dominant matriline. In 1988, Marcos received more grooming than any other female in the group.

Another example of the monkeys' apparent ability to assess each other's relative worth comes from Borgia, a female member of group A. In 1977 Borgia had one juvenile daughter, Leslie, and was ranked third in a group of eight adult females. Over the next 4 years many low-ranking females in group A died (Wrangham 1981), and by 1981 Borgia and Leslie ranked second and third in a group of four females. The dominant female, Escoffier, was young and had no living female kin; her sister had been killed by baboons in 1977 and her mother was killed by a leopard in 1980. Borgia and Leslie soon rose in rank over Escoffier to become first and second ranking. Escoffier, in turn, fell to the bottom of the hierarchy. Despite Escoffier's low status, however, Borgia contined to groom her at higher rates than she groomed any other female, including her own daughter. Borgia behaved as if she remembered Escoffier's former status and recognized that Escoffier might once again rise in rank when she began to produce offspring of her own. In 1986 Leslie died, and Escoffier and her juvenile daughter immediately rose in rank above Borgia, who ended her long life at the bottom of the female dominance hierarchy.

These interactions reflect the tension between those who have power at the moment and those who might become powerful. They suggest that although the dominant female always has the most to offer, losing her female kin makes the prospects for maintaining this status slim. High-ranking females without female kin received less support and grooming than high-ranking females with kin, whereas middle- or low-ranking females with large families received more grooming than might be expected, even before their ranks changed. The animals behaved as if they were hedging their bets, forming bonds with those who had power at the moment but at the same time retaining bonds with those who might be powerful in the future (for further examples see de Waal 1982; Chapais 1988b; Raleigh and McGuire 1989).

Of course, we have no idea what mechanisms—cognitive or otherwise—lay behind Borgia's behavior toward Escoffier or the deference shown by others toward the rising star Marcos. These two anecodotes *suggest* planning, foresight, and an ability to draw distinctions between current and future status, but descriptions alone are obviously insufficient proof that such cognitive mechanisms actually exist. In chapters 3 and 8 we consider in greater detail what monkeys really know about one another.

Reciprocity

One of the central problems in ethology concerns the evolution of altruism (Wilson 1975). Although Darwin's (1859) theory of natural selection predicts that animals should always act selfishly (in their own reproductive interests), naturalists including Darwin have long recognized that individuals can be altruistic, giving up their lives, or at least their reproductive success, for the benefit of another. The theory of kin selection (Hamilton 1964) provides one solution to the problem of altruism by showing that natural selection can favor cooperation between two or more individuals when the animals involved are close genetic relatives. A second theory, developed by Robert Trivers (1971), addresses the evolution of altruistic behavior among unrelated individuals. Trivers argues that natural selection will favor individuals who are selectively altruistic, directing cooperative behavior only to those who cooperate in return. Such reciprocal altruism can evolve, Trivers argues, whenever three conditions are met: when the benefit to the recipient is greater than the cost to the altruist; when individuals have a high probability of encountering each other again; and when individuals can remember who has behaved altruistically to them in the past and adjust their future behavior accordingly (Trivers 1971, 1985; see also Axelrod and Hamilton 1981).

The theory of reciprocal altruism is of particular relevance to work on social intelligence because the evolution of reciprocity seems to require a number of complex mental operations, including individual recognition, memory, calculation of the costs and benefits of different interactions and, perhaps most important, the ability to detect cheaters. Consequently, reciprocal altruism may have exerted strong selective pressure on at least some aspects of social intelligence, both in nonhuman primates and in other species (e.g., Wilkinson 1984; Trivers 1971, 1985). We return to this point in chapter 3.

Evidence of reciprocal altruism in nonhuman primates was first presented by Craig Packer (1977) in a study of alliance formation among male baboons. Packer found that a male involved in an aggressive interaction with another often solicited support from a third, previously uninvolved individual. In many cases, for example, one male solicited help from another, and the two attackers together drove away a third who had been consorting with an estrous female. The female then formed a consortship with the male who had originally solicited help. Packer observed 140 solicitations for help, 97 of which resulted in support. Directly supporting Trivers' theory, individual males tended to give support most often to those who supported them at the highest rates.

Although some later studies of baboons (e.g., Smuts 1985) have replicated Packer's observation of reciprocal, symmetrical alliances, other have not (e.g., Rasmussen 1980; Smuts 1985; Bercovitch 1988). Noe (1986), for example, found that males formed alliances most often with closely ranked males, who were most likely to reciprocate. Rates of reciprocity were not particularly high, and low-ranking males appeared to be making the best of a bad situation. This does not necessarily mean, however, that alliances were not based on reciprocity. Just as is true for females, differences in dominance rank and fighting ability can lead to asymmetries in the benefits that different males can offer each other (Seyfarth and Cheney 1988b). Noe emphasizes that an alliance partnership in which a high-ranking male only helps his partner once for every two times that his partner helps him might still be entirely reciprocal if each individual invests relatively the same amount in terms of benefits and costs.

If vervet monkeys distribute at least part of their friendly behavior among others in the hope of receiving benefits in return, the animals may be engaged in a form of reciprocal exchange. Reciprocity certainly seems to be involved in the close bonds among family members, since the highest rates of grooming and alliance formation occur among kin. Data from unrelated animals, however, also support the view that cooperative behavior is reciprocally exchanged between individuals, even before the asymmetries in rank-related benefits are taken into account. In 1985 and 1986, for instance, when both kin and nonkin were known, we tested whether *unrelated* vervet females formed alliances at the highest rates with those individuals who groomed them most often. Eight females had a sufficient number of unrelated partners to allow this test to be conducted. In seven of eight cases the correlation between alliances formed and grooming received was strongly positive (table 2.1). Moreover, five of the eight females formed alliances at the highest rate with the same individual who groomed them most often.

Table 2.1 The Relation between Grooming and Rates of Alliance Formation among Unrelated Female Vervet Monkeys

Females	Number of Partners	Correlation between Rate of Alliances Given and Rate of Grooming Received
CY*	4	0.950
AO	4	0.750
AC*	5	0.750
LO*	5	0.750
AU	5	0.725
AM*	5	0.750
AF*	5	0.750
TY	6	−0.114

Note: Values given are Spearman rank correlation coefficients. An asterisk indicates a female who formed alliances at the highest rate with that unrelated individual who groomed her most often.

Sexual Attraction

In addition to kinship, dominance, and reciprocity, many features of vervet monkey social behavior arise as a result of the attraction between males and females. There is nothing surprising in this fact, and we will not belabor the point with unnecessary examples.

To illustrate the role of sexual attraction, consider the timing and consequences of group transfer by males. As we mentioned earlier, at around sexual maturity vervet males leave their natal groups and transfer to neighboring groups. Considerable evidence indicates that the proximate cause of male dispersal in vervets (as in other Old World monkeys; see Pusey and Packer 1987) is sexual attraction to females in other groups. For example, when two groups meet at the border of their ranges, their behavior toward each other is usually aggressive. If the intergroup encounter occurs during the breeding season, however, males are less aggressive and frequently attempt to groom, mount, and copulate with females in the other group. Partially as a result, male transfers occur primarily during the breeding season (Cheney 1983b; Henzi and Lucas 1980). By transferring, males usually increase their dominance rank relative to other males, decrease the rate at which females rebuff their mating attempts, and increase their copulation rate (unpublished data).

Once he has entered a new group, a male interacts with other males primarily in the context of competition for access to sexually receptive females. Male-male aggression, for example, occurs at the highest rate during the breeding season, when the probability of changes in male dominance rank is also the greatest (unpublished data). As with our data on the overall characteristics of male transfer, these observations are not unique to vervets. Smuts (1987a, 1987b) reviews evidence that competition for access to mates accounts for a considerable proportion of male-male interactions in a variety of primate species.

Among the Amboseli vervets, bonds between males and females are relatively weak, and individual males and females seldom form strong associations, in or out of the breeding season (Andelman 1985). This is not true, however, of all Old World monkey species, or even of captive vervet monkeys (Keddy 1986). Among baboons and some species of macaques, for example, competition for access to females and strong mutual attraction between males and females (in the form of grooming, foraging, and resting together) occur even during phases of the female's reproductive cycle when there is no copulation (Seyfarth 1978a, 1978b; Collins 1981; and Smuts 1985 for baboons; Chapais 1981 for rhesus macaques; Fedigan 1982 and Takahata 1982 for Japanese macaques). Long-term "friendships" between males and females are particularly well documented in baboons; in these relationships, individuals can maintain high rates of grooming, proximity, and alliances over periods of 2 years or more (Smuts 1985).

Barbara Smuts argues that close bonds between males and females are ultimately related to reproductive success. She provides evidence that females and their offspring are protected by male "friends" from aggression by conspecifics and possibly predators, whereas males gain access to females with whom they otherwise might not be able to mate. Smuts also points out, however, that while the ultimate benefit of friendships may be increased reproductive success, the persistence of such strong bonds for long periods when females are lactating or pregnant demonstrates that sexual attraction is not the only factor affecting the pattern of interactions among male and female primates. It is unclear why vervet monkeys, whose social behavior is otherwise so similar to that of baboons and macaques, do not exhibit long-term friendships between males and females.

Group defense

Not all social interactions among vervet monkeys involve members of the same group. In Amboseli, vervet groups are territorial, and the members of each group range almost exclusively within an area averaging 28 ha in size (Cheney, Lee, and Seyfarth 1981; Cheney and Seyfarth 1987). Vervets spend a considerable amount of time monitoring the movements of neighboring groups and defending their territory against outsiders (Cheney 1987). Typically, when the members of one group first spot another group they give a loud, trill-like vocalization called a *wrr,* which apparently functions to inform members of their own group and to warn members of the neighboring group that they have been spotted. After this initial contact the animals in both groups may simply go back to what they were doing, occasionally monitoring activity among their neighbors for the next few hours until the groups are out of sight of each other. Particularly if there is water, new grass, or a tree with ripe fruit or flowers near the border, however, the groups may come together and the encounter can escalate into threats, chasing, and fighting among adults (fig. 2.11).

Adult females play a major role in group defense (Cheney 1981, 1987), and long-term data suggest that success in territorial defense ultimately affects an individual's reproductive success. Between 1977 and 1988, each group's territory remained in the same general area, but territory *size* changed markedly. Larger groups (that is, groups with more adult females) were more likely than others to make incursions into other territories and were able to increase the size and quality of their territories at the expense of smaller groups. Large groups also experienced slightly greater rates of infant and juvenile survival, suggesting that the ability to dominate smaller groups has important reproductive consequences (Cheney and Seyfarth 1987).

Figure 2.11. Members of groups A and B threaten each other during an intergroup encounter. Females are active participant's in intergroup encounters.

The adaptive consequences of group defense are particularly apparent when we consider cases of group extinction. As noted earlier, the number of vervets in Amboseli has declined over the past 25 years. Since 1984 we have observed the extinction of six groups, four of them following fusion with a neighboring group. (In the two other cases, all the females died before any fusion occurred.) All fusions took place after a group had declined in size to one or no females and a few juveniles, when it simply seemed too small to compete effectively with its neighbors. In each case the females and juveniles joined that neighboring group with the *fewest* adult females (Hauser, Cheney, and Seyfarth 1986; Isbell 1990). This seems an adaptive choice for two reasons. First, in sheer numbers, the smallest neighboring group can mount the least aggression against intruders. Second, females in small neighboring groups may even be receptive to new members, because the recruitment of more females can potentially improve their ability to compete against their neighbors (Hauser, Cheney, and Seyfarth 1986; Cheney and Seyfarth 1987).

Finally, differences among individual monkeys in territorial behavior provide further evidence of a link between intergroup defense and reproductive success. In some groups of vervets (Cheney, Lee, and Seyfarth 1981), wedge-capped capuchins (*Cebus olivaceus*) (Robinson 1988), and rhesus macaques (Vessey 1968; but see Hausfater 1972), high-ranking females, who benefit most from the resources contained within their range, are the most aggressive participants in intergroup fights. By contrast, low-ranking female vervets (who derive less benefit from the resources within their range) are the individuals most likely to initiate friendly interactions with the members of other groups (Cheney, Lee, and Seyfarth 1981). Similar individual differences exist among males. In baboons, for example, dominant males are more aggressive than subordinate males toward the members of other groups (Cheney and Seyfarth 1977). Subordinate male vervet monkeys are also more likely than others to initiate friendly interactions with the members of other groups (Cheney 1983b; Cheney and Seyfarth 1983). As mentioned earlier, such friendly interactions often precede a male's transfer to another group.

Although vervet intergroup interactions, as well as those of other primates, are generally competitive and antagonistic, not all intergroup relations are identical. There is good evidence that monkeys distinguish among different neighboring groups, recognize the individuals within them, and adjust their behavior accordingly (see also chapter 3).

For example, in nonterritorial species such as baboons, macaques, and capuchins, groups in a local population can often be ranked in a dominance hierarchy based on group size, with larger groups consistently supplanting smaller ones (reviewed in Cheney 1987). The maintenance of such relations, however, does not depend on the physical presence of all group members, and the arrival of even one member of a dominant group can be sufficient to cause all the members of a subordinate group to leave the area (e.g., Kawanaka 1973). Occasionally, intergroup dominance depends on even more subtle factors, such as past relations among the male members of different groups (Gabow 1972).

Similarly, the pattern of migration by natal males in Amboseli was usually not random. Although males in group B, for example, had five neighboring groups to which they might transfer, the majority transferred to group A; group A males, in turn, transferred primarily to group B (Cheney and Seyfarth 1983). Females also tended to be less aggressive toward groups with which their own group had exchanged males than toward other neighboring groups (Cheney and Seyfarth 1982b, 1983).

The precise causal relation between a male's "choice" of groups and reduced female aggression is not clear. Males may have selected groups from which they received little aggression because reduced aggression facilitated social integration. Females may also have been less aggressive toward males with whom they were somewhat familiar. Whatever the exact cause, it seems clear that although vervets and other primates may live in discrete social groups, they nevertheless draw distinctions among their neighbors, recognize the members of other groups as individuals, and adjust their intergroup competitive behavior accordingly.

Using Behavioral Principles to Describe and Predict Social Organization

Although the social behavior of vervet monkeys can for the most part be *described* in terms of the general features we have just discussed, to *explain* why interactions take the form that they do one must specify the motives that govern the behavior of individuals. As a first step in our attempt to explain patterns of interaction among vervets, we have concentrated on the motives and behavior of adult females. Like characters in a Jane Austen novel, female vervets attempt to maintain close bonds with kin, defer to those of higher rank, and simultaneously attempt to establish bonds with animals of high status. In terms of evolutionary function, we believe that each of these motives will ultimately be shown to contribute to an individual's lifetime reproductive success and to represent an optimal strategy given the constraints imposed on females by life in a social group. Attraction to kin reflects the evolutionary benefits to be gained through kin selection (Hamilton 1964); giving way to those of higher rank while simultaneously striving to increase one's status reflects the best "mixed strategy" (Maynard Smith 1974) for animals in groups where high rank is correlated with better reproductive success.

The hypothesis that these motives underlie the social behavior of many Old World monkey females has been derived from data on a number of social groups (Seyfarth 1977); obviously, its general validity cannot be tested using the same groups. Nevertheless, the hypothesis can be independently tested in at least three ways. First, if attraction to kin, deference to those of high rank, and status striving really do account for a large proportion of adult female behavior, we should be able, using only these strategies, to create models of social interaction that accurately reflect what is observed in nature. Second, the same strategies should be able to predict the consequences of social change. Third, by considering the interaction between strategies, we should be able to uncover characteristics of social relationships that would otherwise have gone unnoticed.

Simulating Social Structure

Hinde (1976a, 1976b) defines social structure in nonhuman primates as those aspects of the content, quality, and patterning of social relationships that show regularities across individuals and across societies. In vervet monkeys and other primates, certain patterns of social behavior remain constant despite wide variation in group size, group composition, and habitat quality; these patterns define the species' social structure. Strong social bonds, as measured by high rates of grooming and alliance formation, exist among kin, who also share similar dominance ranks. At the same time, weaker social bonds are created when lower-ranking animals attempt to interact with those of higher rank and when attraction to high-ranking individuals, modified by competition, leads to bonds between individuals of adjacent rank even if they are not members of the same kin group.

One way to test these ideas is to create, through computer simulation, an artificial group of monkeys in which individuals adopt different behavioral "strategies." If our hypothesis is correct, two predictions follow. First, monkeys who behave according to the rules we have described should produce a distribution of interactions that closely resembles what we see in nature. High-ranking animals should receive more grooming than others, kin should groom each other at high rates, and unrelated animals who occupy adjacent ranks should groom each other more often than unrelated animals whose ranks are further apart. Second, the distribution of interactions thus produced should be relatively unaffected by variations in the number of kin, group size, the amount of time available for grooming, or the relative strength of an individual's attraction to kin and her attraction to those of higher status. Compared with the three motives we have identified, these factors are assumed to be relatively unimportant in determining the pattern of grooming among females.

A simulation begins by creating, as part of a computer program, a number of adult females arranged in a linear rank order that defines priority of access to resources (Seyfarth 1977). Some of the females are related; others are not. All females, however, attempt to follow the same behavioral strategies: distribute half of all grooming among close relatives and the other half among unrelated animals in direct proportion to their ranks. Although this is the desired strategy, not all females can achieve it. Sixth-ranking female F, for example, cannot gain access to first-ranking female A whenever A is interacting with B, C, D, or E.

Using only these strategies, the females are allowed to interact freely, each pursuing her "goal" within the constraints imposed by competition. After a period of interaction, the computer generates a pattern of grooming, which can then be compared with data from an actual group. Figure 2.12 compares grooming distributions in three simulated groups of monkeys (the data were published originally in Seyfarth [1977]), with the grooming distribution observed in one group of vervets during 1985–86. In each of the simulated groups the social structure generated is similar to that seen in nature, supporting the first prediction given previously. Most grooming is exchanged between kin, high-ranking females receive more grooming than others, and unrelated females who occupy adjacent ranks groom each other more often than unrelated females whose ranks are more widely separated. Moreover, the three simulations themselves share similar features, despite variation in group size, the number of kin, and the relative strength of an individual's attraction to kin and to higher-ranking animals. This supports

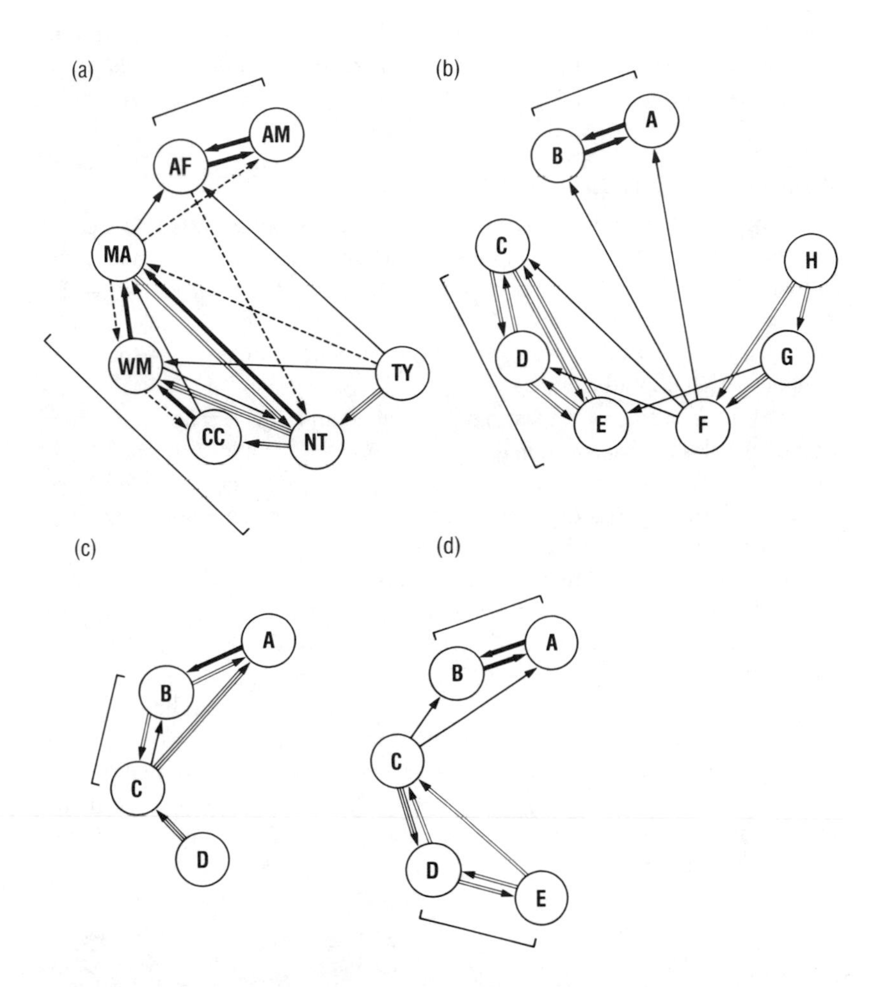

Figure 2.12. Grooming patterns in group B during 1985 to 1986 (A), compared with grooming simulated by a model based on the attractiveness of kin and the attractiveness of high rank (B, C, D). In all simulations, females can give up to four, and receive up to five, units of grooming. In B, the simulation assumes that each female with one relative gives as much grooming as possible to her relative and that each female with two relatives distributes as much grooming as possible equally among her kin. After grooming kin, each female distributes the remainder of her grooming among others according to their ranks and their availability, giving as much grooming as possible to the highest-ranking female available. In C and D, the simulation assumes that each female distributes half of her grooming among her kin and the remainder among unrelated females according to their ranks and their availability (for details see Seyfarth 1977).

the second prediction given previously and explains why similar grooming distributions arise among adult female monkeys despite wide variation in the number of kin present, group size, ecological conditions, or the relative strength of bonds among related individuals.

Predicting the Consequences of Social Change

Occasionally, demographic and social changes such as births and deaths cause changes in the relative ranks or "attractiveness" of particular individuals. These events, whether they occur naturally or are designed by the experimenter, allow us to test our hypothesis, because in each case the hypothesis makes explicit predictions about the sort of changes in social interactions that should occur.

Consider, for example, what happens when an adult female vervet, macaque, or baboon gives birth to an infant. Regardless of her rank, she immediately becomes extremely attractive to other females in the group, who approach, sit near, groom the mother, and groom and handle her infant (fig. 2.13). Although the mother's attractiveness as a social partner declines as her infant grows older (reviewed by Nicolson 1987; Walters and Seyfarth 1987), for a short time attractiveness due to an infant's presence is superimposed on attractiveness that derives from kinship and dominance rank (see also Weisbard and Goy 1976).

Given the observation that new mothers are attractive to others, we can make three predictions about female social interactions. First, when a high-ranking female gives birth, regardless of whether any other females with infants are in the group at the time, she will receive grooming from the same individuals who groom her at other times. This is because high-ranking females are always attractive, and the added attractiveness associ-

Figure 2.13. Like female vervets, female baboons are attracted to infants, and this attraction may temporarily increase the amount of grooming that low-ranking mothers receive from others.

ated with the presence of an infant has little effect on their grooming relations. In contrast, low-ranking females who give birth when there are no other females with infants in the group (for example, at the start of a birth season) will experience a marked change in grooming partners. This is because the presence of an infant temporarily increases the mother's relative attractiveness, increases competition for access to her, and increases the probability that she will receive grooming from high-ranking females who otherwise groom her rarely. Finally, low-ranking females who give birth when there are already other females with infants in the group (for example, in the midst of a birth season) should experience little change in their grooming partners, for reasons that should by now be apparent. Data supporting these predictions are presented in figure 2.14.

Other studies demonstrate the importance of social constraint by means of experiments in which a crucial competitor is removed from the group. Vaitl (1978), for example, tested pairs of squirrel monkeys (*Saimiri sciu-*

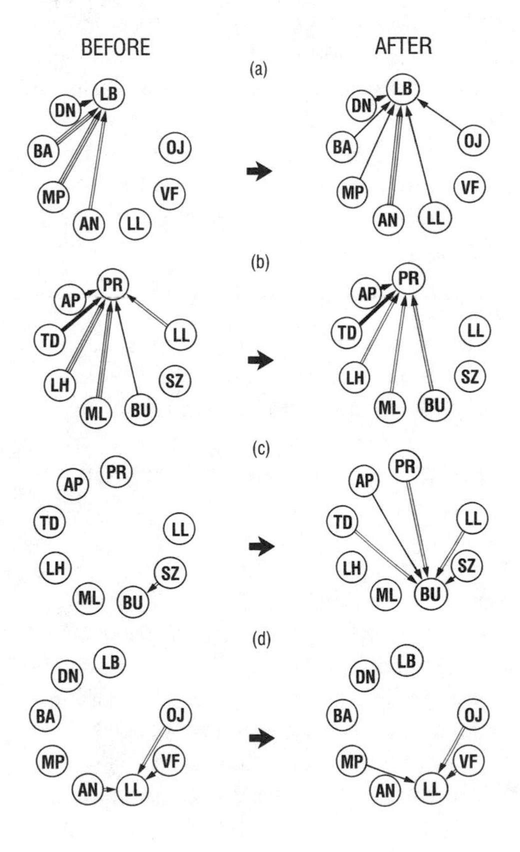

Figure 2.14. Comparison of the grooming received by high- and low-ranking females during the 6 weeks before and the 6 weeks after they gave birth to an infant. A shows grooming received by a high-ranking female who gave birth at the start of a birth season. B shows grooming received by a high-ranking female who gave birth at the end of a birth season. C shows grooming received by a low-ranking female who gave birth at the start of a birth season. D shows grooming received by a low-ranking female who gave birth at the end of a birth season. Redrawn from Seyfarth 1980.

reus) in isolation from their group to determine who preferred to interact with whom. Animals were then returned to the group, and in separate tests each individual's preferred partner was removed. In the absence of their preferred partner, individuals established grooming relations with an animal whom they had previously ignored. These new relationships were inhibited, however, when preferred partners returned.

Similarly, Stammbach and Kummer (1982) compared grooming between female hamadryas baboons when dyads were isolated and when they were part of a larger group. In the group condition, grooming between preferred partners increased significantly, as if these animals sought to intensify their relationship in order to prevent their partner from being taken away by a rival.

Finally, in a study of captive vervet monkeys, Keddy Hector (1989; Keddy 1986) found clear evidence that males preferred high-ranking females over others and that male-male competition constrained the expression of these preferences. Under normal conditions, the alpha male interacted at high rates with the alpha female and her offspring and less often with lower-ranking females and their offspring. Second- and third-ranking males were generally denied access to females and immatures. When the alpha male was removed, however, second-ranking males immediately began interacting with the alpha female, whereas third-ranking males interacted with females of lower rank. Each of these examples demonstrates that behavioral strategies are not always expressed in pure form; the interactions between strategies are as important as the strategies themselves.

Predicting Differences Between Families

As noted previously, in groups where genetically related females occupy adjacent dominance ranks it is often difficult to establish whether frequent grooming between these animals is the result of kin-based preferences or competition for those of high rank. We can, however, test between these hypotheses by examining whether high rates of grooming result from kin preferences alone or whether they result from the simultaneous preference for both kin and high-ranking individuals.

If grooming patterns are determined exclusively by kinship, rates of grooming should be similar from one matriline to the next. If, however, grooming patterns result from *both* kin- and rank-based preferences, behavior within high-ranking matrilines should differ from behavior within low-ranking matrilines. This is because, for the members of high-ranking matrilines, the two strategies reinforce one another: members of one's own matriline are attractive social partners both because they are kin and because they are high-ranking. By contrast, for the members of low-ranking matrilines kinship and status striving counteract one another: low-ranking animals are "torn" between maintaining close bonds with kin and attempting to establish bonds with the members of unrelated high-ranking families. Given the interaction between the attractiveness of kin and the attractiveness of high rank, our hypothesis predicts that high-ranking matrilines should generally be more cohesive than low-ranking matrilines.

Although perfectly straightforward, this prediction is difficult to test in natural groups because matrilines are often small and the strength of bonds within matrilines can be affected by the age and sex of family members. Ties between female vervets and their sons, for example, grow progressively weaker as sons grow older, while ties with daughters remain strong even after the daughters have matured. Only in groups with large matrilines is it possible to make comparisons on the basis of rank while still controlling for offspring age, sex, and family size.

Descriptive data from long-tailed macaques (Fady 1969) suggest that high-ranking matrilines spend more time together, groom more often, and support each other at higher rates than do low-ranking matrilines. Among captive rhesus macaques, high-ranking female kin also reconcile more with each other following fights than do low-ranking female kin (de Waal 1986a, 1989). Similarly, among the rhesus macaques of Cayo Santiago, infants in high-ranking matrilines spend more time on average in proximity to their kin than infants in low-ranking matrilines (Berman 1982). Datta (1983a, 1983c) also found that low-ranking individuals at Cayo Santiago were more likely than high-ranking individuals to "betray" their close kin by supporting a less closely related animal. Studying Japanese macaques, Imakawa (1988) found that cofeeding relationships between mothers and their offspring (both sons and daughters) declined with age, but that this decline was least apparent in high-ranking matrilines. Imakawa (1988: 493) concluded that "middle/low-ranking mothers may not provide their immatures with a secure base for obtaining food" (see also Yamada 1963; French 1981). Finally, in a study designed explicitly to control for other variables, Fairbanks (unpublished) found that adult female vervet monkeys in high-ranking matrilines approached each other more, spent more time together, and groomed each other more than adults in low-ranking matrilines (see also Ehart-Seward and Bramblett 1980). Bonds between adult females and their juvenile daughters were also stronger in high-ranking matrilines, as were bonds between grandmothers and their granddaughters. To date no study has found the opposite result, namely that bonds among family members are stronger in low-ranking matrilines.

These differences would not be predicted by a consideration of kin-based attractiveness alone, but they are a logical consequence of behavioral strategies that are governed, at least in part, by an interaction between preferences based on kin and preferences based on dominance rank. The two strategies are not, as Dunbar (1988 : 225) suggests, mutually exclusive, with kin-based preferences dominant in most groups and rank-based preferences limited "primarily to groups of unrelated animals." Instead the two strategies coexist in many groups, sometimes reinforcing and sometimes conflicting with one another.

As a result, all families are not equal. They recall the first sentence of Tolstoy's *Anna Karenina:* "Happy families are all alike; every unhappy family is unhappy in its own way." In this case, the members of high-ranking families are all alike in their consistently high rates of cooperative behavior, whereas the members of low-ranking families are less cohesive and as a result are less easy to characterize.

A Final Caveat

Before concluding, we should emphasize that *all* of the behavior of female monkeys is unlikely to be explained in terms of kinship and dominance rank. Further research is almost certain to reveal other, equally important behavioral strategies, particularly in species whose social organizations differ from those of baboons, macaques, and vervet monkeys.

In a study of hamadryas baboons, for example, Stammbach (1978) found that the grooming preferences of adult females were related not only to social rank but also to some less tangible aspect of attraction that was independent of dominance. Such individual "preferences" could in some cases override attraction based on rank. Similarly, de Waal and Luttrell (1986) found that age affected the behavior of female rhesus macaques: individuals were attracted not only to their kin and to those of high rank but also to individuals who were the same age as themselves. Colvin (1983a) and Mehlman and Chapais (1988) present similar data on juvenile male rhesus and Japanese macaques, respectively. Finally, our account of the rise of the Marcos matriline, together with Stammbach's (1988a, 1988b) experiments on food-producing "specialists," indicate that animals are attracted not only to those of high rank but also to those who, in the future, might *become* useful or high-ranking. It seems likely that we could account for a larger proportion of grooming (at least in vervet monkeys) if we could assess status striving not just in terms of current rank but also in terms of an individual's probability of advancement.

Summary

Vervet monkeys do not interact at random; their behavior follows particular patterns. To introduce the social behavior of vervets, and to set the stage for future chapters, we have described five general features of vervet social organization: kinship, dominance, reciprocity, sexual attraction, and group defense. Our goal, however, has not been simply to describe what vervets do but to explain why they do it. Focusing on the stable core of vervet monkey groups—the adult females—and using a method originally proposed by Hinde (1976a, 1976b), we have suggested that three motives underlie much of female behavior: attraction to kin, deference to those of high rank, and a desire to increase their own status. The explanatory power of these motives is demonstrated by their ability to account for what we observe, to generate accurate models of social interaction, and to predict features of behavior that might otherwise have escaped our notice. The motives also seem to make evolutionary sense. Attraction to kin reflects the evolutionary benefits to be gained through kin selection (Hamilton 1964); deferring to those of higher rank while simultaneously attempting to increase one's status reflects the best "mixed strategy" (Maynard Smith 1974) for animals in groups where high rank is often correlated with greater reproductive success.

A crucial unanswered question remains, however. Has our analysis really revealed something about the essential nature of vervet monkeys, or has it only told us something about ourselves? After all, the motives we are attributing to monkeys are products of human minds, the minds of those who observe the monkeys. But the ultimate goal of our analysis should not just be social structure as understood by primatologists, but social structure as it appears to the animals themselves. No self-respecting anthropologist would return from 2 years with the Dobu content to report only what he or she thinks of Dobu social structure. The other half of the puzzle is what the Dobu think, and how their view of themselves differs from that of a foreign observer. Have the monkeys, who have certainly seen the same events and more, made the same deductions that we have? Do they *understand* kinship and dominance rank? Or are they just sleepwalking through life, acting out complex strategies without being in any sense aware of what they are doing? To probe further into the minds of our subjects, in the next chapter we present a series of experiments and observations designed to test whether the regularities of social behavior we humans see might also exist in the minds of the monkeys themselves.

CHAPTER THREE | SOCIAL KNOWLEDGE

The vervet monkeys had moved out of their sleeping trees to forage on the ground. While the adults fed, the juveniles played in a nearby bush. Macaulay, the rambunctious son of a low-ranking female, wrestled Carlyle, the juvenile daughter of the highest-ranking female in the group, to the ground. Carlyle screamed, chased Macaulay away, and then went to forage next to her mother. Apparently the fight had been noticed by others, because 20 minutes later Shelley, Carlyle's sister, approached Austen, Macaulay's sister, and without provocation bit her on the tail.

This anecdote sets the stage for what is by now the familiar sort of popular article on nonhuman primates. Read any description of a long-term study of nonhuman primates and you will find an account of complex kinship networks, friendships, struggles for dominance, and shifting alliances. In fact, one of the fascinating things about primates is that their social structure often seems as rich and complex as our own. When we read Shakespeare's account of the blood feud between the Capulets and the Montagues, we take it for granted that members of the two families have a well-developed sense of their own and other peoples' social relationships. If they didn't, there could be no feud and no irony or tragedy in Romeo and Juliet's romance (to the detriment of the play, if not Romeo and Juliet). Monkeys also behave as if they recognize that relations within their own families are similar to relations in other families, and they use this knowledge to retaliate against their opponents. But still a nagging question remains: How much do the monkeys really know about what they are doing? Do they understand concepts like "kinship," or are they simply responding on the basis of associations they have formed between other group members?

There are many examples of animals—including humans—executing complex behavior without really knowing what they are doing. Ants, for example, remove the carcasses of dead conspecifics from their nest. The function of this behavior is to rid the nest of bacteria, but the ants are almost certainly not aware of the relation between corpses and disease; they are simply responding to the presence of oleic acid on the decaying corpse.

Ants will remove *anything* that smells of oleic acid, regardless of whether it is dead or infected (Wilson 1971). Even a live ant dabbed with oleic acid will be dragged, struggling, out of the nest. The ants' behavior is a fixed, unmodifiable response to a particular stimulus. As Mike Schmidt, late of the Philadelphia Phillies, describes fielding line drives at third base, "It's not something you think about: you just *react*."

The fact that we can describe vervet monkey social behavior in terms of relationships and strategies does not, therefore, necessarily mean that knowledge of these principles is what guides the monkeys' actions. They might be following a few relatively simple rules of action: avoid some individuals, groom or copulate with others. Alternatively, they might, like their human observers, recognize the relations that exist among others and even understand concepts like dominance, kinship, and reciprocity. At this point in our discussion we simply do not know.

Research on animal intelligence frequently attempts to distinguish between *knowing how* and *knowing that*, a distinction first drawn by the philosopher Gilbert Ryle (1949; see also Dickinson 1980; Whiten and Byrne 1988a). Knowing how refers to the ability to perform a specific procedural task based on recognition of a particular stimulus. An ant drags out anything that smells of oleic acid; Mike Schmidt reacts to the crack of a bat; a vervet monkey mother runs to her offspring's aid whenever she hears it scream. By contrast, knowing that refers to "declarative representations or knowledge" (Dickinson 1980) and implies an ability to make statements and causal inferences about the world. Rather than simply running whenever her offspring screams, for example, a vervet mother might understand enough about the relation between dominance rank and kinship to recognize that discretion is often the better part of valor and that she should intervene on her offspring's behalf only when the offspring is fighting with a member of a lower-ranking matriline. In other words, because it refers to more general knowledge *about* things and can be divorced from a particular response, knowing that allows greater flexibility in behavior, depending upon changes in the social and physical environment (see discussion by Whiten and Byrne 1988a).

We must distinguish, therefore, between knowledge that can be used only in a limited set of circumstances and knowledge that can be applied more broadly. A monkey may have formed an association between two members of the same matriline because the two animals are often encountered together. As a result, the monkey knows that whenever she approaches one individual she is also likely to be near the other. Such knowledge, however, might be limited to these two individuals or to a small set of animals within the monkey's own group. It would prepare the monkey for some (indeed, many) sorts of interactions but not for those that depended on the recognition of more differentiated relationships—for example, the difference between a relative and a "friend."

Alternatively, the monkey might have interacted with many different kin pairs and she might have inferred, on the basis of her experiences and observations, that such relationships share similar properties regardless of the individuals involved. The monkey might even have labels, like *closely bonded* or *enemies,* that help her order relationships into types. In this case the monkey's knowledge would be less constrained by particular stimuli, more general, and more abstract. It could also be applied in a much wider variety of circumstances.

In chapter 2 we presented evidence suggesting that vervet monkeys' knowledge of their social environment—that is, their knowledge of each other—may be declarative rather than procedural. In competing to interact with the members of high-ranking families, for example, the monkeys act as if they recognize that some animals are useful allies and that bonds with these individuals can potentially help to maintain or even improve their own status. But is this really so? Can we actually provide evidence that the monkeys assess each others' relationships and classify them into types? Or are the animals just following a few simple rules? To answer these questions requires a closer examination of what monkeys actually know about social relationships and how such knowledge affects their behavior. That is the purpose of this chapter.

Probing into the minds of monkeys, however, is not easy. Unlike anthropologists studying humans, we cannot simply interview our subjects and ask them what they think about each other. Instead, we must rely on a variety of indirect methods, including observations, anecdotes, and experiments—each focusing on situations in which the monkeys reveal, by their behavior, some of what they know about the principles that govern their interactions. By using different methods and drawing on data from a number of species, we hope that conceptual or methodological weaknesses in one area can be wholly or partially overcome by work in another. However, no single set of experiments or observations can ever provide the kind of ringing, definitive proof one would like. Instead we circle the problem, trying, from as many different angles as possible, to understand a perspective on social life that is different from our own.

Our argument may be summarized as follows. We begin with evidence that vervet monkeys, like many other nonhuman primates and indeed many other birds and mammals, recognize each other as individuals even when interactions occur at low rates. Monkeys also remember previous interactions, keep track of who has cooperated with them in the past, and adjust their future behavior accordingly. Although such knowledge can lead to extremely complex interactions, such as reciprocal exchange, it represents a relatively simple level of social intelligence, concerned only with knowledge that a monkey possesses about his own relationships.

There is also evidence, however, that primates observe the behavior of others and make judgments about relationships in which they are not themselves involved. They seem to recognize, for example, not only their own status and kin group but also the status and kin groups of others. In this respect, a monkey may be able to step outside his own egocentric world and recognize not only his own relationships but also the relationships that exist among others.

Finally, monkeys may not only recognize bonds among others but also compare and classify relationships, making same/different judgments between them. Beyond recognizing that individual members of the same matriline are close associates, for example, monkeys also seem to recognize some characteristic of relationships among kin that is similar across families. To do so, the monkeys must somehow be able not only to identify individuals that associate together but also to infer general properties of social relationships and compare relationships on the basis of these properties.

Drawing on results presented in chapter 2, our analysis is organized around the major features of social behavior in vervet monkeys, macaques, and baboons. Because these animals act *as if* they recognize kinship, dominance, status striving, reciprocity, and group membership, it seems logical to start by looking closely at such rules of interaction and asking whether individuals really do understand them. We begin with knowledge of group membership, because in asking what a monkey knows about the members of other groups we consider whether monkeys ever acquire knowledge about individuals with whom they interact only rarely. After a brief detour to introduce the method of field playback experiments used in much of our research, we then examine the monkeys' knowledge of social relationships within their own group. Here we begin with reciprocity, where the least is known, then turn to friendship, kinship, and dominance rank, where much more data are available.

Evaluating Other Individuals

Group Membership

The great majority of social interactions in vervet monkey society occur among members of the same group. Data in chapter 2, however, suggest that primates also recognize individuals in other groups and distinguish among their various neighbors. Young male vervets, for example, are most likely to transfer to the adjacent group that has received the most males from their own group in the recent past (Cheney and Seyfarth 1983), while among Japanese macaques the sight of only one individual from a dominant group is sufficient to cause members of a subordinate group to move away (Kawanaka 1973).

To test the hypothesis that vervets recognize the member of other groups, we borrowed an experimental technique used to study recognition of neighbors in territorial songbirds (Brooks and Falls 1975). We focused on our three main study groups (A, B, and C), whose territories were adjacent and slightly overlapping. Our subjects were members of group B. We reasoned that if vervets could recognize animals in other groups by voice alone, we should evoke little response from the members of group B if we played a vocalization given by a member of group A from group A's territory. By contrast, if group B members heard the vocalization of a member of group A coming from territory C, this event would be highly anomalous, and the animals' responses should be much stronger. The hypothesis that vervet monkeys can recognize the members of other groups by voice alone therefore predicts that calls from the same individual, played to the same subjects under two different conditions, will evoke markedly different responses.

Because this is the first playback experiment we will be describing, and because readers may be unfamiliar with the application of this technique to free-ranging primates, we pause here to discuss how such experiments are done, as well as the perverse and ignominious ways in which they can fail. Our method of playback experiments was developed jointly with Peter Marler in 1977 (e.g., Seyfarth, Cheney, and Marler 1980b). Colleagues in ethology will recognize, however, that many of the techniques we describe are not original but borrow extensively from work conducted on birds over the past 30 years.

•

Playback experiments can be designed to reproduce events that occur naturally or to present subjects with a stimulus they normally would be unlikely to encounter. The *neighbor recognition* experiments described previously provide examples of both types. In the "control" condition (vocalization of a group A animal played from territory A), we simply replicated an event that occurred naturally many times each month. Since we could not always anticipate when an intergroup call would occur, we were often caught unprepared, unable to quantify precisely how animals responded when they heard a neighbor's call. Playback experiments allowed us to present a call we knew was coming, to choose a particular animal whose responses we wished to measure, and to film her responses for subsequent analysis. By contrast, in the "experimental" condition (vocalization of a group A animal played from territory C), we presented the monkeys with the same signal but in a context they were unlikely to have experienced before. Such novel and perhaps even alarming material—in this case suggesting a major territorial incursion by a neighboring group—can elicit responses that reveal some of what animals know about their environment.

Whatever their exact purpose, most of our playback experiments were based on the assumption that vervets are able to identify individuals by voice alone. This assumption seemed justified, since evidence for the discrimination of individuals by voice alone had already been well documented in numerous species of birds and mammals when we began our study. We consider the phenomenon of individual recognition further in subsequent sections of this chapter.

To produce meaningful results, playback experiments must occur at low rates and must blend in as inconspicuously as possible with the animals' natural social behavior. If they are carried out too often or become associated with unusual or abnormal events, the animals will not be fooled and will soon cease responding altogether. When conducting the neighbor recognition experiments, therefore, we set a number of conditions that had to be met before a vocalization could be played. First, the designated subject had to be in the center of her group's range, away from any territorial border, and the group must not have interacted with any other group during the previous 2 hours. Further, the subject could not recently have been involved in any large, escalated fight in her group, nor could she have recently encountered a predator.

If our aim was to play a vocalization from a loudspeaker hidden within territory A, the members of group A had to be at the other end of their territory, out of sight of our subject. If this condition was not met, we could have biased our results in a number of ways. If group A members were in the immediate area, for example, they might have responded to the call themselves, and we would have been unable to determine if our subject was responding to the playback or to the responses of the group A animals.

When these preconditions had been met, one of us prepared to film the subject while the other walked into the neighboring territory (on average, 120 m away) carrying a playback tape and loudspeaker. Vocalizations used in experiments had to have been recorded from known individuals during the past 6 months. They also had to be free of background noise and any other distortion.

Some playback experiments elicited strong behavioral responses from the monkeys (see chapter 4). Far more often, however, responses were more subtle and involved simply a change in the direction of the subject's gaze for a few seconds. Since it was crucial to be able to analyze each experiment with some precision, we filmed our experiments with a Super-8 sound movie camera or, in later years, a video camera. At the start of each trial, we filmed the animals for a predetermined time (15 seconds in the group recognition experiments) before playing the call; filming then continued for 1 minute after the call had been played. This allowed us to control for each subject's behavior in the seconds before playback and to define a response in terms of the difference between behavior immediately before and after a call had been heard. Following the completion of each trial, we drew a map to record the location of all trees, bushes, and major landmarks as well as the distances between camera, subject, and speaker.

Lest the animals become suspicious of the experimental procedure, we separated all trials by intervals of at least 2 days. To ensure that the sight of playback equipment would not signal an imminent experiment, we often filmed the animals during the intervening period and (rather absurdly) hid speakers in nearby bushes. More important, since natural intergroup encounters occurred roughly every other day (Cheney 1987), subjects usually had the opportunity to see and hear their neighbors giving intergroup calls between successive trials. By experimenting at low rates we therefore reduced the likelihood that the monkeys would regard our playbacks as abnormal and cease responding to them.

Finally, a word should be said about the many cases in which an experiment had been planned and set into motion, yet could not be completed. Often, for example, by the time we had positioned the speaker and were ready to start filming, our subject had moved into a thick bush. On other occasions group A or group C might suddenly appear and a genuine encounter would begin. Trials could also be aborted because a predator was spotted, an elephant approached to doze under the tree where the monkeys were feeding, or a Maasai tribesman appeared and asked us to help him smuggle cooking oil into Tanzania. Worst of all, with an otherwise perfect *mise-en-scène,* the camera's battery would run down or it would be discovered, too late, that a cable connected to the speaker was not properly in place. Loud, mutual recriminations would follow, with considerable time being devoted to determining exactly who was at fault. Frustrating as they were, these aborted trials were valuable because they meant that on many occasions (for some types of experiments, up to 70% of all attempts), the monkeys saw us go through all the preparations to conduct an experiment without hearing a call. Once again, these failed attempts reduced the likelihood that animals would habituate to the experimental procedure.

Along with these precautions, we took the usual care to ensure that our tests were not biased in some other way. In the neighbor recognition study, for example, we controlled for the order of stimulus presentation by having 9 of 20 subjects hear the "appropriate" call (a member of group A calling from A's territory) first and 11 of 20 hear the "inappropriate" call first. To control for the possibility that calls from particular individuals might be more salient than others, we used calls from six individuals in three groups as stimuli and the same vocalization in a matched-pairs design as both an appropriate and an inappropriate stimulus.

Despite all these precautions, none of our experiments ever achieved the precision and control of some laboratory tests. We simply did not know everything that had happened to our subjects on the day they were tested, nor could we completely control the myriad contextual variables (birds, ungulates, insects, other monkeys) present under natural conditions. Despite our care in selecting equipment, no loudspeaker can ever perfectly duplicate a monkey's voice, and when we say "a speaker was hidden behind a tree or in a bush," this is sheer pretense; a speaker may have looked well hidden to us, but almost certainly the vervets could often see it. Many of these problems were alleviated by allowing different vocalizations to serve as each others' controls. If some aspect of our procedure in the group recognition experiments was biased, for example, it should have been equally biased in both the appropriate and inappropriate trials. In the end, we controlled what we could and hoped that the benefits of experimenting on animals in their natural habitat would outweigh the imprecision of our techniques.

•

We now return to the experiments at hand, designed to examine whether monkeys can recognize the members of other groups by voice alone. We used as a stimulus the intergroup *wrr* vocalization, a loud, relatively long, trilling call given by females and juveniles when they spot another group (chapters 4 and 5). In 14 paired trials, the vocalizers were adult females; in 4 others they were subadult males who had not yet transferred from their natal group.

If vervets recognized the calls of animals in neighboring groups, they should have responded more strongly to a call when it was played from an inappropriate territory than when it was played from the true range of the vocalizer's group. In fact, this is exactly what occurred. In a typical trial, subjects responded to the playback of a call from the appropriate territory by looking briefly toward the speaker and then returning to their former activity. By contrast, when they heard the same call from the inappropriate territory, a significant number (18 of 20) subjects looked toward the speaker for longer durations. In some cases they ran into trees and gave leaping displays normally used in intergroup encounters (Cheney and Seyfarth 1982b). Vervets therefore seemed to associate the vocalizations of individual members of other groups with those groups' territories. Significantly, they did so even though the vocalizers had never lived in the same group with the subjects and had interacted with the subjects only during intergroup encounters.

Other studies have used playback experiments to investigate the ways in which neighboring groups' calls regulate spacing between groups. For example, in territorial species like gibbons (Chivers and MacKinnon 1977) and dusky titi monkeys (*Callicebus moloch*) (Robinson 1981), intergroup calls cause animals to approach the boundaries of their range. By contrast, in species characterized by groups with overlapping ranges, similar playback experiments cause neighboring groups to move away from the area from which the calls are played (e.g., Waser 1976 for gray-cheeked mangabeys (*Cercocebus albigena*); Cheney 1987 for a review). In none of these experiments, however, did investigators attempt to distinguish among responses to *different* neighboring groups. As a result, it is impossible to determine whether the monkeys recognized where specific neighboring groups were located.

Although our experiments demonstrated that vervets associate a particular vocalization with a particular group, they did not directly test whether monkeys can distinguish among the individual members of other groups. We examined this question in another series of experiments, although sample sizes were too small to permit any conclusive results.

In Amboseli, high-ranking female vervets are usually (but by no means always) more active and aggressive participants in intergroup encounters than low-ranking females, and they also give more *wrr* calls (Cheney 1981; Cheney, Lee, and Seyfarth 1981; Cheney 1987). Taken together with reports that rhesus and Japanese macaques apparently recognize dominant males in other groups (chapter 2), this observation suggests that vervets might also distinguish among the intergroup *wrrs* of high- and low-ranking females in neighboring groups.

To test this hypothesis, we played the *wrr* vocalizations of dominant and subordinate females in paired trials to female subjects in neighboring groups. Seven of ten subjects responded for longer durations (by looking in the direction of the speaker) when they were played the *wrr* of the neighboring group's dominant female than when they heard a more subordinate female. Although not statistically significant, these results suggest that vervets do distinguish among the calls of different individual neighbors.

Although results do not, of course, prove that vervets recognize the dominance relations of females in neighboring groups, it is worth emphasizing that when subjects responded more to the more dominant female, they were not simply responding to the female whose *wrr* they had heard most often. When we counted the total number of *wrrs* given by each female over the 8-month period when these experiments were conducted, we found that there was no relation between the strength of subjects' responses and the frequency with which they heard each female in real intergroup encounters. In four cases, the two comparison females gave an equal number of *wrrs*. In only half of the six remaining paired trials did subjects respond more strongly to the female who gave more *wrrs* under natural conditions. There is, therefore, some evidence that females did discriminate among different *individual* members of different groups and that these responses were not simply based on the relative frequency of different females' *wrrs*.

Reciprocity

Young males leaving their natal group generally join that neighboring group with which their natal group has the least aggressive relations (chapter 2; Cheney and Seyfarth 1983). This observation suggests that vervets may remember past interactions and that the nature of these interactions may affect subsequent behavior. Social interactions within a group provide even stronger support for this hypothesis. As noted in chapter 2, male baboons are most likely to form alliances with those who have supported them in the past (Packer 1977), and when vervet monkey females form alliances with unrelated individuals, they do so most often with those who have previously groomed them at the highest rates (fig. 3.1A and B; see also table 2.1). Although such data are suggestive, they do not demonstrate

Figure 3.1. A, Juvenile male Wordsworth grooms juvenile female Acton. B, Several hours later, Wordsworth (center) *solicits the support of Acton* (left) *in a coalition against Macauley. Like many other nonhuman primates, vervets solicit support through specific vocalizations and by rapidly looking back and forth from their rival to their intended ally.*

conclusively that cooperative behavior between two individuals has a direct causal influence on their future interactions.

To gain a clearer picture of the mechanisms that might lead to cooperative, reciprocal exchanges, we designed an experiment that examined whether grooming between two unrelated vervet monkeys increases the likelihood of future support in aggressive disputes (for details see Seyfarth and Cheney 1984a). The playback stimulus used in these trials was the vervets' *threat grunt* (or *woof woof*: Struhsaker 1967a). Under normal conditions, individuals give threat grunts when they are threatening or chasing another group member. The calls also apparently function to solicit support because they often cause other animals to come running to the caller's aid.

We began by selecting two individuals, A and B, for a matched pair of trials. We waited until we observed A grooming B; then, 30 to 90 minutes after the grooming bout had ended, we played A's threat grunt within earshot of B from a concealed loudspeaker. (Clearly, exchanges of cooperative behavior are unlikely to be restricted to the 90-minute period that we used in the experiments, but we were constrained by our ability to control events for longer periods of time.) Some days later we played A's threat grunt near animal B after a period of at least 2 hours in which A had *not* groomed B. If grooming and alliances really did constitute part of a reciprocal system of interactions, prior grooming of B by A should have increased B's willingness to support A in a dispute. B's response to playback of A's call should therefore have been stronger after A had groomed B. We chose subjects from 10 pairs of unrelated animals in three social groups and systematically varied the order in which subjects were played threat grunts in the presence or absence of prior grooming. To test whether reciprocity was also influenced by kinship, we carried out a similar test on nine pairs of close kin, defined as individuals whose degree of relatedness, *r*, was greater than 0.25. Kin pairs were mother-offspring pairs or siblings through the maternal line.

Subjects responded to playbacks by looking toward the speaker. Among kin, prior grooming had no effect on the duration of an individual's response (fig. 3.2). Close relatives responded to each other's solicitations with similar intensity whether or not they had recently groomed. Prior grooming, however, strongly affected the behavior of nonkin. The response duration of unrelated animals was significantly longer after grooming had occurred than in its absence. The effect of prior grooming on the behavior of unrelated animals was so marked that in some cases they responded more strongly than related animals did under comparable conditions (fig. 3.2; Seyfarth and Cheney 1984a). Unlike kin, unrelated animals seemed to be engaged in a form of reciprocal exchange.

We can speculate that when a vervet monkey hears another calling for support, her decision to attend (and potentially to intervene) is affected by at least two considerations: Is the animal a relative? and, if not, What has she done for me lately? The fact that playbacks elicited attentiveness rather than overt action also reminds us that although grooming may increase the readiness of unrelated animals to attend to each other's solicitations, actual physical involvement in an alliance is likely to depend on further assessment of the potential costs of intervention. Such costs are no doubt influenced by the relative dominance ranks of the participants and the intensity of each dispute (Cheney 1983a; Gouzoules, Gouzoules, and Marler 1984; see also Marler 1976b).

In functional terms, the difference in behavior between related and unrelated pairs of monkeys is consistent with evolutionary theory, which argues that an individual can potentially increase its reproductive success by aiding kin regardless of whether prior altruistic behavior has occurred (Hamilton 1964). By contrast, the theory of reciprocal altruism suggests that an individual interacting with nonkin can potentially increase its reproductive success only if its support of another is part of a long-term, reciprocal relationship (Trivers 1971, 1985; West-Eberhard 1975).

In terms of proximate mechanisms, there are a number of explanations for the differing behavior of kin and nonkin, none of them mutually exclusive. For example, bonds formed among kin during development may have long-term consequences that override the short-term effects of social interactions. Relatives might therefore come to each others' aid even if they have not recently interacted. If we think of behavior like grooming as contributing slowly and incrementally to the establishment and maintenance of a relationship (Seyfarth, Cheney, and Hinde 1978), much as conversation,

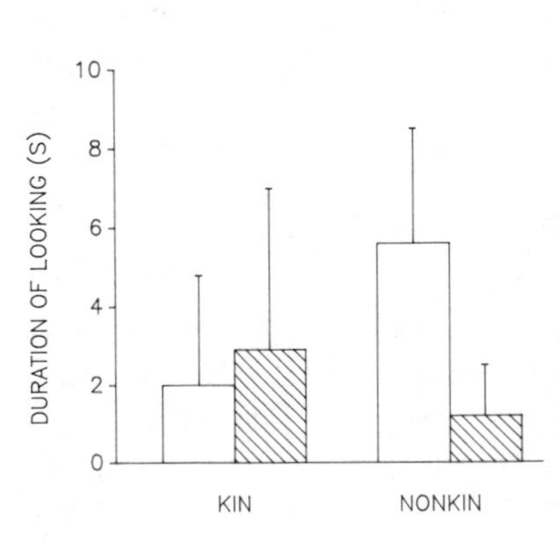

Figure 3.2. Results of experiments testing for reciprocal altruism in kin and nonkin pairs after grooming (open histograms) *and after a period of time when no grooming had occurred* (shaded histograms). *Histograms show means and standard deviations for 9 subjects in kin trials and 10 subjects in nonkin trials. For nonkin, prior grooming significantly increased the number of seconds that subjects looked toward the speaker* (*two-tailed Wilcoxon test, P <0.05*). *For kin, prior grooming had no effect on durations of response. Redrawn from Seyfarth and Cheney 1984a.*

a hug, or a pat on the back helps subtly to reaffirm friendship among humans, then it is easy to imagine how the occurrence (or lack) of a single grooming bout might have little effect on relations among kin, who groom, play, forage, and sleep together often. The same bout of grooming, however, could produce major changes in the relationship of two unrelated animals who interact less often.

Before considering other, more theoretical, problems associated with the theory of reciprocal altruism, we should add a final caveat regarding our experiments, because readers will have recognized that they address only half of the issue. Although we may have shown that cooperative acts *increase* the willingness of others to respond to solicitations for aid, our data do not examine whether noncooperative or even spiteful acts *decrease* it. To test the reciprocity theory completely, we would need also to play A's threat grunt to B either after A had threatened B or after A had failed to support B in an alliance. If playbacks under these conditions decreased B's response compared with her response in the absence of prior interactions, we could conclude with more confidence that B's heightened response following grooming was in fact evidence for reciprocal altruism. Unfortunately, the decline in the Amboseli vervet population prevented us from conducting these experiments.

If vervets do monitor the failure to reciprocate aid, any retaliation is likely to be subtle and to occur only in the form of a decreased likelihood to respond to aid solicitations rather than as overt aggression. Observational studies of rhesus and stumptailed macaques (*Macaca arctoides*), for example, have suggested that animals rarely threaten (or show contra reciprocity toward) individuals who form alliances against them, probably because the relative stability of dominance hierarchies prevents low-ranking animals from attacking previous opponents (de Waal and Luttrell 1988).

This generalization, interestingly, seems not to be true of chimpanzees. Dominance among chimpanzees is more dependent upon alliance partners than it is among Old World monkeys, and individuals do seem to retaliate against animals who have formed alliances against them in the past (de Waal and Luttrell 1988; de Waal 1989). A chimpanzee who forms an alliance against a third can expect his opponent later to form an alliance against himself, to the potential decrement of his own rank. The implied threat of revenge may therefore strongly influence patterns of aggression and alliances in chimpanzee communities.

Although grooming and alliances in vervet monkeys, like mutual cooperation in other animal species (e.g., the vampire bats studied by Wilkinson 1984), may constitute an adaptive system of reciprocal interactions, reciprocity in nonhuman species nevertheless differs in two important respects from reciprocity among humans. First, as Trivers (1971) himself noted, the evolutionary theory of reciprocal altruism takes the altruism out of reciprocity: one monkey grooms another not through genuinely unselfish motives but because the monkey hopes to elicit support in return. Many social scientists, unsatisfied with this approach, have sought an explanation for the evolution of human altruism that does not rest entirely on selfish motives (Fiske 1990; Frank 1988; see also Alexander 1987).

Second, though we have used the word *reciprocal* throughout this section, we do not yet know the extent to which strict notions of reciprocity, involving the exchange of items of roughly equal value, actually underlie the animals' behavior. Are the monkeys really calculating the costs and benefits of a grooming bout or alliances and then computing the difference between them? Or are they using some simpler rule of thumb? Obtaining an answer is complicated by our own inability to weigh the costs and benefits, from a monkey's perspective, of different sorts of interactions with different individuals.

Consider, for example, a low-ranking female who attempts to establish close relations with a female of higher rank. The low-ranking female's ultimate goal may be a relationship in which grooming is traded for support in alliances or tolerance at desirable food resources. To achieve this goal, however, the low-ranking female may have to endure months or even years when she grooms the high-ranking female often but receives little benefit in return. Furthermore, low-ranking females may derive so much benefit from rare alliances with high-ranking partners that they are willing to groom them or support them in alliances 10 times for every bout of support they receive in return, or even gamble on that prospect without actual concrete guarantees (Cheney 1983a).

Asymmetrical relationships are common in primates, and they have sometimes been offered as evidence that the quest for reciprocal exchange plays an unimportant role in the distribution of grooming (e.g., Silk, Samuels and Rodman 1981; Fairbanks 1980; de Waal and Luttrell 1988). Nevertheless, as Boyd (1988) has shown, it is theoretically possible for an apparently asymmetrical relationship to be entirely reciprocal if the imbalance in the costs and benefits that each partner has to offer are taken into account (Seyfarth and Cheney 1988b). Noe (1986) emphasizes this point in his analysis of coalitions among adult male baboons. Indeed, until we find some way to measure the relative costs and benefits that individuals derive from interactions, almost any relationship can be seen as reciprocal (or asymmetrical), depending upon how one chooses to measure the participants' gains and losses. The point here is not to create a tautology or to dismiss Trivers' theory as unworkable. Instead we must be aware that theories of reciprocity, especially when applied to groups in which individuals differ widely in their ability to help one other, are slippery concepts—difficult to state precisely and difficult to translate into rigorously testable predictions.

Recognizing the Relationships of Others: Kinship, Friendship, and Dominance Rank

Knowledge About Other Animals' Companions

Thus far, we have used data on cross-group recognition and reciprocity to illustrate some of the knowledge a monkey acquires about other animals based on his own interactions with them. To understand a rank hierarchy, however, or to predict which individuals are likely to form alliances with each other, an animal must step outside his own sphere of interactions and recognize the relations that exist among others (see discussion by Harcourt 1988). Such knowledge can only be obtained by observing interactions in which one is not involved and making the appropriate inferences. There is, in fact, growing evidence that monkeys do possess knowledge of other animals' social relationships and that such knowledge affects their behavior.

Studies of hamadryas baboons in Ethiopia were the first to suggest that nonhuman primates assess the relationships that exist among others. Under natural conditions, hamadryas baboons are organized into one-male units, each of which contains one fully adult male and two to nine adult females (Kummer 1968; Sigg et al. 1982; reviewed in Stammbach 1987). One-male units frequently come into contact with single, unattached males, and a unit leader must constantly defend himself against attempts by other males to take over his females. Experiments using captive hamadryas have shown that "rival" males assess the strength of an owner's relationship with his females before competing to acquire them. Rival males do not attempt to take over a female if they have previously seen her interact with her owner. Such "respect of possession" holds even when the rival is dominant to the owner in other contexts (Kummer, Goetz, and Angst 1974). This phenomenon seems to be widespread. Among both gelada baboons in Ethiopia and savanna baboons in Kenya, challenges from a rival are less likely to occur if a male has strong grooming relations with a female and more likely to occur if grooming relations are weak (Dunbar 1983b; Smuts 1985).

To test the hypothesis that rivals make judgments about the strength of bonds between a male and his females, Bachmann and Kummer (1980) studied six adult males and six adult females, using choice tests to determine how strongly each male preferred each female and how strongly each female preferred each male. A male-female pair was then placed in a large enclosure and allowed to interact freely. Different rival males were allowed to watch the pair and then given an opportunity to challenge the owner for possession of his female. Bachmann and Kummer found that the probability of a challenge was not correlated with the rival's or the owner's preference for a particular female. The *female's* preference, however, did seem to make a difference: if a female was with an owner whom she groomed at a high rate, this inhibited challenges from middle- and low-ranking rivals. The two highest-ranking males challenged all owners regardless of the females' behavior. Although Bachmann and Kummer could not rule out the possibility that rival males were simply responding to the females' actions rather than their relationships, the experiments suggested that males may assess the strength of attraction between an owner and his female and avoid challenging an owner when the pair's relationship is close. This strategy seems adaptive, because aggressive challenges, which involve potential injury, can be costly if the contested female prefers to remain with her current mate.

Further evidence that monkeys recognize relations among others comes from playback experiments on vervet monkeys. As we noted earlier, many of our experiments, like those on cross-group recognition or reciprocal altruism, rely on the assumption that vervets recognize the calls of different individuals. Early in our study we investigated this assumption with a relatively simple case: maternal recognition of offsprings' screams.

When infant and juvenile vervets scream during rough play (fig. 3.3), their mothers often run to support them (chapter 2). This behavior suggests that females can distinguish among the calls of different individuals. To test this hypothesis, we played the scream of a 2-year-old juvenile to its mother and two control females who also had offspring in the group. So, for example, in a typical trial, we waited until three adult females, Profumo, Teapot Dome, and Maginot Line, were sitting near each other while their

Figure 3.3. Juvenile male Bobby Vee leaps away during play with another juvenile. If their play becomes too rough, juveniles may scream, attracting the attention of their mothers.

offspring were out of sight elsewhere in the group. We then played, through a concealed loudspeaker, the distress scream of Profumo's daugher, Shelley. We found that mothers consistently looked toward or approached the speaker for longer durations than did control females, indicating that they recognized the voice of their offspring (Cheney and Seyfarth 1980). This result was entirely expected, given the many studies that had already shown individual recognition by voice in primates (e.g., Waser 1977; Kaplan, Winship-Ball, and Sim 1978; Hansen 1976) as well as birds and other animals (e.g., Emlen 1971; Petrinovich 1974; Brooks and Falls 1975; Kroodsma 1976).

More interesting, however, was the behavior of control females. When the responses of control females were compared with their behavior before the scream was played, we found that playbacks significantly increased the likelihood that control females would look at the *mother*. By contrast, there was no change in the likelihood that control females would look at each other. In many cases control females looked at the mother before she herself had made any apparent response, suggesting that control females did not simply look at the mother because she in some way cued their behavior (Cheney and Seyfarth 1980, 1982b). Instead, it was as if the females were saying, "That scream goes with Shelley, and Shelley goes with Profumo. What's Profumo going to do about it?" By associating particular screams with particular juveniles, and these juveniles with particular adult females, the control females behaved as if they recognized the kinship relations that existed among other group members.

At this point, it is important to emphasize that whenever we speak of kin recognition in vervets or any other primate species, we define the term operationally, as the recognition of a close social bond. The ability to recognize other animals' kin does not imply that monkeys necessarily have concepts of kinship or genetic relatedness but simply that they recognize the close associates of other group members. In most cases close associates are also kin, and this rule of thumb appears to be the primary mechanism underlying kin recognition in nonhuman primates (Frederickson and Sackett 1984; Gouzoules 1984; Gouzoules and Gouzoules 1987; Waldman, Frumhoff, and Sherman 1988). There is at present no evidence that monkeys discriminate among different kinship relations that are characterized by similar rates of interaction—for example, adult sisters as opposed to mother-daughter pairs. The relevant experiments simply have not been conducted.

Monkeys seem not only to distinguish among the screams of different juveniles but also to differentiate among different types of aggressive interactions. In a study of maternal intervention in the semicaptive population of rhesus macaques on Cayo Santiago, Gouzoules, Gouzoules, and Marler (1984) noticed that the screams of juveniles varied systematically in their acoustic features, that different types of screams were given in different types of conflicts, and that mothers responded differently to different scream types. Mothers reacted most strongly to the screams given to higher-ranking opponents, next most strongly to screams given to lower-ranking opponents, and least strongly to screams given to relatives. Through its screams, in other words, a juvenile effectively classifies its opponents according to kinship and dominance. By her selective responses, an adult female reveals knowledge of both her offspring's voice and her offspring's network of social relationships. We will discuss these experiments further in chapter 4.

For additional evidence that monkeys recognize the kinship relations (or close associates) of other group members, consider the phenomenon of redirected aggression. In many primate species, an animal that has been involved in a fight will "redirect" aggression and threaten a third, previously uninvolved individual (fig. 3.4). Rhesus macaques (Judge 1982) and vervet monkeys (Cheney and Seyfarth 1986, 1989) do not distribute such redirected aggression randomly but direct it toward a close relative of the prior opponent. Vervets in Amboseli, for example, were significantly more likely to threaten unrelated individuals following a fight with those animals' close kin than during matched control periods (fig. 3.5). This was not because fights generally increased aggression toward unrelated animals. Instead, aggression seemed to be directed specifically toward the kin of prior opponents.

Kin-biased redirected aggression was even evident *within* matrilines (fig.

Figure 3.4. Adult female Disney (right) *threatens subadult male Sedaka moments after she chased Sedaka's younger brother. A juvenile female* (center) *forms an alliance with Disney and also threatens Sedaka.*

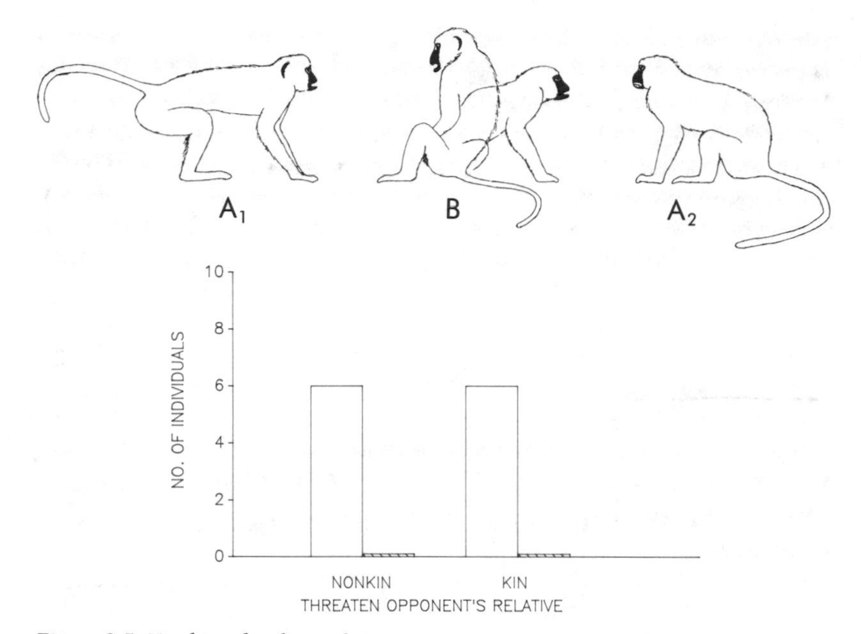

Figure 3.5. Kin-biased redirected aggression in vervet monkeys from 1985 to 1986. Open histograms show the number of individuals who redirected aggression against an individual more after fighting with that individual's relative than during a matched control period. Hatched histograms show the number of individuals who were as likely to redirect aggression after a fight with an individual's relative as during the matched control period. Fights among nonkin are considered separately from fight among kin. Both nonkin and kin were significantly more likely to threaten another animal after fighting with that animal's relative than during matched control periods (two-tailed sign test, $P < 0.02$; data from Cheney and Seyfarth 1989). Drawing by John Watanabe.

3.5). For example, an adult female who was chased from a food source by her sister might later seek out and lunge at her sister's daughter (her own niece). Although it might initially seem surprising that an individual would redirect aggression against her own relative, fights among relatives are in fact quite common in most primate species. In our vervet groups fights within families were as frequent as fights between families, although within-family fights seldom escalated to involve bites or injuries (Cheney and Seyfarth 1989). Anyone who has ever experienced a large family reunion over a long rainy weekend should find it intuitively easy to understand that even minor annoyances with a particular relative can easily expand to include that relative's children or parents. In the case of vervet monkeys, the data suggest that a previous fight with a close relative was an important context for fights within matrilines.

Figure 3.6. Adult female Carlyle handles female Austen's newborn infant shortly before grooming her. Females sometimes reconcile after fights by grooming, touching, or handling each others' infants.

Similar kin-biased patterns of interaction were evident in a behavior that is the mirror image of redirected aggression; namely, reconciliation (fig. 3.6). In many species of monkeys and apes, animals sometimes reconcile after fights by approaching their former opponents and touching, hugging, or grooming them (e.g., Cords 1988 for long-tailed macaques; de Waal and Yoshihara 1983 and de Waal 1989 for rhesus macaques; de Waal 1989 for stump-tailed macaques; York and Rowell 1988 for patas monkeys (*Erythrocebus patas*); de Waal and van Roosmalen 1979 and de Waal 1989 for chimpanzees). It is not only the primary antagonists who reconcile, however. Monkeys will also reconcile with the *kin* of their former opponents. Studying reconciliation among captive patas monkeys, York and Rowell (1988) found that unrelated animals contacted the kin of their former opponents almost twice as often following a fight as during matched control periods. Similarly, vervet monkeys were significantly more likely to groom or initiate a friendly interaction with an unrelated animal following a fight with that animal's kin than in the absence of such a fight (Cheney and Seyfarth 1989; fig. 3.7).

Unlike redirected aggression, however, kin-biased reconciliation was not evident *within* matrilines. For vervets, patterns of reconciliation among members of the same extended family differed in two important respects from reconciliation among nonkin. First, although unrelated animals were more likely to reconcile with their opponents' kin than with their opponents themselves, related animals were slightly more likely to reconcile directly with their opponents (fig. 3.7). Second, reconciliation appeared to be a more important context for affinitive interactions among nonkin than among kin. Nonkin were significantly more likely to initiate friendly interactions both with their opponents and with their opponents' kin following a fight than during control periods. In contrast, related individuals were as likely to interact with their opponents and their opponents' kin (who were also their own kin) during control periods as they were following a fight. Apparently, the generally high rates of grooming and friendly interactions among kin swamped the effect of affinitive interactions in the context of reconciliation.

These differences in patterns of reconciliation among related and unrelated vervets are similar to those reported by Cords (1988), who found that juvenile male long-tailed macaques reconciled at higher rates with unrelated opponents than with related ones. Relationships among unrelated animals are typically less predictable and stable than those among relatives, and Cords suggests that postconflict affinitive interactions may function as a repair mechanism for relationships among nonkin. Such conciliatory interactions may be less important for kin, who interact at high rates in any case.

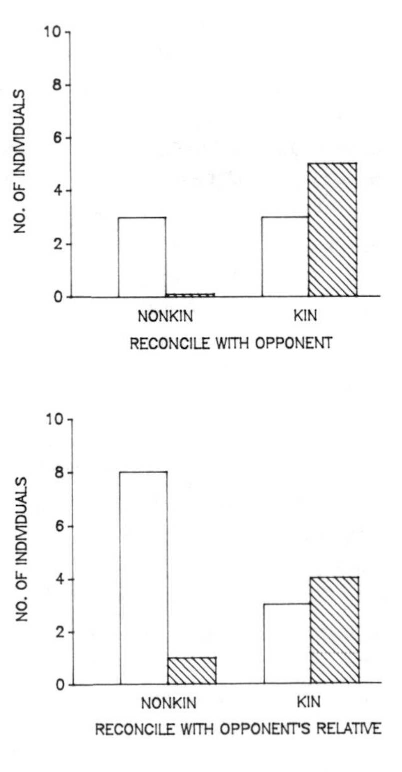

Figure 3.7. Reconciliation and kin-biased reconciliation in vervet monkeys from 1985 to 1986. Open histograms show the number of individuals who reconciled with an opponent (top graph) *or with an opponent's relative* (bottom graph) *more after a fight than during a matched control period. Hatched histograms show the number of individuals who were as likely to reconcile after a fight as during the matched control period. Nonkin were significantly more likely to reconcile with an opponent's relative after a fight than during matched control periods (two-tailed sign test, P* <*0.02; data from Cheney and Seyfarth 1989).*

The fact that unrelated vervets reconciled with their opponents' kin as well as with their opponents themselves suggests that conflict resolution extends beyond individual opponents to their entire families. There is a good reason for this. Over 20% of all aggressive interactions among female vervets involved alliances of two individuals against a third, and vervets formed approximately 65% of their alliances with family members (Cheney and Seyfarth 1987). Since an aggressive interaction with a particular individual is likely to expand to include other members of that individual's matriline, it may be as important to reconcile with the opponent's family as with the opponent herself, particularly if the opponent is a member of a higher-ranking matriline (see also Judge 1983). Reconciliation with an opponent's relative may have the added advantage of establishing affinitive contact with a relevant, yet uninvolved, individual while nevertheless avoiding the opponent.

We should emphasize that these results are preliminary. They suggest that a fight with a particular individual is an important *context* for subsequent interactions with that individual's kin, but they do not conclusively prove a close *causal relation* between the two events. In most of the studies cited here, it has been difficult to show that individuals interact with their opponent or their opponent's kin more than with any other group member following a fight (see e.g., de Waal and Ren 1988). Perhaps this is because at any given time an individual is dealing with several social agendas: within a half-hour period a female might not only chase a low-ranking individual from food but also groom her own sister, be threatened away from a dominant female's infant, and form an alliance with her daughter. Each of these interactions may influence her subsequent interactions, enormously complicating our analysis. The female might indeed subsequently threaten her low-ranking opponent's kin, but if any of her *other* interactions also affects her behavior, we may be unable to show that the female's interactions with her opponent's kin increase relative to all her other contacts. The only way to resolve this problem is with experiments on captive primates that more precisely control for and limit an individual's interactions.

Knowledge about other animals' relationships may not be restricted to the recognition of matrilineal kin bonds. Consider, for example, the pattern of redirected aggression among pairs of male-female "friends" in savanna baboons (Smuts 1983, 1985; 1987b). As we mentioned in chapter 2, baboon males and females sometimes form long-term pair bonds, or "friendships," in which proximity and cooperative behavior are maintained throughout the female's reproductive cycle (see also Seyfarth 1978a, 1978b; Altmann 1980; and Kaufmann 1965 for rhesus macaques). In some baboon groups, friendships are maintained for years at a time (Smuts 1983, 1985; Strum 1984). In the best documented study of friendships, Smuts (1985) found that females and males often redirected aggression against their opponents' friends. Following a fight with another male, for example, a male frequently appeared to seek out his rival's female friend and chase her. The baboons, in other words, seemed to recognize friendships.

In sum, monkeys in a number of different species appear to observe interactions in which they are apparently not involved and to recognize the relationships that exist among others. In this respect, monkeys make good primatologists. A male considers how strongly a female prefers her partner before he attempts to take her away; juveniles and adult females take note of their opponents' kin as they plot retaliation or reconciliation; and adult females, upon hearing a juvenile's cry for help, learn to expect a response from the mother.

Knowledge About Other Animals' Ranks

Dominance relations in vervets and many other primates are typically transitive (chapter 2). This allows a human observer to assemble, from data on interactions between pairs of individuals, a rank hierarchy that orders the behavior of a large number of animals. The fact that we can derive such hierarchies does not, however, prove that they also exist in the minds of monkeys. It is certainly possible that monkeys attend to each others' dominance interactions and that they recognize rank orders (and transitive relations) among others in their group. Alternatively, each monkey may simply know who is dominant and who is subordinate to herself, having derived this knowledge from personal experience. In the latter case, a dominance hierarchy would occur as an incidental outcome of paired interactions.

Some of the data on status striving presented in chapter 2 suggest that monkeys do indeed recognize the dominance ranks of others. When competing over grooming partners, for example, both female vervets and female baboons supplant each other, on average, most often for access to the highest-ranking individual, next most often for access to the second-highest-ranking individual, third most often for access to the third-highest-ranking individual, and so on (fig. 2.9; see also Seyfarth 1976, 1980). This pattern does not occur simply because high-ranking females spend more time grooming and are therefore more likely to be available as objects of competition; females of different ranks spend roughly equal amounts of time in grooming interactions. The observed pattern, moreover, is consistent across many different individuals. In other words, adult females seem not only to rank one another but also to "agree" on their ranking of the most preferred grooming partners. Similarly, in both pigtailed macaques (*Macaca nemestrina*) (Gouzoules 1975) and olive baboons (Scott 1984) the intensity of male-male competition for mates is directly related to the rank of the female involved.

Additional hints that monkeys are able to judge the ranks of others emerge when we consider the details of social behavior when adult female vervet monkeys compete for access to grooming partners. As noted earlier, competition over access to a grooming partner occurs whenever one female approaches two females who are grooming, supplants one of them, and then grooms or is groomed by the remaining individual. In a small proportion of all cases such competition takes a form that is especially interesting for our present purpose: a high-ranking female (ranked 2, for example, in a group of six adult females) approaches two groomers who are both lower ranking than she is (say, females ranked 4 and 5). Though females 4 and 5 are both subordinate to female 2, they are not equally likely to depart. From 1985 to 1986, in 29 out of 30 interactions that took this form, the higher-ranking female (female 4, in our generic example) did nothing, while the lower-ranking female (female 5) moved away (see fig. 3.8). This result was independent of kin relations among the individuals involved.

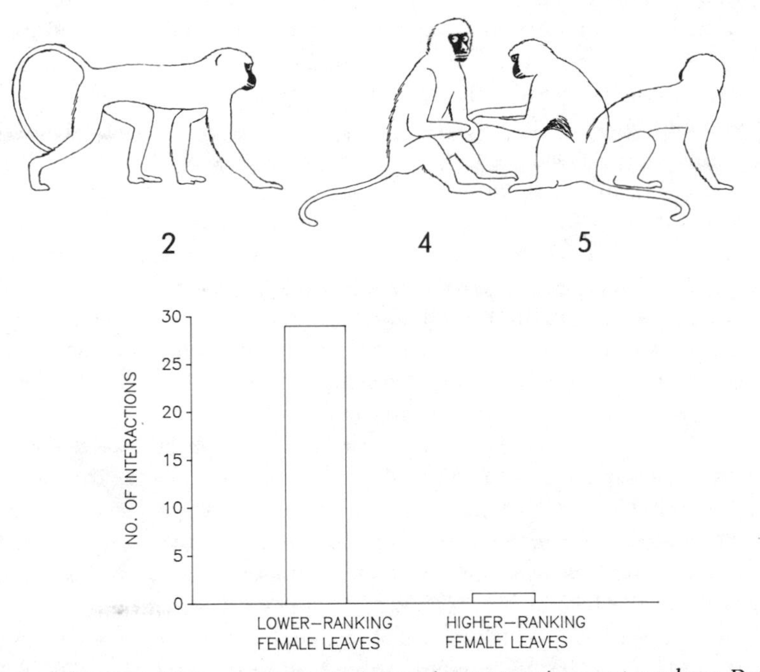

Figure 3.8. Competition over access to a grooming partner in vervet monkeys. Data are taken from all cases from 1985 to 1986 in which a high-ranking female (for example, female 2) approached two lower-ranking females (for example, females 4 and 5) and supplanted one individual and groomed the other. Of the two females who were approached, the lower ranking was significantly more likely to be supplanted (two-tailed sign test, $P < 0.01$). Drawing by John Watanabe.

It is female 4, of course, whose behavior is most interesting. She acts as if she has made the following computation: "We're both subordinate to female 2, so *someone* has to move away. However, female 5 is more subordinate than I am, so I can stay put." Female 4's behavior, therefore, suggests that she recognizes the ranking:

Female 2 > female 4 > female 5

To do this, she must not only know her own status relative to females 2 and 5 but also their status relative to each other. In other words, she must recognize a rank hierarchy.

An alternative explanation might argue that female 2's approach simply has a greater effect on female 5 than it does on female 4. If the probability of a supplant depends on the magnitude of the difference between two individuals' ranks, the result could be explained without positing that the females know each others' ranks. Data gathered in 1985 and 1986, however, do not support this argument. For example, when dominant females approached others who ranked two, three, or four steps beneath them in the hierarchy, the subordinate was supplanted in 61%, 54%, and 63% of all cases, respectively ($N = 101$, 61, and 48 approaches).

It is also possible that, unnoticed by us, female 4 gave a subtle glance or shrug in female 5's direction as female 2 approached. This would certainly have caused female 5 to move away, and again, it would explain our observations without requiring any knowledge of ranking on the monkeys' part. Clearly, we will never be able to exclude this possibility entirely. If these gestures do occur, however, they are extremely subtle and certainly do not resemble any other form of threats or supplants.

Nonhuman primates may not be the only species to rank each other. Linear, transitive dominance hierarchies are common, for example, in wild dogs, hyenas, and a variety of birds (e.g., Frame et al. 1979; Yasukawa 1979; Rowher 1982; Dufty 1986; Frank 1986). In a study of captive goldfinches (*Carduelis tristis*), Popp (1987) observed competitive interactions among individuals at a feeding site that contained two perches. He found that when a dominant bird flew into a site that was already occupied by two subordinate animals, it usually approached and supplanted the more subordinate of the two, as if it recognized the birds' relative ranks (see also Marler 1955, 1956a). As in the case of monkeys, however, simpler explanations are possible. In this instance, rather than recognizing the other birds' relative ranks, the dominant bird may simply have distinguished individuals whose latency to fly off in past interactions was different. The subordinate birds' behavior might reveal more about goldfinches' understanding of dominance hierarchies. Did the bird that flew away recognize that it was more subordinate than the one that stayed?

In the absence of experiments designed specifically to test for animals' understanding of dominance hierarchies, no one set of observational data can ever prove decisively that monkeys recognize each others' ranks. For the moment, we can only conclude that a variety of data from a number of different species suggest that monkeys can rank one another.

We turn now to the question of how they might do it. Consider first some experiments by Michael D'Amato and his colleagues (D'Amato and Colombo 1988; see also D'Amato and Salmon 1984; D'Amato et al. 1985). In these tests, captive brown capuchins (*Cebus apella*) were trained to respond to five stimuli (a circle, a plus sign, a dot, a vertical line, and an hourglass (hereafter A, B, C, D, and E) in a specified order: first AB, then ABC, then ABCD, and finally ABCDE. To test the animals' knowledge of the sequential position of each item, subjects were given pairwise tests (for example, BC or DA) and rewarded for responding only to pairs that appeared in the correct sequential order. The monkeys performed well. In addition, their latency to respond was shortest when the first item in the test series was A, longer when it was B, longer still when it was C, and so on. Their latency was also shortest when the two items in the test series were adjacently ranked, longer when they were separated by one item, and longer still when they were separated by two items. D'Amato and Colombo believe that these results demonstrate "an internal representation of the sequential order of the five items" (p. 136; see also D'Amato and Colombo 1989; and 1990).

D'Amato and colleagues argue that the representation of rank order in capuchins is based on *associative transitivity*, which they contrast with *transitive inference*. In associative transitivity no inference is involved because there is nothing in the intitial conditional discrimination that demands a particular pairing of stimuli on the test trials. There is no underlying rule, in other words, that is common to the pairs AB and BC. As a result, in the absence of prior association, the subject has no way of inferring that in the test trials A should be linked with C. In many respects, the experiments test only whether monkeys are capable of ordering stimuli sequentially.

By contrast, experiments that test for transitivity in children (Bryant and Trabasso 1971), squirrel monkeys (McGonigle and Chalmers 1977), and chimpanzees (Gillan 1981) have all involved identification of a *relation* between the training stimuli: for example, A is longer than B, B is longer than C, and so on. This may have allowed transitivity to be inferred on subsequent tests (D'Amato and Salmon 1984). Gillan, for example, taught chimpanzees that container E had more food than container D, D had more food than C, C more than B, and B more than A. He then tested individuals on novel pairs like BD, BE, and CE. The animals consistently chose the container in each pair that was associated with the greater amount of food. In this and other tests, it seems possible that subjects inferred the relation *greater than* and solved test problems according to this relational rule rather than according to the prior association of particular stimuli (for alternative explanations see McGonigle and Chalmers 1977; Breslow 1981; D'Amato and Salmon 1984).

Although socially living monkeys seem to recognize the dominance ranks of others, we know very little about how these ranks are learned or how ranks are represented in the animals' minds. One means by which a monkey might acquire information about other animals' ranks is simply through *brute force*, a method similar to D'Amato's associative transitivity. By this method, a monkey just observes and remembers all possible dyadic interactions among other group members until he is able to conclude that A is dominant to everyone, B is dominant to everyone but A, C is dominant to everyone but A and B, and so on. The brute force method does not require the ability to make transitive inferences, but it does demand that a monkey observe at least one interaction between all other group members before constructing a dominance hierarchy. In contrast, a monkey who could make transitive inferences about rank relations among other group members could construct a linear dominance hierarchy on the basis of partial information without having to observe interactions among all pairs of individuals.

At present no data allow us to choose between these alternatives, though tests on captive squirrel monkeys (McGonigle and Chalmers 1977) and chimpanzees (Gillan 1981) suggest that transitive inference is at least possible. In some cases, it is difficult to explain the behavior of monkeys in large groups without assuming that the animals are using the more efficient method of transitive inference. Although vervet monkeys typically live in groups of fewer than 30 individuals, macaque and baboon groups commonly exceed 100 members. Observers often report spending months with a group without ever seeing some individuals interact. Yet when data on social interactions within such groups are analyzed (e.g., Scott 1984) there is still evidence that the animals construct rank orders of their fellow group members. Since these rank orders include individuals who interact only rarely, it seems probable that their places have been calculated through observation of a subset of dyadic interactions, with the additional assumption that ranks are transitive.

Recognizing Other Animals' Relations

We can think of social groups of monkeys and apes as being composed of many different long- and short-term alliances among related and unrelated animals. To gain a social and reproductive advantage over others, individuals must be able not only to predict each others' *behavior*, but also to assess each others' *relationships*. It is not enough to know who is dominant or subordinate to oneself; one must also know who is allied to whom and who is likely to come to an opponent's aid. For this reason, we might predict individuals to be sensitive to other animals' relationships in any species in which alliances are common (Harcourt 1988).

Monkeys do seem to recognize the social relationships that exist among other group members, and judgments about these relationships seem to underlie much of their behavior. Males assess the closeness of bonds between other males and their females before attempting a takeover; females assess the ranks of others when competing for grooming partners; and females and juveniles, apparently recognizing the ways in which matrilineal kin act in unison, direct reconciliation or retaliation to their opponents' kin as well as to their opponents themselves.

Knowledge of other animals' social relationships can only be obtained by observing the behavior of others and making the appropriate deductions. Primates may not be limited to an egocentric view of the world but may be able to step outside their own immediate experience to make judgments about the experiences of others. Monkeys, in this respect, seem to differ from Anthony Powell's infamous character Widmerpool, who was "one of those persons capable of envisioning others only in relation to himself." The assessments that monkeys make of one another, moreover, are not simple, but seem to occur along at least two dimensions simultaneously. Classification of individuals on the basis of kinship or close association, on the one hand, and on the basis of dominance rank, on the other, means that two individuals may sometimes be lumped together as members of the same family, while at other times they may be considered separately, one ranking higher than the other.

Clearly, we can make no precise statements about the mechanisms that underlie the monkeys' knowledge of each others' relationships. Although there are hints that certain "higher" cognitive processes may sometimes be involved, we cannot rule out the possibility that most social knowledge is based on relatively simple associative learning (Dasser 1985). For example, even if monkeys do have abstract concepts like *closely bonded*, their knowledge of these bonds may derive principally from associations formed between individuals who interact at high rates. Similarly, although there is evidence that monkeys infer transitive relations when constructing a dominance hierarchy, it is also possible that they simply memorize the outcome of every possible dyadic interaction.

Although we cannot state conclusively that monkeys recognize each others' close bonds and dominance ranks through inference rather than brute memory, the ability to classify others into abstract categories like *closely bonded* would have at least two functional advantages. First, it would allow individuals to identify types of relationships quickly and to predict the behavior of others based on partial information. As a result, a monkey who joins a new group, or whose group receives an influx of migrants, could make accurate predictions about behavior without having to observe interactions between each pair of individuals. Second, as group size increases, the ability to form categories and to make judgments based on these categories would provide an increasingly efficient method for recognizing the characteristics of relationships and predicting what specific individuals are likely to do next.

The Representation of Social Relationships

The Problem

Nonhuman primates classify other individuals according to their patterns of association and seem to recognize the bonds and enmities held by individuals other than themselves. Humans, though, go several steps further, to classify different types of relationships into superordinate categories that are independent of the particular individuals involved. If a friend mentions a sister, an uncle, or a husband, we immediately have some idea of the nature of her relationship with the other person, even if we have never met the individual in question. And if the friend tells us that her uncle wrecked her new car and her husband closed her bank account and left town, we are shocked at least in part because their behavior is at variance with what we typically expect of people in these categories. In fact, it could easily be argued that humans are overly eager to classify relationships. "The friend of my enemy is also my enemy" is, cognitively speaking, a delightfully complex concept, redolent of all sorts of inference, transitivity, and classification. It can, however, lead to awkward overgeneralizations and less than adaptive behavior. Is there any evidence that monkeys, too, classify social bonds into higher-order units that allow relationships to be compared independent of the individuals involved?

To examine this problem we begin with a brief discussion of the evidence for concept formation in animals. We then describe the crucial distinction between judgments based on the elements of a stimulus (its shape or size, for example) and judgments based on the relation between two stimuli. We then discuss the evidence for a concept of social relationships in monkeys.

Animal "Concepts"

Many animals appear to classify objects according to "concepts"—relatively abstract criteria that are not based on any single perceptual feature (Lea 1984). In a classic test of concept formation in birds, Herrnstein and Loveland (1964) showed slides to pigeons (*Columba livia*) and rewarded them for pecking only when they saw a slide containing one or more trees. When the pigeons had achieved a certain level of performance, they were tested with hundreds of novel slides. Slides containing trees—tall trees, leafless trees, even parts of trees—were more likely to receive pecks than stimuli without trees. The authors concluded that pigeons could generalize from specific stimuli to recognize the members of a general class. In short, the pigeons had apparently formed a concept of *tree*. Similar results were obtained when pigeons were rewarded for classifying pictures of people, fish, and human-made breeds of pigeons (Herrnstein 1979, 1985).

What was the nature of the pigeons' concept of tree? Although the question is difficult to answer precisely, the pigeons did not seem to be responding on the basis of perceptual similarity, since leaves and bark do not resemble trees. Indeed, no single set of perceptual criteria was necessary or sufficient to account for the birds' behavior. The general case for a mental representation seemed strong; the exact basis of this representation, however, was unclear.

Similar experiments conducted on monkeys produced mixed results. Schrier, Angarella, and Povar (1984), for example, trained stump-tailed macaques to respond to a series of slides of monkeys or humans. Although all subjects then transferred to novel slides, they showed less accuracy than pigeons. In further experiments that required rhesus macaques to discriminate people from a large set of items, Schrier and Brady (1987) found stronger evidence for classification of novel slides after training (see also Sands, Lincoln, and Wright 1982).

Convincing as these tests might at first appear, the notion of concepts in pigeons and even in monkeys has been questioned on several grounds. First, there is evidence that the pigeons' performance is at least partially the result of rote memory. Vaughan and Greene (1984) arbitrarily divided slides into positive and negative exemplars by tossing a coin. Pigeons correctly discriminated the two classes, even when tests involved over 100 slides. Apparently, the acquisition of a "concept" in pigeons is as rapid when the stimuli are *randomly* sorted as it is when the stimuli have some feature in common (see also Herrnstein 1985, 1990 for reviews). Many species of birds and mammals have the capacity to remember vast numbers of food storage sites and previously visited food patches (chapter 9), so it is perhaps not surprising that some aspects of category classification might be based on memory. Nevertheless, as Herrnstein (1990) points out, even a huge capacity to remember arbitrarily classified exemplars is unlikely to be sufficient under natural conditions, where the number of exemplars of a given class (for example, acorns for a foraging squirrel) is likely to be open ended.

Animals could also classify items by generalizing some component or collection of components from previous instances to novel exemplars. In an effort to address this issue, D'Amato and van Sant (1988) trained capuchin monkeys to distinguish between slides that did or did not contain a person or part of a person. They found that the monkeys transferred easily to novel slides, suggesting that they had some concept of person. But what was the discrimination based on? A number of factors suggested that the monkeys might not have formed an abstract concept of person but might instead simply have generalized a number of features (including, oddly, the presence or absence of the color red), which, if present, were sufficient to identify humans. In fact, as D'Amato and van Sant (1988:52) argue, it is extremely difficult in principle to distinguish between stimulus generalization and the formation of an abstract concept, because both are based to at least some degree on physical similarity. As a result, "Transfer to new positive examplars might arise from the animal's abstracting a variety of relevant features from previous positive instances and assembling them into an abstract representation, such as a prototype, which would qualify as concept-mediated transfer. Or it might be due to 'mindless' generalization from a specific relevant or even irrelevant feature of previously encountered positive examplars." D'Amato and van Sant also point out that their results are somewhat paradoxical, because "Anyone who has worked with monkeys . . . would have difficulty believing that their responses to humans are totally devoid of a conceptual basis" (1988:54).

Premack (1983b) also suggests that class discrimination in pigeons probably depends on the recognition of a small set of relevant features. He argues that only language-trained apes probably make regular use of abstract categories; other animals rely on a more concrete, "imaginal" code. In an explicit test of this hypothesis, Roberts and Mazmanian (1988) investigated the ability of pigeons, squirrel monkeys, and humans to discriminate three classes of increasing abstraction. At the most concrete level, subjects were asked to discriminate kingfishers from other birds. At a more abstract level, they were required to discriminate birds from other animals. Finally, at the most abstract level, they had to discriminate animals from nonanimals. Humans, not surprisingly, had no difficulty making any of these discriminations. The pigeons and squirrel monkeys performed well on the kingfisher/other bird problem. After some training, they also learned to identify animal pictures correctly, although they continued to have difficulty classifying birds and other animals. Generally, squirrel monkeys performed better than pigeons on the animal/nonanimal discrimination, which Roberts and Mazmanian argue must have been based on *some* abstract concept, since it is difficult to see any shared common feature between, say, a buffalo and a spider. The critical features used by pigeons and monkeys to identify animal pictures, however, remain unknown (see also the discussion by Medin and Smith 1984).

The ambiguous results obtained from these experiments suggest that the notion of a concept, at least when applied to animals, is inherently vague. We may be able to agree that a concept is something more than a collection of individual exemplars or even a prototype, but in many respects its precise features cannot help but remain elusive. D'Amato and van Sant (1988) argue that further efforts to identify concepts in animals through the discrimination of photographs may be futile. Monkeys might well have a highly developed concept of humans that simply cannot be revealed by lifeless, two-dimensional photographs. D'Amato and van Sant recommend that future research concentrate on identifying the mechanisms used by different species to classify stimuli in their environment. Different species clearly rely on different means to achieve the same result. Just because an animal classifies slides in the same way that a human might does not mean that it has formed the same concept.

Humans classify objects not only according to physical similarity but also according to function. Moreover, they label categories and can identify single items within a category while simultaneously recognizing them as part of a general class. To date, the only nonhuman species that have demonstrated similar abilities in the laboratory are language-trained chimpanzees. Chimpanzees can learn to classify objects such as toys or tools not just according to perceptual criteria but also according to their function. The chimpanzees trained by Premack (1976, 1986), for example, easily identified different parts of fruit, such as seeds, as *fruit,* even when they were using symbols for fruit. Similarly, the chimpanzees Austin and Sherman not only grouped different items into superordinate functional classes, like *toy* or *food* but also grouped the symbols for different toys and fruit into these superordinate classes (Savage-Rumbaugh et al. 1980; chapter 5). In all cases, however, the classification of objects into superordinate classes was preceded by a period of time when human trainers taught the chimpanzees the appropriate labels for these classes. We simply do not know whether chimpanzees ever classify items naturally in the absence of human intervention.

Judgments Based on Elements, Relations, and Relations Between Relations

Even if pigeons do recognize resemblances among stimuli belonging to the same class, it is difficult to train them to manipulate their internal representations of stimuli to make, for example, same/different judgments between them (Premack 1983b; Herrnstein 1985). Monkeys, in contrast, can readily be taught to solve problems that require recognition of a *relation* between objects rather than a specific physical attribute. In oddity tests, for instance, a subject is presented with three objects, two of which are the same and one of which is different. He receives a reward only if he chooses the different object. If a small number of stimulus objects are recombined trial after trial, the subject could, of course, achieve above-chance performance through associative learning. Many monkey species, however, achieve scores of 80 to 90% correct even when new stimuli are used for each problem and a given set of stimuli is presented for only one trial (e.g., Harlow 1949; Strong and Hedges 1966; Davis et al. 1967). Such levels of performance suggest that animals are using an abstract hypothesis—*pick the odd object*. The hypothesis is called abstract because *odd* does not refer to any specific stimulus dimension, as does *red* or *square*. Instead, oddity is a concept that specifies a relation between objects independent of their specific attributes (Essock-Vitale and Seyfarth 1987; Roitblat 1987).

Judgments based on relations among items have been demonstrated more often in nonhuman primates than in other taxa, and primates seem to recognize abstract relations more easily than at least some other animals. Although it is possible, for example, to train pigeons to recognize such relations, the procedural details of the test appear to be far more critical for pigeons than they are for monkeys, and the relational distinction can easily be disrupted (Wright et al. 1983; Herrnstein 1990). Other studies, however, suggest that the dichotomy between primates and other animals is less distinct than the tests with pigeons imply. Pepperberg (1983, 1987), for instance, taught an African grey parrot (*Psittacus erithacus*) to make same/different judgments about the color, shape, and material of objects (see also Schusterman 1988 for data on sea lions). The same parrot was also able to generalize numerical discriminations from training sets to novel items. Comparable numerical abilities have been demonstrated for rats (Church and Meck 1984; Capaldi and Miller 1988) and chimpanzees (Matsuzawa 1985; Boysen and Berntson 1989), suggesting that many species may have a concept of numerosity that is based on relatively abstract criteria (see also Gallistel 1989a and 1989b).

Premack (1983b, 1986) contends that tasks like oddity tests require only judgments about relations between elements, not relations between relations. By contrast, judgments about relations between relations are involved in tasks like analogical reasoning. They are less fundamental and universal than judgments about relations between elements, and thus far have been demonstrated only in language-trained chimpanzees.

In his study of analogical reasoning in chimpanzees, Premack (1976, 1983b) trained an adult female subject, Sarah, to make same/different judgments between pairs of stimuli (see chapters 5 and 8 for further discussion). Once Sarah could use the words *same* and *different* correctly even when she was confronted with entirely new stimuli, she was shown two pairs of items arranged in the form A/A′ and B/B′. Her task was to judge whether the relation shown on the left was the same as or different from the relation shown on the right. Alternatively, Sarah was given an incomplete analogy like A/A′ *same as* B/? Her task then was to complete the analogy in a way that satisfied this relation. Sarah performed these analogy problems with apparent ease.

In the most complex test, the objects presented to her shared no obvious physical similarity. For example, Sarah was asked "lock is to key as closed paint can is to ——"; the options for completing the analogy were a can opener and a paint brush. Here the identity between two such relations is not based on physical similarity (in fact they look quite different), but on the underlying relation *opening*, which both cases instantiate. Hence it is not the stimuli themselves but this relation that must be represented in the mind of the chimpanzee. To solve an analogy, the chimpanzee must infer the appropriate relation for each stimulus pair and then compare these two relations to see if they are the same (Gillan, Premack, and Woodruff 1981; Premack 1983b). In other words, she must somehow form a representation of the concept instantiated by each pair and then compare these representations. Again, Sarah solved the problems correctly.

Premack (1983b) argues that the ability to form such abstract representations is enhanced by, and may require, language training. His claim is not that chimpanzees naturally lack the ability to reason abstractly. Instead, he believes that all primates possess the potential for such skills but that only chimpanzees given language training are able to realize this potential.

The Assessment of Social Relationships by Monkeys

Premack's tests prompt one to ask whether group-living primates might use abstract criteria to make relational judgments about their social companions. A comparable problem in the social domain might concern the judgment of relations within different kin groups: is the relation mother A/infant A the same as or different from the relation mother B/infant B (Cheney and Seyfarth 1982c)? Premack's analogy tests therefore bring us back to the central question of this section: is there any evidence that primates, in their assessment of each others' behavior, ever classify relationships using criteria that are not restricted to the particular individuals involved?

The data that most directly address this question come from a study conducted by Verena Dasser (1988a) on a group of 40 captive long-tailed macaques. With considerable effort, Dasser trained two adult females so that they could be temporarily removed from the group and placed in a small test room to view slides of other group members. In one test that used a simultaneous discrimination procedure, the subject saw two slides. One showed a mother and her offspring and the other showed an unrelated pair of group members. The subject was rewarded for pressing a response button below the mother-offspring slide. Having been trained to respond to one mother-offspring pair (five different slides of the same mother and her juvenile daughter), the subject was tested using 14 novel slides of different mothers and offspring paired with 14 novel pairs of unrelated animals (all of which included at least one adult female). The mother-offspring pairs varied widely in their physical characteristics. Some slides showed mothers and their infant daughters; others showed mothers and juvenile sons or mothers and adult daughters. Nonetheless, in all 14 tests the subject correctly selected the mother-offspring pair.

In a second test that used a match-to-sample procedure (or, more accurately, a match-to-a-relation-of-the-sample), the mother was represented as the sample on a center screen, while one of her offspring and another stimulus animal of the same age and sex as the offspring were given as positive and negative alternatives, respectively. Having learned to select the offspring during training, the subject was presented with 22 novel combinations of mother, offspring, and an unrelated individual. She chose correctly on 20 of 22 tests ($P < 0.001$).

Finally, to test whether monkeys could recognize other categories of social affiliation, Dasser (1988b) trained a subject to identify a pair of siblings and then tested the subject's ability to distinguish novel sibling pairs from mother-offspring pairs, pairs of otherwise related group members, like aunts and nieces, and pairs of unrelated group members. The subject correctly identified the sibling pair in 19 of 27 tests ($P = 0.05$). Seven of the eight errors occurred when she was asked to compare siblings with a mother-offspring pair; one occurred when she compared siblings with two less closely related members of the same matriline.

Data on redirected aggression and reconciliation in vervet monkeys provide additional evidence that animals classify social relationships into types that are not restricted to the particular individuals involved. Recall that in some monkey species redirected aggression and reconciliation are kin biased, so that animals often interact with the kin of their prior opponents. In vervet monkeys, redirected aggression and reconciliation can extend even to the previously uninvolved kin of prior opponents. Data gathered in two social groups over two different time periods showed that an animal was more likely to threaten another individual if one of her own close relatives and one of her opponent's close relatives had recently been involved in a fight (see fig. 3.9; Cheney and Seyfarth 1986, 1989). We call this behavior *complex redirected aggression* for brevity. The same was true of reconciliation: two unrelated individuals were more likely to engage in an affinitive

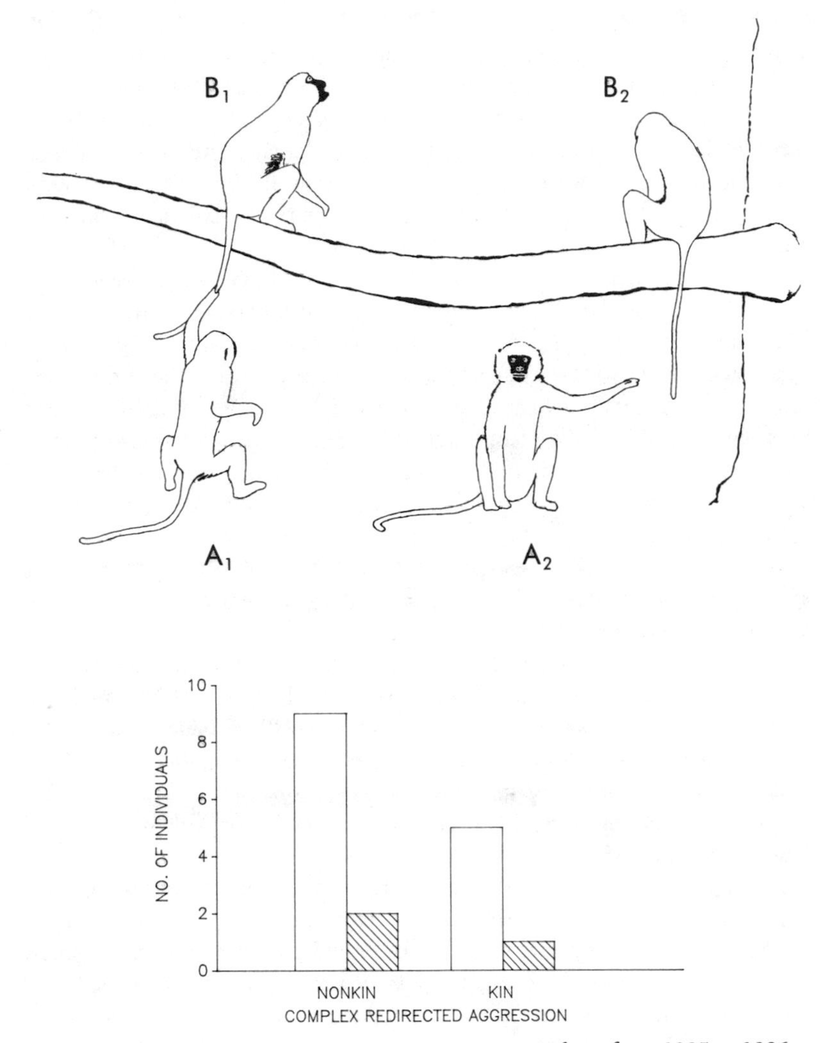

Figure 3.9. Complex redirected aggression in vervet monkeys from 1985 to 1986. Open histograms show the number of individuals who behaved aggressively toward an opponent more after a fight between one of their own relatives and one of their opponent's relatives than during a matched control period. Hatched histograms show the number of individuals who were as likely to act aggressively after a fight as during the matched control period. Nonkin were significantly more likely to act aggressively toward an opponent after a fight between their own relatives and their opponent's relatives than during matched control periods (two-tailed sign test, P <0.05). Among kin, the difference approached significance (P = 0.11; data from Cheney and Seyfarth 1989). Drawing by John Watanabe.

interaction following a fight between their close kin than during matched control periods. So, in the example given at the beginning of this chapter, the fight between Macaulay and Carlyle apparently caused Shelley, Carlyle's sister, to attack Austen, Macaulay's sister. Of course, the parallel is not exact: if the prior opponents were both adult females this did not necessarily mean that the subsequent opponents would both be their daughters. Vervet families are simply too small for such perfectly balanced analogies even to arise.

Once again, these results are preliminary. They have not yet been replicated in other populations, and possible confounding factors have not been eliminated entirely. In particular, we have not yet ruled out the possibility that vervets do not redirect aggression specifically toward the kin of their relatives' opponents but instead simply generalize across *all* members of a given family. As we have mentioned, many aggressive interactions involve coalitions among related animals, and vervets frequently threaten *both* their relatives' opponents and their relatives' opponents' kin. So, for example, when A1 threatens B1, A2 may threaten both B1 and B2 within the next few hours. It therefore remains possible that animals do not so much seek out specific members of a given family as retaliate against the entire family. This *stimulus generalization* argument seems unlikely, however, given the observation that vervets behaved differently following fights with their *own* kin than following fights with unrelated opponents. Recall that unrelated animals were more likely to reconcile with their opponents' kin than with their opponents themselves, while related animals were more likely to reconcile directly with their opponents. Had monkeys simply generalized their behavior across entire matrilines, there should have been no difference within and between matrilines.

Bearing in mind the speculative nature of these results, this more complex form of redirected aggression in vervet monkeys supports Dasser's experiments in suggesting that monkeys recognize that certain types of social relationships share similar characteristics. When a vervet monkey (say, A2) threatens an unrelated animal (B2) following a fight between one of her own relatives (A1) and one of her opponent's relatives (B1), A2 acts as if she recognizes that the relationship between B2 and B1 is in some way similar to her own relationship with A1. In other words, we may think of A2 as having been presented with a natural problem in analogical reasoning:

A1/B1 *same as* A2/?

A2 correctly completes the analogy by directing aggression to another member of the B family.

Definitive proof that monkeys are capable of solving social analogies, and that language training is not a necessary prerequisite, can only come from laboratory tests. We can imagine, for example, an experiment in which a monkey is asked the following question of its fellow group members: mother A is to infant A as mother B is to infant B, juvenile B, or infant C? Dasser's results and our less rigorous observational data suggest that monkeys would solve this problem with ease. The relevant tests, however, have not yet been attempted.

We have no idea how monkeys might complete social analogies, much less how they might represent social relationships in their minds. One possibility is that they use physical resemblance as a cue, since members of the same matriline often, but not always, look alike. Note, however, that vervets and long-tailed macaques treat bonds between kin as similar even when they involve pairs of animals whose within-family resemblances, at least to a human observer, are markedly different. In Dasser's (1988a) study, for example, subjects generalized to a diverse array of mother-offspring pairs (mothers and very young infants; mothers and juvenile sons; mothers and adult daughters), even though they had been trained with only one example from this category. Similarly, male and female baboon "friends" do not resemble each other, yet other baboons nevertheless seem to recognize that certain pairs of individuals associate at high rates (Smuts 1985).

There is no hard evidence that vervets or any other monkey species recognize kinship in any sense other than as a close association between two individuals (e.g., Frederickson and Sackett 1984). Association rates, however, do not entirely explain differential treatment of kin and nonkin, because kin do not always interact at higher rates than nonkin. Even the same types of kinship bonds are not always characterized by similar kinds and rates of interactions. Some mother-offspring pairs, for example, are close and interact often, whereas others are more distant (e.g., Hinde 1974; Altmann 1980). All, however, are placed in the same category of social relationship (Dasser 1988a). Furthermore, while bonds within matrilineal kin groups can be extremely variable (depending, for example, on the ages and sex of family members), monkeys nevertheless treat competitive interactions as pitting one family against another (Dunbar 1983a; Walters 1987; Cheney and Seyfarth 1986, 1989).

In sum, monkeys seem to use a metric to classify social relationships that cannot be explained simply in terms of physical features or the number and type of interactions. Instead, their classification seems based on an abstraction that includes all of these elements. These observations raise the possibility that monkeys recognize some similarity among kinship bonds across different families. To recognize that certain sorts of bonds share similar characteristics independent of the particular individuals involved, monkeys must compare each other not according to physical features or a specific type of interaction, but according to an underlying relation that has been abstracted from a series of interactions over time. Monkeys take note of the elements that make up a relationship (grooming, alliances, and so on). They then make judgments of similarity or difference not by comparing specific elements but by comparing the different relationships that these elements instantiate.

Throughout this section, we have drawn attention to the parallels that exist between the solution of analogical reasoning problems by a language-trained chimpanzee and the recognition of social alliances by vervet monkeys. Although we believe the parallels are significant, we still know very little about the cognitive mechanisms used by the monkeys, and it does not necessarily follow from our results that vervets employ an analogical format to represent and classify social relationships.

Among humans, categories of social relationships are often articulated by labels such as friends, lovers, or enemies. Labeling allows us to characterize and compare relationships even between unfamiliar people, and it makes transfer to new stimuli immeasurably easier. Accurate use of a category like *friend* implies that we recognize the category's characteristics and can use the category in novel situations. In some of the ape language studies animals have been taught symbols that allow them to label relations like *greater than* or *less than* (Premack 1976) or classes of items like *tools, food,* or *color* (Savage-Rumbaugh et al. 1980). As a result, their ability to deal with novel stimuli improves markedly (Matsuzawa 1988).

Under natural conditions, however, monkeys do not seem to use labels to identify superordinate classes or their constituent parts. Vervets appear to have no vocalization, for example, that refers to family or matriline, much less to any of the particular individuals that comprise a family. Moreover, there is no evidence that monkeys recognize relations between relations in any context other than a social one. We consider this issue in more detail in chapters 6, 8, and 9.

Summary

In chapter 2 we argued that the social behavior of vervet monkeys and many other nonhuman primates is best explained if we assume that individuals are pursuing a number of strategies. The monkeys act as if they are striving to maintain close relations with kin and to establish, whenever possible, bonds with the members of high-ranking families. When interacting outside their immediate family they strive for reciprocity and are more willing to cooperate with another animal if he or she has recently cooperated with them. Our explanations of primate social behavior therefore imply that monkeys possess a sophisticated knowledge of their social companions and the relations that exist among them.

In this chapter we have explored the extent and the limitations of primate social intelligence. Monkeys do, indeed, seem to recognize the members of other groups and to associate these individuals with particular areas, even though they interact with them only rarely. Within their own group, monkeys recognize others as individuals, remember who has cooperated with them in the past, and adjust their future cooperative behavior accordingly. In this respect, the monkeys' behavior is consistent with theories that predict reciprocal social exchange. It is difficult, however, to determine whether monkeys possess a concept of reciprocity like our own, because it is difficult for us, as observers, to determine precisely what costs and benefits the monkeys attach to different sorts of interaction.

Although individuals of many animal species seem to know a great deal about their own social relations, the social knowledge of primates is most striking when we consider what a monkey knows about the social relations of *others* in her group. A variety of evidence indicates that monkeys recognize the close associations that exist among others, both within matrilineal kin groups and in the long-term pair bonds of particular males and females. Monkeys may also recognize the dominance relations of others—knowledge that allows them to construct a dominance hierarchy of their fellow group members.

Finally, monkeys seem not only to recognize the relationships of others but also to compare relationships, judging some to be similar and others to be different. The criteria for such judgments appear to be independent of the particular individuals involved. Monkeys seem to recognize, for instance, that bonds within families are similar even when the individuals involved are of widely different ages and sex. Monkeys therefore seem capable of classifying relationships according to one or more abstract properties. In other words, they represent social relationships in their minds and compare relationships on the basis of these representations. Whether they are at all aware of what they are doing, or whether they can employ this ability outside the context of social interactions, remains to be determined.

CHAPTER FOUR | VOCAL COMMUNICATION

During the 1981 All-England tennis championships at Wimbledon, officials found themselves confronted with a dilemma. Some players, particularly men like Jimmy Connors, were regularly giving loud grunts as they hit the ball. The grunters' opponents protested to the officials, asking that this behavior be stopped. Opponents claimed that grunts were distracting and were given on purpose to throw them off their timing. But when the officials confronted players like Connors, they received a slightly different explanation. "Sure," explained Connors, "some players do grunt on purpose—but not me. I really have no control over my grunting, it just *happens* when I hit the ball hard." Like Connors, most of the other grunters were willing to admit that *some* players did grunt intentionally, but each denied that he himself had any conscious control over these particular vocalizations. When the officials tried to determine, by observing different players, which grunts were intentional and which were not, they found the distinction virtually impossible to make. The only conclusion they could agree on was that grunts were indeed distracting, regardless of whether they were given on purpose or just happened as part of the exertion of hitting a ball hard.

The Wimbledon referees' dilemma captures many of the problems confronted by anyone studying the vocalizations of monkeys and apes. Vervet monkeys, for example, vocalize to one another in a variety of circumstances: they give loud alarm calls when they see a predator, *wrrs* and chutters in encounters with other groups, threat grunts and a different kind of chutter in fights with members of their own group, and quiet, guttural grunts during relaxed social interactions. In each case it is, quite simply, impossible to tell whether a monkey deliberately intends to communicate to another, or whether the monkey has no control over his vocalizing and calls simply come out as part of his ongoing behavior.

control?

The two interpretations of grunting at Wimbledon, that they are given deliberately with an intent to distract and that they just emerge involuntarily, also exemplify the dichotomy drawn during the 1960s and 1970s between human speech and the vocalizations of monkeys and apes (reviewed in Marler 1977a, 1978). Human speech, everyone agreed, was under voluntary control, could be detached from emotion (we can talk about fear without actually being afraid), and involved activity in certain higher cortical areas of the brain. In contrast, laboratory experiments and the descriptions of vocalizations that were beginning to emerge from early field studies suggested that nonhuman primate vocalizations were relatively involuntary, occurred only in highly emotional circumstances, and were under limited higher cortical control (e.g., Myers 1976; Ploog 1981). In addition, although human words were known to represent, or stand for, objects and events in the external world, the calls of nonhuman primates were thought never to represent anything other than an individual's emotional state or his imminent behavior (e.g., Lancaster 1975; Premack 1975). At the time, these conclusions seemed entirely reasonable, since a strict dichotomy between speech and the calls of monkeys and apes was supported by data from at least five quite different but complementary sources.

First, early attempts to condition primate vocalizations were unsuccessful. Even when they were rewarded for doing so, monkeys apparently could not learn to vocalize selectively, producing calls at certain times and withholding them at others. Since vocalizations were defined as voluntary only if their delivery was not obligatory and if they could be extinguished or reinforced through conditioning, investigators concluded that calling in primates was an involuntary reflex (e.g., Myers, Horel, and Pennypacker 1965; Yamaguchi and Myers 1972; Myers 1976).

Second, neurophysiological studies of squirrel monkeys suggested that seemingly normal vocalizations could be elicited through electrical stimulation of subcortical areas in the brain, like the anterior limbic cortex, an area long known to be associated with emotional expression in humans (e.g., Robinson 1967; Jurgens et al. 1967; Jurgens and Ploog 1970). The elicitation of apparently normal vocalizations by stimulating the limbic system led some researchers (e.g., Washburn 1982) to conclude that higher cortical areas are not involved when animals vocalize under natural conditions.

Third, reviews of natural communication among birds and mammals (e.g., Smith 1965, 1969, 1977) found little evidence that animal vocalizations ever denoted, or referred to, objects and events external to the signaling individual. Instead, borrowing a term from linguistics (Abercrombie 1967; Lyons 1972), animal vocalizations were assumed to be largely *indexical*, that is, concerned with the signaler or his subsequent behavior. The dance "language" of the honey bee (von Frisch 1967) was considered a notable exception. When this ethological view was considered in conjunction with the existing neurophysiological data, some investigators concluded that the close link between a vocalization and the signaler's subsequent behavior was not accidental but could be traced to common physiological mechanisms. Like Jimmy Connors' grunts on the tennis court, animal calls were good predictors of subsequent behavior because the same events that produced the behavior also produced the vocalization (e.g., Myers 1976; Ploog 1981).

Fourth, research on the perception of speech revealed that humans perceive an acoustically graded continuum of speech sounds as a series of relatively discrete categories (Liberman et al. 1967; Liberman 1982). Many linguists believed that such categorical perception was a necessary precursor of language. On the basis of their own experience and acoustic analysis using sound spectrographs, field biologists noted that monkeys and apes also had *discrete* and *graded* calls in their repertoires. Discrete calls were easy to distinguish by ear in the field and were acoustically different from one another. Other call types graded into one another and were generally difficult to distinguish—by ear or when displayed on a sound spectrogram (e.g., Marler 1965; Struhsaker 1967a). Repertoires of discrete signals were thought to be common among many forest monkeys, who live in a habitat where visibility is poor and long-distance communication predominates. Such calls allow animals to communicate unambiguously, even in the absence of supporting visual signals. By contrast, monkeys and apes living in open country were described as having a repertoire in which signals grade acoustically into each other, and they were thought to use these signals at close range in combination with visual cues (Marler 1965). Peter Marler (1976a) hypothesized that a crucial step in the evolution of language occurred when early humans began to perceive graded vocal signals in a discrete manner.

A final dichotomy between human language and nonhuman primate vocalizations concerned learning. Humans obviously learn many aspects of language, and language learning can be severely disrupted by early deafness (Feldman, Goldin-Meadow, and Gleitman 1978) or social isolation (Curtiss 1977). In contrast, research on the ontogeny of calling in squirrel monkeys revealed neither gradual development nor dependence on environmental cues. Winter and colleagues (1973), for example, found that squirrel monkeys could apparently produce their species' entire vocal repertoire, without anything resembling practice, 6 days after birth. Similarly, Talmadge-Riggs and colleagues (1972) found no vocal abnormalities in squirrel monkeys deafened shortly after birth or in animals raised with muted mothers. Third, as recently as 1984 Herzog and Hopf demonstrated that squirrel monkey infants who had been reared in isolation were nevertheless able to distinguish and respond appropriately to the different alarm calls given by members of their own species.

In recent years, these dichotomies drawn between human and nonhuman primate vocalizations—voluntary versus involuntary, referential versus indexical, graded versus discrete, learned versus unmodifiable—have gradually broken down, at least partly as a result of new techniques employed in field and laboratory research. In the well-known ape language projects, for example, investigators have achieved striking success in their attempts to teach some elements of human language to chimpanzees, gorillas, and orangutans (see Gardner and Gardner 1975; Terrace 1979; Miles 1983; Premack 1986; and Savage-Rumbaugh 1986 for reviews). The results of these studies suggest that at least some of the cognitive mechanisms thought to underlie human language are present in the great apes.

Furthermore, as a result of recent work it now appears that primate vocalizations can be conditioned (Steklis and Raleigh 1979). In one study, for example, rhesus macaques sitting in a chamber with red, green, blue, or white lights learned to vocalize when only a light of one color was illuminated (Sutton et al. 1973). In a second study, subjects learned to give *coos* or barks, but not other vocalizations, when a certain light was on (Sutton, Samson, and Larson 1978; see also Wilson 1975). Third, Sinnott, Stebbins, and Moody (1975) trained macaques to give a *coo* vocalization while sitting in a chair. Then, through headphones, the monkeys heard either low-frequency noise (200 to 500 Hz) that masked the frequency of their calls or high-frequency noise (8 to 16 kHz) that fell outside the range of their vocalizations. Like humans, the monkeys immediately increased the amplitude of their calls in the former condition but showed no change in the latter. Taken together, such results strongly suggest that nonhuman primate vocalizations are under voluntary control.

As a final challenge to the dichotomy drawn between nonhuman primate vocalizations and human speech, research on primates in their natural habitat has revealed parallels between the way monkeys use vocalizations and the way humans use words. There are also striking similarities between the development of vocalizations among monkeys and the earliest use of words by human infants. In this chapter we review these results, focusing in particular on the vocalizations of vervets and certain other Old World monkeys. We describe how calls are used, how they develop, and the social context in which they are embedded. Our aim is not to prove that monkeys possess language. Instead, following Hockett (1960), we use language as a point of departure for a comparative study of communication in nonhuman primate species. Our analysis, we hope, will reveal as much about the *differences* between human and nonhuman primate communication as it does about the similarities.

This chapter, however, contains more than just a descriptive account of vervet monkey vocal communication. A second goal is to link the monkeys' communication with their knowledge of the world around them. Once it can be shown that monkeys use vocalizations to communicate about objects or events, communication—like language—becomes a tool for understanding how animals *think*. Already, in chapter 3, we have described how playback experiments using one individual's call can shed light on other animals' ability to recognize group members, remember past interactions, and even assess social relationships in which they are not themselves involved. Now, in this and the next chapter, we use the vocalizations of vervets to study how the same individuals recognize and classify predators, social situations, and other features of their environment. Like their social behavior, the monkeys' vocalizations offer a glimpse of how they see the world.

Alarm Calls

In 1967, Tom Struhsaker reported that vervet monkeys in Amboseli give different-sounding alarm calls in response to at least three different predators: leopards, eagles, and snakes. Each alarm call elicits a different, apparently adaptive escape response from other vervets nearby. A loud, barking alarm call is given to leopards (*Panthera pardus*; fig. 4.1) and other cat species like caracals (*Felis caracal*) and servals (*Felis serval*). Hereafter we refer to this call as the vervets' *leopard alarm*. When vervets on the ground hear a leopard alarm, they run into trees. Leopards in Amboseli typically hunt vervets by concealing themselves in bushes and pouncing on a monkey as it walks by (Altmann and Altmann 1970; pers. obs.). Apparently, vervets are

Figure 4.1. The leopard is one of the primary predators of vervet monkeys in Amboseli. Photograph by G. Anzenberger.

safe from leopards when they are in trees because the vervets' small size and agility make them difficult to catch.

In contrast, vervets give an acoustically different alarm call—a short, double-syllable cough called an *eagle alarm*—in response to the two large species of eagle that prey on them—the martial eagle (*Polemaetus bellicosus* fig. 4.2) and the crowned eagle (*Stephanoaetus coronatus*). Both species can take vervets of all age-sex classes. They hunt monkeys from the air, attacking from a long stoop at great speed (Brown 1966; Brown and Amadon 1968). Both raptor species seem skilled at taking monkeys in trees and on the ground (Brown and Amadon 1968; pers. obs.). Vervets on the ground respond to eagle alarms by looking up in the air or running into bushes. Vervets in trees respond to eagle alarms by looking up, and occasionally running down, out of the tree, and into a bush.

Finally, when vervets encounter pythons (*Python sebae*) or poisonous snakes like mambas (*Dendroaspis* spp.) and cobras (*Naja* spp.), they give a third, acoustically distinct alarm call, onomatopoetically termed a *chutter*. We refer to this as the vervets' *snake alarm*. Pythons hunt vervet monkeys primarily on the ground, by hiding in tall grass. The monkeys' best defense against a python is to be constantly aware of where the snake is. Upon hearing a snake alarm, vervet monkeys on the ground stand bipedally and peer into the grass around them (fig. 4.3). Once they see the snake, the monkeys often approach and mob it, repeatedly giving snake alarms from a safe distance.

Figure 4.2. The martial eagle is the largest of the African eagles and preys on birds, small mammals, and monkeys. Photograph by G. Anzenberger.

Figure 4.3. Members of group A stand bipedally and give alarm calls at a python as it emerges from a bush. Photograph by Richard Wrangham.

Based on Struhsaker's (1967a) description, vervet monkeys certainly seemed to be using calls to denote different external referents (e.g., Altmann 1967; Marler 1977a, 1978), an interpretation that directly contradicted views of primate vocalizations held at the time. Nevertheless, legitimate doubts about this interpretation were raised. W. John Smith, for example, described vervet alarms as "referring to different escape tactics" and with "no referents external to the communicators" (1977: 181), while the psycholinguist John Marshall claimed, "Even the alarm calls of the vervet monkey which seem, superficially, to be 'naming' the type of predator are more plausibly regarded as expressing no more than the relative intensity of the fearful and aggressive emotions aroused by the various predators" (1970: 234). Given the information available at the time, Smith and Marshall were appropriately conservative in their interpretation of the mechanisms underlying vervet alarms. Each distinguished between calls that provide information only about the signaler's emotional state or subsequent behavior (a relatively simple, straightforward explanation) and calls that denote a specific external referent (an explanation that implied more complex cognitive processes). There seemed no need to attribute sophisticated mental processes to the vervet monkeys when simpler mechanisms could adequately account for their behavior. Not discussed was the possibility that calls might *both* denote external referents and signal other sorts of information.

In 1977, working in collaboration with Peter Marler, we began a study of vervet monkey alarm calls with the goal of testing these different hypotheses. Our research was conducted in the same area where Tom Struhsaker had carried out his original study. We began by tape recording alarm calls and observing the behavior of vervet monkeys in actual encounters with leopards, eagles, and pythons. We then analyzed our recordings by playing them into a sonagraph, a machine that filters (or, in recent years, digitizes) the incoming signal and then displays this signal on a sound spectrogram. Spectrograms (like those in fig. 4.4) show, for each instant in time, the amount of energy present at different frequencies.

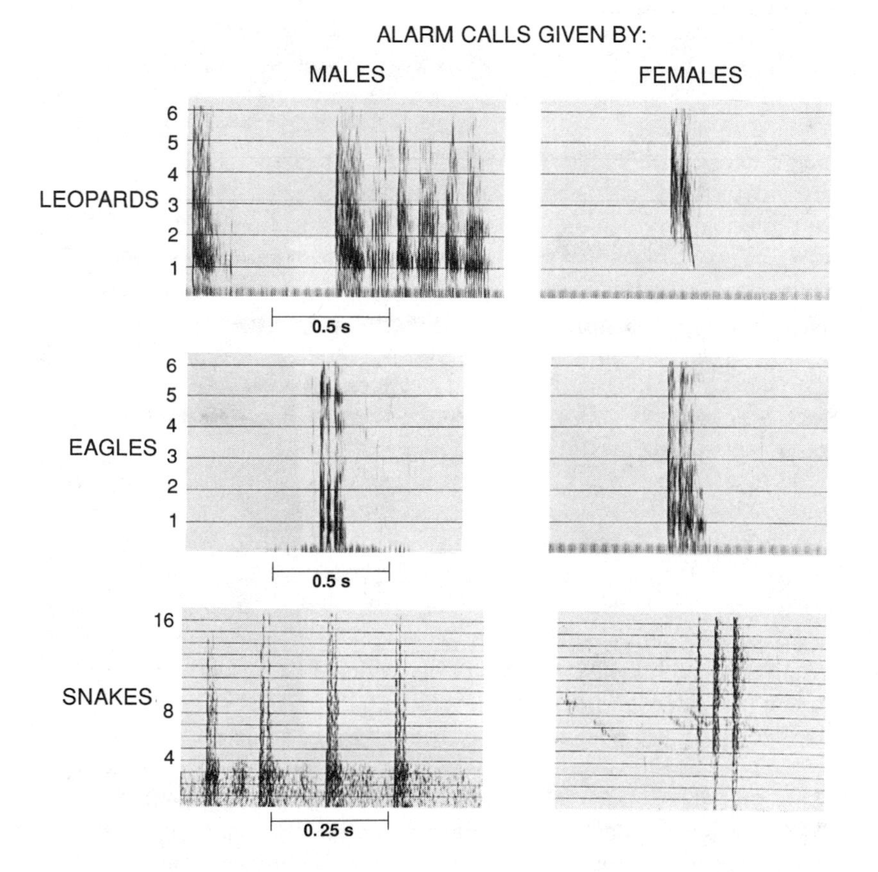

Figure 4.4. Spectrograms of alarm calls given by adult male and female vervet monkeys to leopards, martial eagles, and pythons. In each spectrogram, X-axis indicates time: Y-axis indicates frequency in units of 1 kHz (1,000 cycles/sec).

As Struhsaker had originally reported, we found that leopard, eagle, and snake alarm calls are easily distinguished by ear and easy to tell apart when displayed as sound spectrograms (fig. 4.4). Each type of alarm call also exhibits consistent acoustic features from one individual to the next (Seyfarth, Cheney, and Marler 1980b). In some cases there are sex differences in the vocalizations of adults; in others there are not. Adult males and adult females give acoustically similar eagle alarms and acoustically similar snake alarms; however, the leopard alarms given by adult males and adult females are acoustically quite distinct. Male leopard alarm calls consist of a repeated series of barks, whereas female leopard alarms consist of only a single, high-pitched chirp (fig. 4.4). Like Struhsaker, we found that each alarm call type elicits different escape responses.

Although our observations were consistent with Struhsaker's suggestion that the monkeys' different alarm calls denote different classes of predators, there were, as noted earlier, alternative explanations. For example, each call might have been a general "alerting" signal that caused animals to look all around them. Then, once the monkeys had spotted the predator, they could have responded on the basis of what they had seen. If this were the case, acoustic differences among alarm call types would be largely irrelevant. Alternatively, as Marshall (1970) had suggested, different alarm calls might simply reflect different levels of fear or excitement in the presence of different predators. Supporting this view, we found that leopard alarm calls (particularly those given by adult males) are generally louder and longer than eagle alarm calls, which are louder (but not longer) than snake alarm calls (Seyfarth, Cheney, and Marler 1980b).

Playback experiments allowed us to test these alternative explanations because they provided us with a means of examining separately how variation in a call's acoustic features and variation in other contextual events affected the monkeys' responses to a particular vocalization. When conducting an experiment, we first waited until no predator was in the area and a number of individual vervets were visible on the ground or in a tree. We then began filming our subjects' behavior. After 10 seconds we played, from a concealed loudspeaker, a leopard, eagle, or snake alarm call that we had previously recorded from an individual in the subjects' own group.

If vervet alarm calls were simply general alerting signals, the monkeys should have responded in similar ways to all of the acoustically different calls. If call meaning was determined primarily by context, then the response to each alarm should have varied depending on the context in which it was presented. By contrast, if each call's meaning was determined largely by its acoustic features, a given call type (leopard, eagle, or snake alarm) should have elicited a functionally consistent set of responses regardless of the context in which it was presented. Finally, if calls conveyed information primarily about the emotional state of the caller and only secondarily about the type of predator that had been seen, it should have been possible to blur the distinction among responses to the different call types by varying acoustic features associated with a signaler's level of excitement.

Alarm call playbacks produced two sorts of responses. In response to all three types of alarm, subjects looked toward the speaker and scanned the surrounding area. They behaved as if they were searching for additional cues, both from the source of the alarm call and elsewhere. More important, however, each type of alarm call also elicited a distinct set of responses. When subjects were on the ground, leopard alarms caused a significant number to run into trees. Eagle alarms caused a significant number of subjects to look up in the air or run into bushes, and snake alarms caused them to stand bipedally, looking down at the ground around them. When subjects were in trees, eagle alarms caused them to look up, and, in some cases, to run down, out of the tree and into a bush. Snake alarms caused monkeys to look down (Seyfarth, Cheney, and Marler 1980a, 1980b).

These qualitatively different responses demonstrated that the different alarm calls alone, even in the absence of a predator, provide the monkeys with sufficient information to make distinct and apparently adaptive responses. We also found that varying the length and amplitude of alarms—two features that would presumably mirror a signaler's level of fear or excitement—had no effect on the responses alarm calls evoked. Variation in the acoustic structure of different call types was the only feature both necessary and sufficient to explain differences in response (Seyfarth, Cheney, and Marler 1980b).

This is not to suggest that there is no emotional, or affective, component to the vervets' alarm calls. Encounters between vervet monkeys and their predators are often emotionally charged events, and monkeys responding to an alarm call almost certainly attend to such features as loudness, length, rate of delivery, and the number of individuals alarm calling in order to assess how close a predator is and whether it poses an immediate danger. Over the years we have noticed that if a predator is actually attacking, more animals give alarm calls, alarm calls are louder, and (to our ears at least) calls sound more "urgent" and the signalers seem more "distressed" than when a predator is not engaged in an attack.

Two other, admittedly rare, events illustrate the complex relation between alarm calls and the emotional state of a caller. In fewer than 1% of all vervet encounters with eagles ($N > 1{,}000$), an eagle attacked monkeys who were on the ground. As the bird began its stoop, with only seconds left before it might have grabbed a victim in its talons, one or more adult males gave *leopard* alarm calls. One interpretation of this rare event is that, under conditions of imminent danger, the monkeys give the alarm call that is, for them, associated with the greatest fear. Alternatively, one need not assume that emotions alone are guiding the vervets' judgments: when an eagle has committed itself to an attack on the ground, in open country, the best warning may indeed be one that causes others to scatter and to run for the nearest tree. It certainly is not advisable, under conditions of immediate attack, simply to look up into the air.

Similarly, during fewer than 2% of the 502 intergroup encounters that we observed between 1977 and 1986, a male gave leopard alarm calls. Again, while this suggests that leopard alarm calls may reflect emotional arousal and simply "spill out" during highly charged events, three observations suggest that this cannot be the sole explanation. First, most of the leopard alarm calls that occurred during intergroup encounters were given by the same male, Kitui, whose behavior may not be representative of vervets in general. Second, Kitui and other males gave these alarm calls in two specific contexts: when their own group was being driven back toward the center of its territory and hence was losing the encounter and when a new migrant male was attempting to approach their group. Giving leopard alarm calls allowed the males to grasp at least temporary victory from the jaws of defeat, since the alarms caused all animals to run into trees, ending the dispute and keeping the migrant male at bay.

Although these observations are by no means conclusive, they do at least suggest the possibility that alarm calls given during intergroup encounters represent attempts to deceive the members of other groups. We discuss these possibly deceptive signals further in chapter 8. Finally, intergroup encounters are noisy, conspicuous aggregations of many monkeys, all of whom are intent on their opponents. For this reason such encounters may actually attract predators; we have seen predators (a caracal, a serval, and a martial eagle) attack fighting groups of vervet monkeys on three separate occasions. It therefore seems possible that Kitui and other males were not simply excited, or even trying to deceive, but had actually seen a predator.

Cases like these, which raise the possibility that great excitement may override the relationship between a specific call and the predator for which it stands, were extremely rare. Far more often, vervet monkeys gave alarm calls in more relaxed circumstances that suggested a referential, denotative function. In a typical encounter with a martial eagle, for example, one of us was following an adult female, Philby (a member of group 3), who casually looked up in the air as she foraged. Philby stared intently at the sky for a few seconds and then, with no sign of imminent flight, gave an eagle alarm call. Using binoculars, we could see a martial eagle soaring far above us. Other monkeys nearby, in this case the adult females Burgess, MacLean, and Blunt, responded to Philby's call by looking up. Burgess also gave an eagle alarm call, but the other two females did not. No one got up to run. During the next 10 minutes each female checked the sky periodically, keeping track of the eagle as it soared overhead and eventually disappeared. Watching their behavior, our strong impression was that Philby's alarm call had served simply to alert others that an eagle was nearby.

On another occasion we played a snake alarm call to three individuals foraging in tall grass. Two subjects, the juveniles Leslie and Sedaka, responded by standing on their hind legs and peering at the ground around them, but a third subject, adult female Borgia, did nothing. Disappointed, we completed the experiment, put away the camera, and began collecting data on social behavior. A few hours later, one of us was following Borgia as she foraged toward the area where our experiment had been conducted. Entering the area, Borgia stood on her hind legs and scanned the ground around her. Clearly, our experiment had conveyed quite specific information to Borgia even though she had not chosen to act on it at the time.

Here again, we do not mean to argue that alarm calls provide information *exclusively* about external referents; a referential function by no means rules out the possibility that calls also convey information about the caller's subsequent behavior or about the conditional probability that a caller will act in certain ways given certain other events (e.g., Smith 1977; Hinde 1981). Obviously, the fact that different predators evoke both acoustically distinct alarm calls and different escape responses means that in many cases there will be a close link between alarm call type and behavior. Even in human language, where the referential function of signals is not in doubt, it is often difficult if not impossible to distinguish whether someone's use of a particular word refers to a specific object or to the likelihood that she will, at some time in the future, behave in a certain way (e.g., Marler 1961). Our experiments, then, are not meant to disprove the notion that calls provide information about subsequent behavior. Instead, we suggest that information about the caller's subsequent behavior is not the only information transmitted by the vervets' alarms, nor is it invariably the most important; vocalizations may also denote objects and events in the external world. In chapter 5 we discuss the meaning of vervet vocalizations in greater detail.

Further evidence of the monkeys' ability to classify alarm calls into types independent of other cues comes from research on captive vervets by Michael Owren (Owren and Bernacki 1988; Owren 1990a, 1990b). Using alarm calls that we had tape recorded in Amboseli, Owren trained two vervet monkeys to distinguish between the eagle and snake alarms given by one adult female, Alaska Pipeline. One vervet subject had been born in the wild and captured at approximately age 2; the other had been raised in a captive group. During the experiments, each female was seated in a restraining chair, wearing headphones. When she heard an eagle alarm she was rewarded for pushing a lever to the right; when she heard a snake alarm she was rewarded for pushing the lever to the left. Both vervets learned to distinguish between Alaska Pipeline's alarms. This was perhaps not surprising, since all calls came from the same individual, and variation other than between alarm call type was minimal. Once the animals had completed training, however, they were tested with 48 eagle and snake alarms that had been recorded from 17 different animals, including adult males, adult females, and juveniles. These calls exhibited considerable acoustic variation due to individual differences, the caller's age, sex, level of excitement, and the quality of our recordings. Despite such variation, the subjects immediately classified these novel stimuli with an accuracy that was significantly above chance. In the face of considerable idiosyncratic variation from one vocalization to the next, the monkeys still sorted calls into the same acoustic classes that were functionally important to vervets in the wild.

Given these results from both field and laboratory, we suggest that "semantic" and indexical information are combined in vervet alarm calls, much as they are combined in human speech. Each type of alarm refers to, or denotes, a particular type of predator. Different types of alarm are distinguished by their different acoustic properties and convey a meaning that is relatively independent of the context in which they are given. Supplementing and enriching this semantic information are indexical features that provide information about, for example, a caller's identity, her level of fear and anxiety, or the probability that she is likely to flee. Semantic information is of primary importance, but it is by no means the only sort of information conveyed. We return to the meaning of vervet alarms in chapter 5.

Other Vervet Alarm Calls

As Struhsaker (1967a) originally reported, leopards, eagles, and snakes are not the only species that elicit acoustically distinct alarm calls from vervet monkeys. The vervets also give what Struhsaker termed a *minor mammalian predator alarm* to carnivores that rarely prey on vervets, such as jackals (*Canis mesomelas*), hyenas (*Crocuta crocuta*), lions (*Panthera leo*), and cheetahs (*Acinonyx jubatus*). When monkeys hear this call they immediately become vigilant and either watch the predator as it moves through the area or move relatively slowly toward the safety of a tree. The minor mammalian predator alarm is a quiet vocalization that, unlike any other call in the vervets' repertoire, has its major energy concentrated at an extremely high frequency, around 32 kHz (fig. 4.5). At frequencies this high the vervets' hearing, otherwise very humanlike, is far superior to our own (Owren et al. 1988). In our experience, whether or not vervets give a minor mammalian predator alarm call depends upon what the predator is doing. For example, if a lion or hyena is chasing or killing a wildebeest, vervets may initially give an alarm call to it, but their alarm calling ends very quickly. The minor mammalian predator alarm shown in figure 4.5 was given in response to a jackal as it brought down an impala calf.

Another of the vervets' alarms is given in response to *unfamiliar humans* (fig. 4.5). Vervets give this alarm call most often in response to Maasai who come into the park to graze their livestock. When Maasai approach, the monkeys first call and then silently move off to nearby bushes or the tops of trees, where they sit until the Maasai and their cattle have passed (fig. 4.6). We are not sure exactly why monkeys dislike Maasai so much, because the Maasai express no particular antipathy toward them. In Amboseli, local people explain the vervets' behavior as a reaction against Maasai children, who sometimes throw rocks at the monkeys. The vervets' alarm calling, however, is no louder or more intense when children are present than when they are not. And Maasai are not the only humans who elicit the unfamiliar human alarm. When we began our study and the monkeys were not used to seeing humans at close range, they gave unfamiliar human alarm calls in response to us when we arrived each morning. Months later, after the animals had become habituated to us, we evoked a different vocalization, the *chutter to an observer* (Struhsaker 1967a), if we came upon a monkey suddenly or approached too closely to an infant. This chutter, along with a number of other, acoustically similar chutters, is described in greater detail on page 120.

Finally, vervet monkeys give a sixth alarm call in response to *baboons*, which occasionally prey on vervets. The call (fig. 4.5) apparently functions to alert others to the baboons' approach, but vervets seldom leave the immediate area unless the baboons actually enter their trees or chase them. Like minor mammalian predator alarms and unfamiliar human alarms, the baboon alarm call is a very quiet call, difficult to tape record and difficult

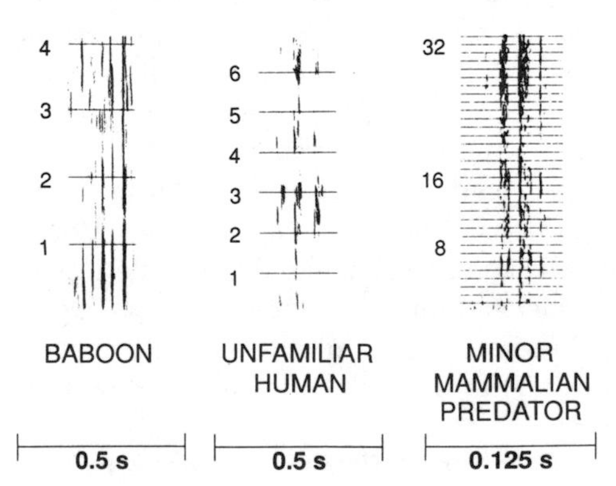

Figure 4.5. Spectrograms of alarm calls given by vervet monkeys to a baboon, an unfamiliar human, and a minor mammalian predator (in this case, a jackal).

Figure 4.6. Adult female Charing Cross watches Maasai herdsmen bringing goats to water. She has just uttered an unfamiliar human *alarm call.*

even to hear if there is the slightest amount of background noise. Although superficial examination of spectrograms suggests that minor mammalian predator alarms, unfamiliar human alarms, and baboon alarms differ from each other and differ from the somewhat similar alarm calls given to snakes, we do not have a sufficiently large sample of recordings of these calls to test this impression.

Alarm Calls in Other Species

The alarm calls of vervet monkeys are by no means unique among birds and mammals. Struhsaker (1975), for example, describes acoustically distinct alarm calls given by red colobus monkeys (*Colobus badius*) to birds and terrestrial predators. Similarly, several species of ground squirrels (genus *Spermophilus*) and prairie dogs (genus *Cynomys*) give acoustically different alarm calls in response to ground and aerial predators. Different alarm call types also elicit different responses (Melchior 1971; Turner 1973; Sherman 1977, 1985; Hoogland 1983). California ground squirrels, for instance, give "whistles" in response to eagles and hawks, plus three acoustically distinct "chatter-chat" vocalizations in response to snakes, terrestrial mammalian predators, and aggressive conspecifics (Owings and Virginia 1978; Leger, Owings, and Gelfand 1980; Owings and Leger 1980). When they hear a whistle, animals become vigilant in the "down alert" posture (making them less conspicuous) and run toward their burrows. Upon hearing a chatter-chat, animals become vigilant in the "up alert" posture (making them more conspicuous). They do not necessarily run toward a burrow (Leger and Owings 1978).

There are, however, important differences between the alarm calls of ground squirrels and the alarm calls of vervet monkeys. In contrast to vervet alarms, acoustically different alarm calls by ground squirrels are not tightly linked with a particular sort of predator. Both California and Belding's ground squirrels, for example, give whistle alarms when a predator arrives suddenly and there is little time to escape. Most sudden attacks come from raptors, but occasionally squirrels are surprised by a mammalian carnivore. When this occurs, the mammalian carnivore elicits whistles. Similarly, chatter-chat alarms are given to predators that have been spotted at a distance. Typically, such predators are mammalian carnivores, but it is not unusual for the squirrels also to give chatter-chat alarms to a distant hawk (Leger, Owings, and Gelfand 1980; Robinson 1981).

To explain these observations, Owings and Hennessy (1984) suggest that rather than denoting different classes of predators, ground squirrel alarm calls signal different levels of "urgency." Whistles are high-urgency calls that signal imminent danger (often, but not always, associated with raptors); chatter-chats are low-urgency calls that require vigilance but not immediate flight. Compared with the alarm calls of vervet monkeys, ground squirrel alarm calls correspond more closely with the appropriate escape strategy and degree of urgency than with the class of predator involved.

This difference, moreover, makes adaptive sense. Among ground squirrels, the type of predator is less important than the immediacy of danger, because squirrels evade different predators in roughly similar ways. Ground squirrels can escape from *any* predator by entering their burrows. In contrast, the type of predator makes a great deal of difference to vervet monkeys, who must escape from terrestrial and aerial predators in qualitatively different ways. As a result of different selective pressures, vervet alarm calls denote the type of predator involved, leaving the exact response up to each individual, while ground squirrel alarm calls denote levels of urgency and are more ambiguous about the exact source of danger.

Many birds also have acoustically distinct alarm calls for terrestrial predators and raptors (e.g., Daanje 1941; Nice 1943; Marler 1956b; Ryden 1978; Thielke 1976; Latimer 1977; Walters 1990). At present, however, we know very little about the stimuli that elicit such calls or the precise responses that they evoke (but see Gyger, Marler, and Pickert 1987; Walters in press). On page 136 and in chapter 5, we describe the alarm calls given by one East African bird, the superb starling, in response to ground and aerial predators. We also present the results of playback experiments designed to test how much the vervets know about the starlings' different alarm calls.

Other Vocalizations

We have used data on vervet monkey alarm calls to argue against traditional interpretations of primate vocal communication. Our argument would be considerably weakened, however, if it could be shown that alarm calls were somehow atypical and that no other vocalizations in the vervets' repertoire ever functioned to denote objects and events external to the signaler. Indeed, if one is interested in comparing the natural vocalizations of monkeys with language, or with the "linguistic" communication of captive apes, the most important data would seem to come not from alarm calls but from the many vocalizations exchanged among primates during social interactions, when animals are resting, grooming, foraging, or playing. In the past, these vocalizations were assumed to be general purpose *contact calls* that identified the caller, announced his location, and kept the group together (e.g., Rowell 1972). Recent research, however, reveals considerably greater complexity.

Vervet Monkey Grunts

Like the players at Wimbledon, vervet monkeys frequently grunt to each other during normal social interactions. The vervets' grunts are harsh, raspy signals that sound like a human clearing his throat with his mouth open. As Struhsaker (1967a) originally noted, grunts are given in at least four distinct social circumstances. First, a monkey may grunt as she approaches a more dominant individual; second, a monkey may grunt as she approaches a subordinate. Third, monkeys often grunt as they watch another animal, or as they themselves, initiate a group movement across an open plain. Fourth, grunts may be given when a monkey has apparently just spotted the members of another group (fig. 4.7). Even to an experi-

Figure 4.7. Adult females Marcos and Charing Cross look at a neighboring group just after Charing Cross has given the grunt to another group.

enced human listener, there are no immediately obvious audible differences among grunts, either from one context to another or across individuals. When grunts are displayed on sound spectrograms, there are also no consistent differences in acoustic structure from one context to the next (fig. 4.8). Although grunts are occasionally answered by other group members, in most cases grunts evoke no salient behavioral responses. Changes in the direction of gaze, which are difficult to measure in the wild, seem the only obvious response when one individual grunts to another.

Vervet monkey grunts, therefore, are strikingly different from alarm calls. While alarm calls given in response to different predators are easily distinguished acoustically, grunts given in different social contexts sound very much alike. Unlike alarm calls, grunts occur in quiet, relatively relaxed circumstances and evoke no obvious response from those nearby. From an observer's perspective, watching monkeys grunt to each other is very much like watching humans engaged in conversation without being able to hear what they are saying: the creatures *seem* to be exchanging some sort of information, but we have no idea what it is. As Quine's (1960) imaginary linguist soon learned, if one relies solely on observation, the only way to

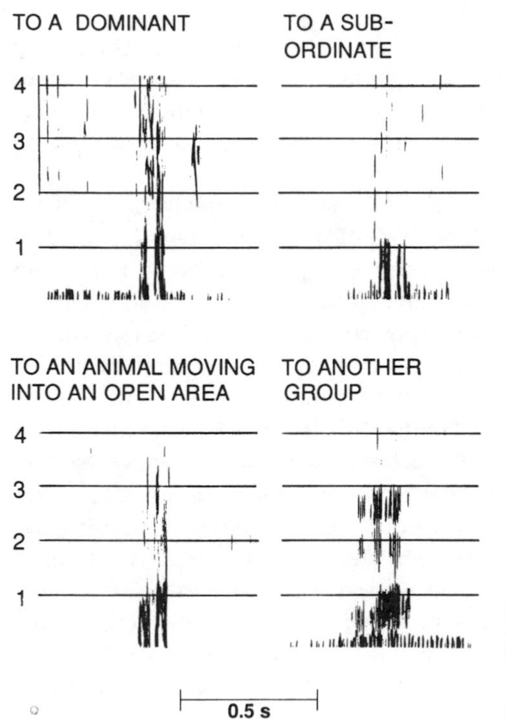

Figure 4.8. Spectrograms of grunts given by adult female Teapot Dome in four different social situations.

measure a call's meaning is through the responses it evokes in others. And since these responses are hardly discernible, we seem to be faced with an intractable problem.

Grunts also recall the Wimbledon referees' dilemma. Like tennis referees, we cannot through observation alone distinguish between calls that provide information exclusively about the caller's behavior or emotional state and calls that also function to denote objects or events in the environment.

Given these methodological dilemmas, the most obvious starting place is to adopt Smith's (1977) influential view that the meaning of an animal's vocalization (that is, the information it conveys to others) is a function of its message (in this case, its acoustic properties) and of the context in which it is given. Animals, Smith argues, have relatively small repertoires of signals, each of which conveys a broad, general message. A small repertoire of general signals can nevertheless elicit a variety of responses because of variation in the contexts in which calls are given. Applied to vervet grunts, this hypothesis would predict that vervets are using a single vocalization in a variety of circumstances. The grunt itself is a manifestation of a particular level of arousal and provides general information about the vocalizer's identity, location, or subsequent behavior. Variation in the responses evoked by different grunts is accounted for by variation that the receiver perceives in the context in which they are given.

In contrast, consideration of the vervets' alarm calls suggests an alternative explanation: namely, that what seems to a human listener to be one grunt is in fact a number of different grunts. Each grunt type conveys specific information that depends more on a call's acoustic properties than on the context in which it is given.

To test between these hypotheses, we designed the following set of experiments. First, grunts from the same individual were tape recorded in each of the four social contexts described above. Then, over a number of months, we played each grunt to subjects from a concealed loudspeaker and filmed their responses. For example, we might play Bokassa's *grunt to a dominant* to Duvalier on one day and then, 3 or more days later, play Bokassa's *grunt to another group*. Throughout these trials we allowed social context to vary freely. Tests were conducted, for instance, when there were dominant or subordinate animals nearby, when the group was foraging or resting, or when animals were at the center or the edge of their range. We reasoned that if the grunts were really one vocalization whose meaning was largely determined by context, subjects should show no consistent differences when responding to different calls. Instead, responses should be a function of the variable contexts in which calls were presented. On the other hand, if each of the grunts was different, and if each carried specific information that was relatively independent of context, we should find consistent differences in responses to each grunt type, regardless of the varying circumstances in which it was played.

Our overall method thus parallels that adopted by Quine's imaginary linguist, who, unable to determine what his subjects' words mean, poses hundreds of yes-no questions in an attempt to clarify one word's meaning relative to that of another. In our case, unable to determine what the vervets' grunts mean through observation, we asked subjects, "Is grunt A different from grunt B? If so, is grunt A different from grunt C? Do grunts B and C differ?" and so on.

As an example of the results we obtained, consider the comparison between grunts that had originally been given to a dominant and grunts that had originally been given to another group. Here we used as stimuli one grunt of each type from three different individuals. Eighteen subjects heard first one call and then, a few days later, the other. Subjects responded in many ways, but two responses appeared consistently and were consistently different across the two grunt types. Grunts to a dominant caused subjects to look toward the loudspeaker, whereas grunts to another group caused subjects to look out, toward the horizon, in the direction the loudspeaker was pointed (Cheney and Seyfarth 1982a). Grunts to another group therefore directed the listener's attention away from the speaker and in the direction toward which, under normal conditions, the vocalizer would have been facing.

Consistent differences in responses to different grunt types appeared in many, but not all, of our paired comparisons. Grunts to a subordinate, grunts to a dominant, grunts to an animal moving into an open area, and grunts to another group all elicited responses that were consistently different from each other. There was, however, no difference between the responses elicited by grunts to a dominant male and grunts to a dominant female (Cheney and Seyfarth 1982a).

By their behavior, then, the monkeys seemed to be saying that although their grunts sound more or less the same to us, to *them* each grunt transmits a specific sort of information. In many cases, this information can include events external to the signaling individual, such as the approach of another group or the movement of animals into an open area. Although the vervets' grunts are, in many ways, different from their alarm calls, the two sorts of vocalizations function in a similar manner.

Since monkeys responded in consistently different ways to different grunt types despite variation in social context, we conclude that the information conveyed by grunts—like the information contained in human words—depends as much, or more, on a particular call's acoustic properties than on the circumstances in which it is given. Of course, this does not mean that contextual variables are irrelevant. Common sense suggests that context must be important for vervets, just as contextual cues can enrich and modify the meaning of words for humans. Results suggest, however, that in many cases monkeys make less use of contextual cues than they do of acoustic features when interpreting the meaning of a particular call.

To follow up our field experiments, we analyzed the acoustic features of vervet grunts in an attempt to determine the cues that might be used by the monkeys to distinguish one grunt type from another. To begin, we considered 32 grunts that had been given by one adult female, Teapot Dome, in the four social contexts described above. For each call we measured a total of 16 acoustic features, including call length, number of acoustic units, peaks in the frequency spectrum, and changes in frequency peaks over time (Seyfarth and Cheney 1984b). We then searched for any acoustic measure that differed consistently from one context to the next; in other words, any acoustic measure that might allow us (or a vervet) to determine what Teapot Dome was signaling about even if she could not be seen.

Once we had a list of these purportedly distinctive features, we tested our hypothesis using 216 grunts from 13 other individuals, 3 of whom (Profumo, Alaska Pipeline, and Maginot Line) were adult females in Teapot Dome's own group. Grunts were first classified according to social context, then classified according to the acoustic features that had proved successful with the grunts of Teapot Dome. For 82% of all calls (177/216) the two types of classification agreed (Seyfarth and Cheney 1984b).

Establishing the cues that vervets might use to distinguish different grunts is not just an exercise in acoustic analysis; it can also provide insights into how monkeys perceive themselves and each other. Consider, for example, the vocalizations given and received by three juvenile male vervets who transferred from one study group to another during the 1980 breeding season. Males generally transfer to a group with an adjacent home range, and the process of transferring usually takes 4 or 5 days before it is completed (Cheney and Seyfarth 1983). During the early stages of transfer, a male often spends the day foraging, feeding, and interacting with his new group, then returns at dusk to sleep with his old group.

In 1980 the three young males, all members of group A, grunted at high rates whenever they saw the dominant male, Pius, in group B. Acoustic analysis indicated that these calls were all grunts to another group. When the males began their attempt to transfer into group B, they continued to give grunts to another group to Pius, and their use of this particular call persisted even after their transfer had been completed and they were sleeping each night with group B. During the same period, the *females* in group B also treated the three juvenile males as outsiders, giving grunts to another group whenever the males came near. Between 12 and 15 days later, however, all of the animals' vocalizations changed. Now the young males' grunts to Pius became grunts to a dominant, and females vocalized toward the new arrivals with grunts to a dominant or grunts to a subordinate.

Even after the young males had joined their new group, therefore, they still seemed to perceive themselves (and were perceived by others) as outsiders. Information of this sort—which concerns not only what monkeys do but also what they think—would not be accessible to us unless we had some way of recognizing a mismatch between the vocalization that would have been expected under a particular set of circumstances and the vocalization that was actually given.

Finally, a brief comment on the apparent function of vervet grunts. Reviewing the results of our experiments, together with data on naturally occurring grunts, we find that the different calls are often associated with different beneficial social consequences. When subordinate animals grunt to a dominant, or dominant animals grunt to a subordinate, grunts decrease the probability that the subordinate will move away from the dominant. Grunts increase the likelihood that subordinate and dominant individuals will forage, feed, or sit together (unpublished data). Grunts to a subordinate, therefore, allow dominant animals to interact with those of lower rank when they choose without frightening them away, whereas grunts to a dominant give subordinate animals the opportunity to interact with those of higher rank. The grunt given by animals as they themselves move into an open area or as they watch other animals do so directs listeners' attention both toward the caller and outward, in the direction the caller is facing. By so doing, these particular grunts may decrease the risk of predation by increasing the number of vigilant animals. Finally, grunts to another group also direct attention outward, in the direction the caller is facing, and therefore serve as the initial warning that another group is nearby.

There are, then, ample reasons why vervets need more than one grunt and why natural selection may have favored individuals who can use at least four acoustically different grunts in these four different circumstances. This post hoc explanation, however, begs an important question: why stop at four? Having watched vervets for many years, and having seen them die at high rates, we can think of many situations in which the monkeys could make good use of a vocalization but have not apparently developed one. Vervet mothers, for example, have no call that conveys the information *follow me*. Mothers often leave their infants in what seem to be vulnerable positions, and they make no apparent attempt to let the infants know that they are moving off or where they are going. On one occasion, for example, a mother walked away, leaving her infant in a tree just as a group of baboons was approaching. When the infant suddenly became aware that her mother was gone she vocalized loudly, attracting both the mother's and the baboons' attention. The mother looked toward her infant but made no sound; she apparently had no way to signal that the infant should simply follow her.

It has been suggested that a crucial evolutionary transition occurs whenever a species begins to divide a graded stream of acoustic sounds into discrete categories (Marler 1976a). Implicitly, such arguments assume that once categorical signaling has been achieved *in principle,* the most difficult problem has been surmounted and individuals will be free to develop a large number of discrete, highly specific signs. Data from vervets, however, suggest that other constraints are at work. Vervets may be able to divide a graded series of sounds into discrete categories, but their repertoire of calls, compared with human language, is still not very large. In terms of evolutionary function, we can easily explain why the vervets have so many grunts, but we cannot explain why they have so few.

Other Close-Range Vocalizations

The system of grunts used by vervet monkeys is by no means unique. Broad classes of vocalizations within which monkeys distinguish subtly different calls are found elsewhere in the vervets' repertoire and have also been elegantly documented in Japanese macaques, rhesus macaques, marmosets, and capuchins. In some cases these calls, like the vervets' grunts and alarm calls, function (among other things) to denote objects or events external to the signaler. In other cases the precise referents of a call are unclear, and it is entirely possible that the call's message (Smith 1977) is concerned primarily with the caller's behavior or state of arousal.

Consider, for instance, a number of similar-sounding chutters given by vervet monkeys. As Struhsaker (1967a) originally noted, vervets give chutters to snakes, to members of their own group, to members of other groups, and to familiar human observers (Cheney 1984). As with the vervets' grunts, all chutters sound alike to human ears. From Struhsaker's work, however, we know that intergroup chutters are consistently different from intragroup chutters, and we have some evidence that both are acoustically different from the chutter given to human observers. Even though we have not compared the vervets' responses to acoustically different chutters in playback experiments, it seems that vervets use them, as they use grunts and alarm calls, to denote different features of their environment.

Another example comes from research on the screams given by rhesus macaques during aggressive interactions. When these calls were first studied (Rowell 1962; Rowell and Hinde 1962), observers noted that their acoustic features were highly variable. Rowell and Hinde explained this variability by suggesting that screams formed a graded system of signals and that each scream variant reflected the caller's level of excitement or distress. Like other vocalizations, screams were assumed to supplement and enrich visual signals, which were the monkeys' primary means of communication.

Studying juvenile rhesus macaques on Cayo Santiago, however, Sarah Gouzoules, Harold Gouzoules, and Peter Marler (1984) noticed a relation between the acoustic properties of screams and the circumstances in which they were given. Some screams, for example, were acoustically noisy (fig. 4.9) and were given most often when the vocalizer was interacting with an individual higher ranking than itself and in situations involving physical contact. "Arched" screams (fig. 4.9) were given to lower-ranking individuals in the absence of physical contact. Both "tonal" and "pulsed" screams

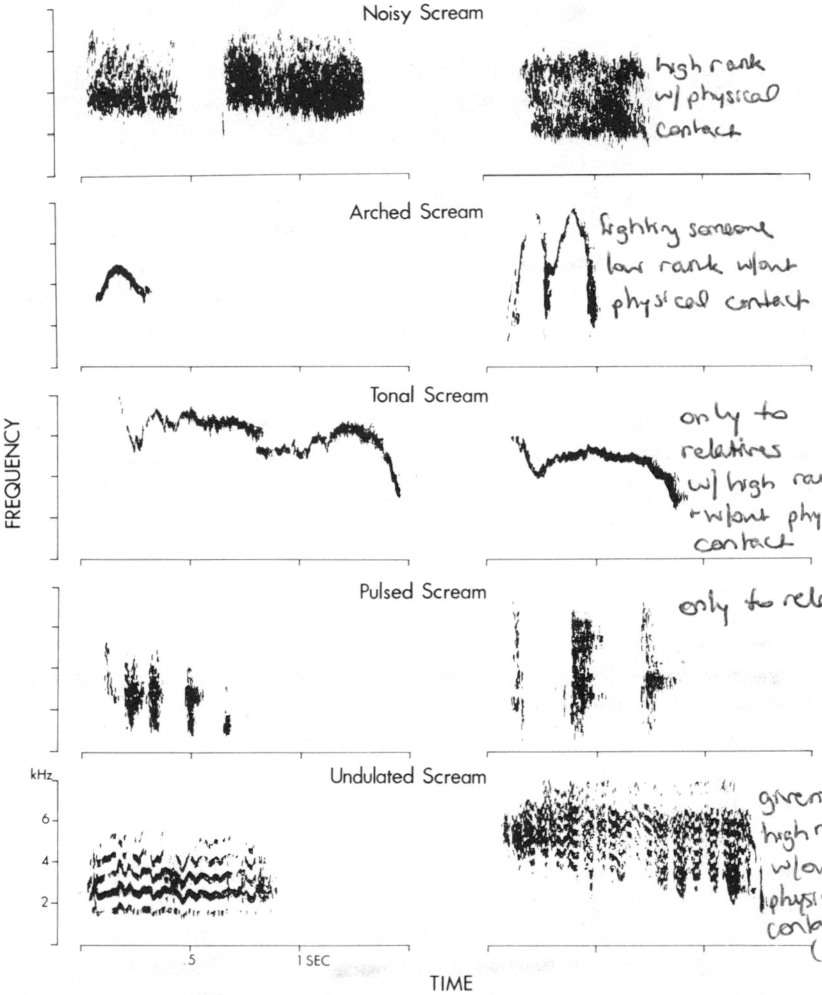

Figure 4.9. Two examples each of five classes of screams given by juvenile rhesus macaques. From Gouzoules, Gouzoules, and Marler 1984, with permission.

were given more often than expected to relatives, and "undulated" screams were given almost exclusively to higher-ranking individuals when no physical contact occurred. Observation revealed no relation between the different screams and the vocalizer's subsequent behavior, arguing against the hypothesis that screams were simply manifestations of arousal or that they provided information primarily about what the vocalizer would do next. Playback experiments, however, showed clearly that mothers responded differently to the different types of screams from their offspring. Mothers responded most strongly to playbacks of noisy screams (physical contact with higher-ranking opponents) and next most strongly to playbacks of arched (lower-ranking opponents), tonal, and pulsed screams (genetically related opponents), in that order. Acoustic differences among scream types, therefore, function to convey information about different external referents (the juveniles' opponents), and mothers make use of this information when responding to their offsprings' cries.

Following up their research with rhesus macaque screams, Harold and Sarah Gouzoules (1989) have shown that pigtailed macaques also give acoustically different screams when interacting with different classes of opponents. It is interesting to note, however, that within essentially identical social contexts the two macaque species use scream types that are quite different acoustically. During contact aggression, for example, rhesus macaques use atonal, noisy screams while pigtails use tonal, frequency-modulated screams. This suggests that there is no simple, direct relation between the physical structure of a monkey's call and his underlying motivation (cf. Morton 1977). In addition, although the acoustic structure of a rhesus macaque's scream depends on the presence or absence of physical contact plus the opponents' kinship and dominance relations, the acoustic structure of pigtailed macaque screams depends only on physical contact and dominance. This suggests that fights among kin occur relatively more often in rhesus than in pigtailed macaques.

Like vervet grunts, macaque screams illustrate these animals' knowledge of their own and other individuals' social relationships. When he screams, a juvenile rhesus macaque effectively classifies his opponent according to kinship and dominance rank, and he classifies the interaction according to the severity of aggression. A juvenile pigtailed macaque classifies his opponent and the interaction only according to the latter two features. By her selective response, the juvenile's mother reveals knowledge of her offspring's voice and her offspring's network of social relationships (Gouzoules, Gouzoules, and Marler 1986). Like the data presented in chapter 3, these results suggest that concepts like kinship and dominance rank, devised by humans to explain what monkeys do, exist not only in the minds of human observers but also in the minds of their subjects.

Subtle discriminations among acoustically similar vocalizations can also be found in the "trills" of pygmy marmosets (*Cebuella pygmaea*) (Pola and Snowdon 1975; Snowdon and Hodun 1981), the close-range vocalizations of wedge-capped capuchins (Robinson 1982), and the *coo* calls of Japanese macaques, first studied in detail by Steven Green (1975). Japanese macaques, Green found, give *coos* to each other in a variety of social situations: for example, when a male is separated from his group, when a dominant approaches a subordinate, when a subordinate approaches a dominant, or when a female is in estrus. At first hearing, *coos* in all these circumstances sound the same. With experience, however, differences begin to emerge: certain *coos*, for instance, have a frequency peak near the start of the call, with steadily falling frequency thereafter (fig. 4.10). Green labeled these *smooth early highs*. Other *coos* have a steadily rising frequency, reaching a peak in the call's latter half (*smooth late highs*, fig. 4.10). These variations in acoustic structure, moreover, are correlated with variations in social situation. Smooth early highs are most commonly used by infants sitting apart from their mothers, whereas smooth late highs are used most frequently by sexually receptive females.

Green's results set the stage for a series of experiments that have produced major changes in our thinking about nonhuman primate vocalizations and the neural mechanisms that underlie them. When humans discriminate between phonemes, two phenomena are apparent. First, when distinguishing different acoustic stimuli we attend to some physical features while ignoring others (e.g., Ladefoged 1975). When distinguishing between the phonemes *ba* and *pa*, for instance, we pay attention to differences in voice onset time, ignoring differences in other acoustic features like amplitude, speech rate, or the fundamental frequency of a speaker's voice

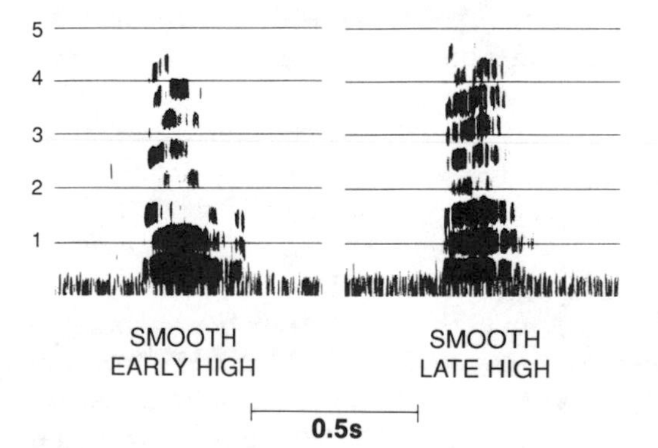

Figure 4.10. Spectrograms of two coo *calls recorded from Japanese macaques. In the classification used by Green 1975, the call on the left is a* smooth early high, *and the call on the right is a* smooth late high. *Calls recorded by M. J. Owren and used with permission.*

(Lisker and Abramson 1964). Humans are apparently predisposed to find some cues perceptually more salient than others.

Second, for most humans the perception of many linguistically relevant sounds is specialized in the left cerebral hemisphere. As a result, subjects commonly show a right ear advantage when asked to distinguish one speech sound from another (reviewed by Bradshaw and Nettleton 1981). For many years, comparable lateralization in the control of behavior was not found among nonhuman primates (e.g., Hamilton 1977; Warren 1977), despite the existence of anatomical differences in the left and right hemispheres of primate brains (reviewed in Nottebohm 1979). To test whether these two features—species-specific sensitivity to particular auditory cues and neural lateralization—might be involved in the processing of *coos* by Japanese macaques, the following experiments were performed.

First, a number of *coos,* both smooth early and smooth late highs, were selected as test stimuli. These calls were played to Japanese macaques and three control species (a rhesus macaque, two bonnet macaques, and a vervet) while the subjects were seated in restraining chairs and wearing headphones. The monkeys' first task was to divide calls into two classes on the basis of peak position, a feature that Green's work had suggested was meaningful for Japanese macaques. All of the *coo* stimuli also differed in terms of onset frequency: some had relatively high and others relatively low initial frequencies. A second task of subjects was to divide stimuli into two classes on the basis of initial frequency, presumably a feature that was less biologically important than peak position, at least for Japanese macaques.

Intriguingly, when subjects were rewarded for discriminating *coos* on the basis of peak position, Japanese macaques learned to do so significantly faster than control species. In contrast, when a second group of subjects was rewarded for discriminating *coos* on the basis of initial frequency, Japanese macaques learned to do so at a slower rate than did control species (Zoloth et al. 1979). In a manner that parallels the data from human subjects, Japanese macaques seemed predisposed to classify *coos* on the basis of peak position and to pay less attention to variation in other acoustic features.

Moreover, throughout these experiments Japanese macaques exhibited a right ear advantage, whereas other species did not (Petersen et al. 1978, 1984). This finding parallels results from human subjects, who display a right ear advantage when processing certain speech sounds but not when processing nonlinguistic sounds like pure tones. Finally, to determine the neuroanatomical basis of the macaques' behavior, the subjects' ability to discriminate different types of *coos* was tested before and after ablation of the left or right temporal cortex. Again, paralleling the data from humans (e.g., Penfield and Roberts 1966), damage to the left temporal cortex resulted in a significant, though temporary, impairment in the monkeys' ability to distinguish different *coo* types. Damage to the right temporal cortex had no such effect (Hefner and Hefner 1984).

Syntax

Given the widespread use of many subtly different, acoustically distinct vocalizations in different social situations, it seems logical to ask whether nonhuman primates or any other species ever combine vocalizations into compound utterances, and, if they do, whether they do so in accordance with a particular set of rules, or grammar.

Chomsky (1972:71) defines syntax as a "system constituted by rules that interact to determine the form and intrinsic meaning of a potentially infinite number of sentences." Though perfectly appropriate for research on human speech, this definition is inappropriate for studies of animal communication because it assumes a priori that syntax can exist only in a system of communication that exhibits the formal properties of human language. To get around this problem, scientists studying the calls of nonhuman species have searched instead for any "system of rules that will allow us to predict sequences of signals" (Snowdon 1982:231). This definition of syntax leaves open the question of whether such sequences constitute sentences, or whether they could ever result in an infinite number of messages.

Sequences of animal vocalizations can be of two types (Marler 1977b). In *phonological syntax*, callers take elements from their repertoire of acoustic signals and recombine them in orderly and predictable ways to make new vocalizations. The ordering of elements is important, in the sense that listeners respond differently to the same elements when these elements are presented in different sequences. Phonological syntax does not require that the acoustic elements being combined ever be used in isolation or that they have any meaning when presented on their own. Further, it does not specify any relation between the meaning of elements and the meaning of calls created by their combination. By contrast, in *lexical syntax* the meaning of the compound call results from the sum of meanings of its constituent units (Marler 1977b). To date, many studies of communication in animals have found evidence for phonological syntax; the existence of lexical syntax in nonhuman species is, however, much more problematical.

Two studies suggest the existence of phonological syntax in the vocalizations of at least some nonhuman primates. Studying wedge-capped capuchins in Venezuela, John Robinson (1984) found that certain types of calls, although given singly in one circumstance, were also combined into mixed doublets, triplets, or even quadruplets in other circumstances. The ordering of different call types was predictable, and compound calls seemed to be used in circumstances that were intermediate between those in which the constituent calls had been used singly. Robinson suggested that each call type, given alone, reflected a different emotional state on a continuum from contact seeking to contact avoiding or from affiliation and submission to aggression. By contrast, compound calls reflected "intermediate" emotional states. If this interpretation is correct, however, compound calls might occur simply because, in intermediate social circumstances, monkeys beset with conflicting motives give more than one signal. Such combinations might qualify as phonological syntax but would be considerably simpler than the compound words found in human language, where two words like *foot* and *ball* can be combined to make a third, *football,* whose meaning is more than just the sum of meanings of its constituent parts.

As a second example of rule-governed ordering in primate vocalizations, consider the territorial "songs" of male gibbons, studied by John Mitani and Peter Marler (1989). To construct their songs, male gibbons use a number of acoustically distinct elements, or "notes." These notes, as far as we know, are used in isolation only rarely. Their inherent semantic content when used on their own is therefore not known. Notes are combined according to a set of rules that apparently limits the number of possible songs. Moreover, the songs of different species consist of different note types. When a male gibbon hears the song of a conspecific male, either naturally or in a playback experiment, he responds more strongly than when hearing the song of a male from another species, indicating that note type is an important determinant of song meaning (Mitani 1987). In addition, a male's response to conspecific song is qualitatively different if the notes in that song have been rearranged. The song of male gibbons thus provides an example of phonological syntax. To determine whether gibbon song also involves lexical syntax, however, one would need to determine the meaning of individual notes and establish whether any of these notes brings a unique meaning to the songs in which they are found (Mitani and Marler 1989).

A different approach to the study of sequential ordering has been taken by Charles Snowdon and Jayne Cleveland (1984) in their study of "conversational rules" among pygmy marmosets. In the South American rainforest, marmosets exchange "trills" while foraging for insects out of sight of one another. In the laboratory, Snowdon and Cleveland observed behavior analogous to the turn taking found in human conversation. In their group of three marmosets, trills followed a highly predictable pattern. Animal 1 would call first, then remain silent until animal 2 called, after which both would remain silent until animal 3 called. Individuals differed in the likelihood that they would initiate a bout of calling, and one individual violated the conversational rules more often than others. Snowdon and Cleveland suggest that marmosets use a conversational rule system in their antiphonal calling and that giving calls in a specified order allows each individual to monitor the behavior of many other group members simultaneously. Though the authors are appropriately cautious on this point, the marmosets' adherence to a particular order in the exchange of vocalizations raises the possibility that single individuals might also use predictably ordered sequences of calls in their own sequences of vocalizations.

Evidence from vervet monkey studies is less impressive. Consider first the data on combinations of grunts given by the same individual. From 1985 to 1986, in 14.3% of all cases when a grunt occurred ($N = 678$), the vocalizer grunted two or more times in sequence. Thirty-three of these sequences were tape recorded, and grunts were analyzed to determine their acoustic structure. In every case, the string of grunts was simply the same call, repeated over and over.

Second, on 43 occasions a grunt from one animal elicited a grunt from another. On 15 of these occasions we were able to tape record the exchange and to analyze the calls' acoustic features. Under these conditions, interestingly enough, the second animal always replied by repeating what the first animal had said. If a subordinate, for example, approached a dominant, giving the grunt to a dominant, the dominant individual disregarded her partner's rank and replied with the same call. Perhaps the function of these replies was to communicate "message received." Whatever their purpose, however, they provide no evidence that vervets followed a specific set of rules when combining calls themselves or in vocal exchanges with other group members.

Close-Range Vocalizations and Their Significance

Four conclusions emerge from this brief review of close-range vocalizations in vervets and other nonhuman primates. First, alarm calls are not the only vocalizations that denote objects and events external to the signaler. The chutters and grunts of vervet monkeys, screams of rhesus macaques, and *coos* of Japanese macaques all provide nearby individuals with information about particular features of the social environment. Like the information contained in alarm calls, the information contained in these vocalizations is quite specific and concerns objects and events external to the caller. Although vocalizations may also transmit information about the caller's subsequent behavior, this is by no means the only information they convey.

Second, playback experiments and acoustic analysis demonstrate clearly that the size of vocal repertoires—in primates or any other animal—cannot be assessed by the human ear alone. Where once it was thought that vervets had a grunt, rhesus macaques a scream, and Japanese macaques a *coo,* we now know that the monkeys themselves perceive many different variants of these signals, each with a different meaning. There is, undoubtedly, an upper limit on animal vocal repertoires when compared with the infinite number of messages that can be conveyed through human language. However, the size of vocal repertoires, at least in nonhuman primates, is considerably larger than initially believed and the information conveyed by each call is less general than we had first imagined. Clearly, before we can claim to have documented a species' entire repertoire we must carry out experiments that permit the animals themselves to tell us how many calls they have and how specific call meaning can be.

Third, while it may be heuristically useful in the initial stages of research to divide nonhuman primate vocal repertoires into graded and discrete classes, this distinction breaks down on closer inspection since monkeys, like humans, often perceive a graded continuum of sounds as a number of different, acoustically distinct calls (Snowdon and Pola 1978; Hopp 1985; Snowdon, French, and Cleveland 1986; Owren, Hopp, and Seyfarth 1990; May, Moody, and Stebbins 1989).

Finally, nonhuman primate vocalizations exhibit parallels with language not only in social function but also in the neural mechanisms that underlie call perception. In the one study carried out thus far, Japanese macaques tested with their own species' vocalizations show the same right ear advantage and localization of control in the anterior temproal cortex as do humans.

Development

The study of vocal development must deal with three interrelated areas: the development of vocal production (correct "pronunciation"), the development of vocal usage (how animals come to give specific calls in particular circumstances), and the development of appropriate responses to the vocalizations of others. To date, research on two well-studied animal groups, nonhuman primates and songbirds, has concentrated almost exclusively on vocal production. Results have suggested that although song learning in birds is strongly affected by auditory experience (e.g., Marler 1981), nonhuman primate vocalizations are under stronger genetic control, appear fully developed at birth, and are relatively unaffected by what a young monkey hears (see p. 100). In the sections that follow, we summarize the results of our work on vocal development in vervets, which has placed equal emphasis on call production, usage, and response. The data we report are taken from Seyfarth and Cheney 1980 and 1986.

Vocal Production

To test whether an infant vervet needs practice before it can "pronounce" an alarm call correctly, one would ideally record alarm calls from infants at various ages, beginning shortly after birth, and then see whether the acoustic properties of their alarm calls became more adultlike over time in ways that could not be explained simply by the maturation of the vocal tract. For two reasons, however, it proved difficult to tape record the alarm calls given by infants. First, alarm calls in general occurred unpredictably, and, following Murphy's law, were given at the highest rates when our recording equipment was turned off. Second, infants gave alarm calls far less often than adults. We never heard infants under 1 month of age giving alarms, and, despite our best efforts, we were able to tape record only a few alarms from infants younger than 6 months of age.

The alarm calls we did obtain from young infants were, in every case, acoustically similar to the alarm calls given by adults. Infant alarms also elicited responses from animals nearby that were similar to the responses elicited by adult alarms. Such data suggest that infants are born with an ability to give alarm calls that are acoustically "correct." However, we cannot rule out the possibility that learning also plays a role. Though they are disinclined to give alarms themselves, infants might spend the first 3 or 4 months of their lives learning about alarm calls by listening to adults. Alarm calls are therefore not very helpful in clarifying how experience might affect the development of vocal production.

Grunts provide a more detailed picture of the development of vocal production, since infants begin grunting at high rates from the day they are born. By many acoustic measures, the grunts of infants between 1 and 8 weeks old are different from those of adults. As infants grow older, however, the acoustic features of their grunts gradually come to resemble those of adults. Some features become adultlike by 12 weeks, others by the time infants have reached their first year, and still others not until the infants are between 2 and 3 years old.

Vocal Usage

Vervet monkeys in Amboseli regularly come into contact with over 150 species of birds and mammals, only some of which elicit alarm calls. When infant vervets first begin giving alarm calls, they often make "mistakes" and give alarms to species like warthogs, small hawks, or pigeons that pose no danger to them. These mistakes, however, are not entirely random.

Figure 4.11 compares the species that elicited eagle alarm calls from infant, juvenile, and adult vervets over two 9-month periods in 1980 and 1983. Confronted with the same array of actual and potential predators, adults were most selective, giving eagle alarms almost exclusively to raptors (family *Falconidae*), a group whose members are distinguished from other birds by their relatively large size, curved beaks, and talons. Within this class, adults gave alarm calls most often to the vervets' two confirmed predators, martial and crowned eagles. Juveniles were less selective but more likely to give alarm calls to raptors than to nonraptors. Infants were the least selective and did not distinguish between these two broad classes of birds.

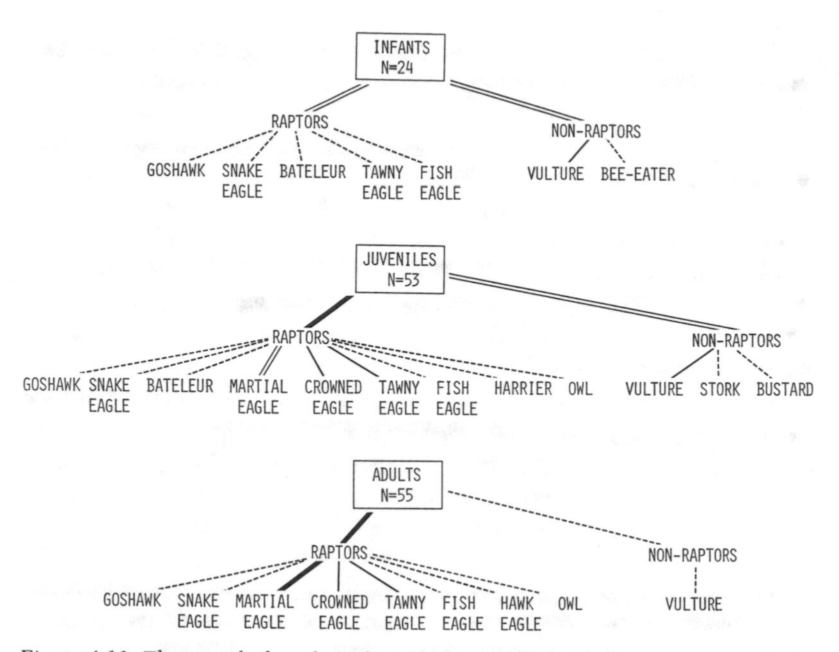

Figure 4.11. The stimuli that elicited eagle alarm calls from vervet monkeys of different ages. Data were collected over two 9-month periods in 1980 and 1983. Infants are animals less than 1 year old; juveniles are 1 to 4 years old; and adults are over 4. N = *number of alarm calls from animals in each age class. Broken lines indicate* <*5 alarms, single lines 6 to 10 alarms, double lines 11 to 15 alarms, and thick solid lines* >*15 alarms. From Seyfarth and Cheney 1986, with permission.*

Nevertheless, infants gave eagle alarm calls *only* to birds and things in the air (for example, a falling leaf).

Similarly, although infants gave leopard alarm calls to a variety of species that posed no danger to them, they restricted their leopard alarms primarily to terrestrial mammals. Moreover, they gave snake alarms exclusively to long, snakelike objects. In other words, infants behaved as if, from a very early age, they were predisposed to divide other species into different classes: predators versus nonpredators and, within the former class, terrestrial carnivores, eagles, and snakes.

This pattern of development in the vervets' use of vocalizations was not unique to alarm calls. When infant vervets first began to grunt, they used many of the acoustically different grunt types found in adult communication. For each of these calls, the relation between grunt type and social situation was imprecise but still not entirely random. Although adults, for example, gave the grunt to an animal moving into an open area only when they themselves or other individuals were moving into an open area, infants between 1 and 4 months of age used this call as they moved into a new area, followed their mothers, or followed a juvenile playmate. Although adults used the grunt to another group exclusively in the presence of another group or a new male member of their own group, infants used the same call as their mother moved away, as they climbed a tree, as they were pushed off the nipple, as they looked at a long-time male member of their own group, and as they looked at another group. Infants behaved as if they were inclined to divide social situations into broad categories, such as *movement or group progression* and *distress, or the proximity or approach of another animal.* Over time, the latter category was further divided to distinguish among *proximity of a dominant, proximity of a subordinate,* and *proximity of another group.* Similar gradual development in vocal usage occurs among pigtailed macaques, who need experience before they can use the appropriate scream in interactions with particular opponents (Gouzoules and Gouzoules 1989).

The use of grunts by infant and adult vervets illustrates the complex learning that must occur if a young monkey is to begin using vocalizations appropriately. Correct grunt usage requires that an animal distinguish between those who are dominant and those who are subordinate to itself. Unlike ground and aerial predators, however, dominant and subordinate animals are not grossly different morphologically, and there is considerable evidence that young primates need social experience before they know which members of their group rank above and below them (e.g., Cheney 1977; Berman 1982; Datta 1983b). Similarly, correct use of grunts requires that an immature vervet distinguish between males that are long-term group residents (to whom he gives the grunt to a dominant) and those who are recent immigrants (to whom he gives the grunt to another group). Again, males in these two classes are not necessarily morphologically distinct, and learning through observation or interaction would seem the most likely means by which an infant would come to make the appropriate discriminations.

learning through observation + social experience

What causes infant and juvenile vervets to make mistakes and give, for example, alarm calls to species that pose no danger to them? One hypothesis argues that such alarms are not mistakes at all but instead reflect the vulnerability of young vervets to a much wider variety of predators. This argument, however, cannot explain why infants give alarm calls to herons, geese, bustards, and even a falling leaf (Seyfarth and Cheney 1980, 1986), which pose no threat at all to them.

hyp

rejects hyp

A second hypothesis, which is supported by observations, argues that infants' mistakes are determined by a potential predator species' proximity and behavior. Infants and juveniles are more likely to give leopard or eagle alarm calls to species that seldom elicit alarms from adults—for example, lions, jackals, or small hawks—if the predator is encountered at close range or if it is hunting.

A third hypothesis derives from studies of object classification in humans. Human categories, Eleanor Rosch claims (1973, 1977), are internally structured in the mind, containing one or more "prototype" objects surrounded by objects of decreasing similarity to the prototype (see also Anglin 1977; Smith and Medin 1981; and Armstrong, Gleitman, and Gleitman 1983 for an opposing view). If Rosch is correct, decisions about the categorization of a new object are based on the object's similarity to a prototypical member of the category. Applying this argument to vervet monkeys, one would predict that when the distance and behavior of different species are held constant, immature vervets would be more likely to give alarm calls to species that look like genuine predators than to species that do not.

The data do not support this hypothesis. For example, among nonpredators the tawny eagle (*Aquila rapax*) was much more likely to elicit alarm calls from immatures than was the black-chested snake eagle (*Circaetus pectoralis*), even though, when seen from below, the black-chested snake eagle bears a close resemblance to the vervets' most common avian predator, the martial eagle (see below; fig. 4.12).

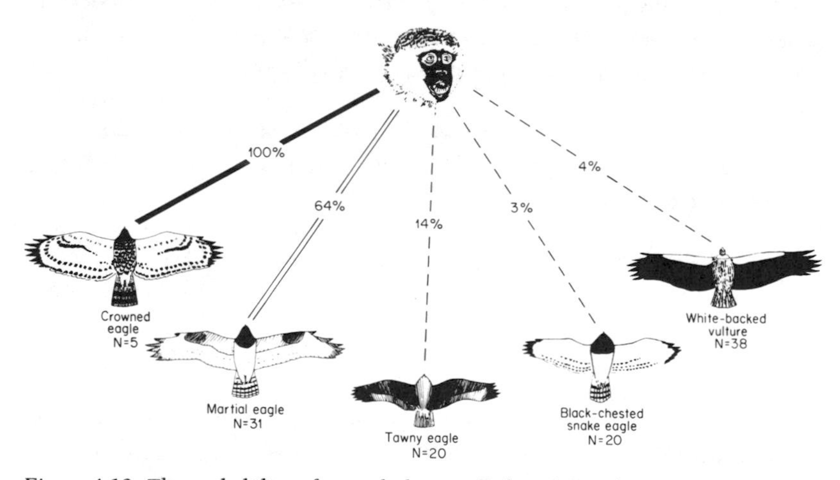

Figure 4.12. The probability of second alarm calls by adults after an infant has given an eagle alarm to an avian species. N = *number of cases in which an infant was the first member of its group to give an eagle alarm to a particular species. Lines and percentage values indicate the frequency with which infant alarms were followed by one or more alarm calls by adults. Drawings by Margaret H. Searcy. From Seyfarth and Cheney 1986, with permission.*

Apparently, then, infants make mistakes because they cannot distinguish predators from nonpredators at a distance and because they are surprised or frightened by a nonpredator that is hunting. Although infants do seem predisposed to distinguish among terrestrial, avian, and snakelike predators, these classes are apparently not coded in the monkeys' minds in terms of specific prototypical species.

How, then, do infants come to recognize the correct association between a particular alarm call and one or two predator species? One answer may lie in the responses shown by adults when an infant gives an alarm call.

As noted earlier, alarm calls by infants and juveniles elicit many of the same responses from others as do alarm calls by adults. If an infant is the first member of his group to give an eagle alarm, for example, adults nearby look up. If the infant has made a mistake and has alarmed at a harmless raptor, adults typically return to what they were doing. However, if the infant has spotted a martial eagle, adults are very likely to give an alarm call themselves (fig. 4.12). These second alarm calls might serve as reinforcers, guiding the infant's developing recognition of the relation between different alarm calls and their referents.

Although the responses of adults may affect the course of infant development, we find no evidence that adults explicitly teach infants. Adults are as likely to give second alarm calls following a correct alarm by an infant as they are following a correct alarm call by another adult. In other words, adults do not behave as if they can attribute ignorance to infants or as if they recognize that infants are particularly in need of encouragement, correction, or instruction. The existence or lack of pedagogy among nonhuman primates raises important questions about the animals' ability to attribute states of mind to others. Pedagogy and attribution are discussed further in chapter 8.

Response to Vocalizations

Adult vervet monkeys respond differently to different alarm calls. Given what we know about the hunting behavior of the vervets' three main predators, these different responses appear to be adaptive. To test whether adultlike responses appear fully developed in infants or whether responses are modified during development, we carried out a series of playback experiments using infants and their mothers as subjects (fig. 4.13). At monthly intervals we played to infants between 3 and 7 months of age a leopard alarm call, an eagle alarm call, and a snake alarm call that had been recorded from members of their own group. We filmed the infants and their mothers and divided responses into three categories: run to mother, a response we had observed frequently in previous years; "adultlike" responses, defined as the typical responses of adults to the playback of each alarm call type; and "wrong" responses, defined as those likely to increase an infant's risk of being killed, given what we know about the hunting strategy of each predator. For example, since leopards hunt vervets by hiding in bushes and eagles are skilled at taking vervets in trees, running into a bush at the sound of a leopard alarm or running into a tree at the sound of an eagle alarm may actually increase an infant's risk of predation.

Figure 4.14 summarizes infant responses. At 3 to 4 months of age, infants typically ran to their mothers. Few showed adultlike responses. Between 4 and 6 months of age, running to the mother decreased and a higher proportion of subjects showed adultlike responses. Many infants, however, also responded in wrong, potentially dangerous ways. Among infants over 6 months old, running to the mother was rare, wrong responses decreased, and most infants behaved like adults.

As with call usage, there were hints that learning from adults might have played a role in the infants' development. When we examined films of infant behavior in the seconds immediately after an alarm call had been played, we found that infants who responded only after first looking at an adult were much more likely to respond correctly than were infants who responded on their own, before looking at another animal (fig. 4.15). This observation does not prove that exposure to older animals is *necessary* for the development of correct responses, since infants might develop normal behavior in the absence of cues from others. It does, however, indicate that helpful cues are available from those nearby and that infants may take advantage of them.

Once again, we found no evidence of active pedagogy on the part of

Figure 4.13. To investigate the ontogeny of infants' responses to alarm calls, alarm calls were played to infants and their mothers at various infant ages when the infants were off their mothers. Here, adult female Amin feeds next to her 2-month-old-son.

adults. When we compared a mother's behavior toward an infant who had responded correctly with the same mother's behavior toward an infant whose response increased his risk, we found no indication that mothers pay particular attention to infants who have behaved inappropriately.

Just as vervets need experience before they can respond appropriately to

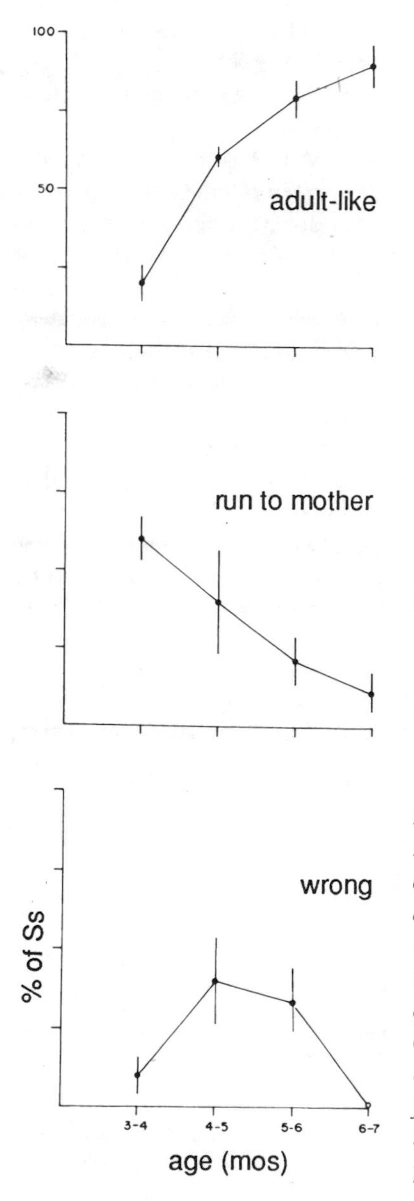

Figure 4.14. The responses of infant subjects to playback of alarm calls at different ages. Classes of response are defined in the text. Seven infants were tested at 3 to 4 months of age, four of these retested at 4 to 5 months of age, and three of these retested at 5 to 6 and 6 to 7 months of age. Values shown at each age represent proportion of subjects (means and standard errors). From Seyfarth and Cheney 1986, with permission.

their own species' vocalizations, experience is also necessary before they know how to respond to the calls of other species. As we describe in chapter 5, vervets in Amboseli inhabit the same areas as a brightly colored songbird, the superb starling (*Spreo superbus*). Like vervets, superb starlings give acoustically different alarm calls in response to different predators. The birds give a harsh, raspy chatter in response to terrestrial predators and a clear, rising or falling whistle in response to hawks and eagles (see fig. 5.7). If adult vervets hear the starling's terrestrial predator alarm call, they often run toward trees; if they hear the starling's raptor alarm call, they typically look up into the air (Cheney and Seyfarth 1985a; see also chapter 5).

Working in Amboseli from 1983 to 1985, Marc Hauser (1988a) noticed that vervets encountered starlings (and heard their alarm calls) at different rates in different habitats. Groups living exclusively in dry woodland areas heard starling terrestrial alarm calls roughly twice every hour, whereas groups whose territories were closest to the swamp heard the same calls more than twice as often. To test whether such differences in exposure affected the age at which infants began responding appropriately to the bird's alarms, Hauser played starling alarm calls to infants at successively older ages and filmed their responses. The starling's song was used as a control. Infants inhabiting ranges near the swamps, where starling alarms were heard more often, responded appropriately at significantly earlier ages than did infants inhabiting the dry woodland, where starling alarms were less frequent. Auditory experience seemed to be the crucial variable, since there was no evidence that infants in the swamp groups were developmentally more advanced than infants in the woodland groups on any other behavioral measure (Hauser 1988a).

Factors Affecting Vocal Development

Vocal development in vervet monkeys occurs gradually during the first 3 years of life. In varying degrees, call production, the use of calls in appro-

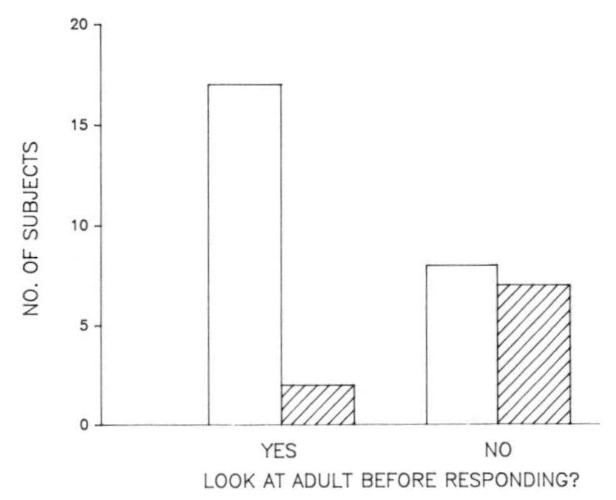

Figure 4.15. The number of infants who responded correctly (open histograms) or incorrectly (shaded histograms) to playback of an alarm call after either first looking at an adult or without looking at an adult before responding. Redrawn from Seyfarth and Cheney 1986.

priate contexts, and responses to the calls of others seem to be affected by both innate predispositions and experience.

For example, from their very first day infant vervets produce a call whose acoustic features are broadly similar to an adult's grunt. Even naive infants seem to "know" that this vocalization is given during social interactions and that grunts in different circumstances have subtly different acoustic properties. Despite these predispositions, infant grunts are "sloppy"; they are badly pronounced and often given in inappropriate situations. Considerable time is needed before a young monkey's grunting is, both acoustically and socially, as well defined as the grunting of adults.

Similarly, from a very early age infants are predisposed to give acoustically different alarm calls to different classes of predator. Large, terrestrial mammals are treated differently from raptors, and both are distinguished from snakes. At the same time, the infants' understanding of the relation between predator species, alarm call type, and response is imprecise. Infants need experience before they can recognize those species within a class that are genuine predators or respond appropriately to their own or other species' alarm calls.

The presence of nearby adults is probably helpful. When an infant hears an alarm call it can look at others before responding, and infants who *do* look first at adults are more likely to respond correctly. If an infant gives an alarm call to a genuine predator, adults will alarm call too, and this cannot help but be encouraging. All the same, we have no evidence that adults recognize their infants' needs or go out of their way to help infants learn how to communicate.

In vervet vocal development, as in language, comprehension precedes production. Among 6 to 7 month-old vervets, for example, responses to alarm calls are indistinguishable from the responses of adults, suggesting that infants of this age can correctly associate each alarm type with a different source of danger. In contrast, the acoustic properties of the same infants' grunts, as well as their ability to use alarm calls and grunts in the appropriate contexts, are far from adultlike and take another 18 months to develop fully.

Summary

Early studies of primate communication drew sharp distinctions, in neural control, development, and function, between nonhuman primate vocalizations and human language. Subsequent research paints a more complicated picture.

The vocal repertoires of nonhuman primates, when assessed by the animals themselves, are far larger than scientists initially perceived them to be. The information each call contains is also more specific and less dependent on context than previously imagined. Free-ranging monkeys also use calls—

alarm signals and close-range vocalizations—in a manner that effectively represents, or denotes, objects and events in their environment. The use of these calls seems to be under relatively voluntary control, since call production can be conditioned in the laboratory and animals in the wild routinely give specific vocalizations only in particular circumstances. Primates make subtle acoustic discriminations when distinguishing between calls, and one well-studied case provides evidence of left hemispheric specialization in vocal perception.

Vocal development in primates exhibits many parallels with the early stages of speech development in young children. Monkeys begin by using certain vocalizations—some clearly pronounced, other more garbled—in particular social situations. They behave as if predisposed to divide events in the world around them into broad categories that require a grunt, a scream, an alarm call, or no vocalization at all. Over time, pronunciation improves and infants sharpen the relation between a call and the objects to which it refers or the context in which it is used. Older infants and juveniles begin to recognize that within each broad context there is a further subdivision into circumstances that call for a specific kind of grunt, a particular sort of scream, or an acoustically different alarm. Throughout development, comprehension precedes production.

Because nonhuman primates use vocalizations to signal *about* things, research on communication offers a glimpse of how they see the world. Studies of vocal communication reinforce the conclusion, already obtained from studies of social behavior, that monkeys classify members of their own species according to group membership, dominance rank, and behavior. Through vocal communication monkeys also reveal their knowledge of other individuals' social relationships and their knowledge of other species.

But what, in the end, do monkeys really *mean* when they vocalize to one another? Can we actually define leopard alarm or grunt to a dominant in the same way we define words like anarchist, bordello, or sycophant? Thus far, we have been purposely vague where questions of meaning are concerned in order to present background information and to summarize the results obtained from the past 10 years' research. This done, we consider what it all means in chapter 5.

CHAPTER FIVE

WHAT THE VOCALIZATIONS OF MONKEYS MEAN

We have called the vocalizations of vervet monkeys semantic signals and have drawn an analogy with human words because of the way these calls *function* in the monkeys' daily lives. When one vervet hears another give an eagle alarm call, the listener responds just as he would if he had seen the eagle himself. It is tempting to suppose that in the monkey's mind the call "stands for" or "conjures up images of" an avian predator, even when the bird itself has not yet been seen. The same is true of leopard alarm calls, grunts to neighboring groups, and many other vocalizations in the monkeys' repertoire.

Clearly, however, descriptive evidence of this sort does not elevate animal signals to the status of human words. Consider first the problem of whether a signaler intends (wants or desires) to communicate with others. In Pavlov's classic experiments, dogs heard a bell every time they were given meat. After a while they began salivating whenever they heard the bell, even if meat did not appear. To Pavlov's dogs, bells evoked the same response, or "stood for," or "conjured up images of," meat, even when the meat wasn't there. This hardly proves that the bells intended to communicate to the dogs in the way that a human intends to communicate when he says, "Dinner is served," or "Your understanding of semantics is really very superficial."

One issue that we must confront, therefore, concerns the factors that cause one animal to vocalize in the presence or even in the absence of another. Given that listeners treat vocalizations as conveying a particular sort of information, is it also the case that signalers *intend* for them to do so? Do animals, for instance, take note of who their audience is or adjust their communication to make sure the message has been received? Do signalers modify their calls if the audience is already aware of their information? As we will discuss, there are intriguing hints that animals do take into account the nature of their audience when they call, but the issue of what animals actually intend to communicate when they signal to others remains unclear. intent unknown

There is another reason for caution in drawing parallels between monkey vocalizations and human words. When humans use words like *apple* or *eagle,* we recognize the referential relation that holds between such signs and the things for which they stand. Referential relations can, for instance, be distinguished from causal relations: the word eagle does not cause a particular bird to appear or result in a particular pattern of behavior. Instead, the word stands for, or represents, an object even when it cannot be seen. We know, moreover, that there is no obligatory relation between the acoustic features of a word and its referent. Although some words, like *choo-choo* or *meow,* do sound like the thing they represent, in most cases the noise that we make when we speak a word provides no clues about its meaning. Words that sound different, like treachery and deceit, are judged to be similar if they mean the same thing, whereas similar-sounding words, like treachery and lechery, are judged to be different if they have different meanings. It is not quite as Lewis Carroll's Humpty Dumpty would have it—that a word means just what we choose it to mean—because there obviously has to be some consensus about the relation between a sound and its referent for communication to take place (admittedly, not a major concern to Humpty Dumpty). Nevertheless, when Alice asks him how he can make words mean so many different things, Humpty Dumpty is not far off the mark in claiming that we are masters of our words.

Humans therefore recognize the referential relation *X means Y* and, for the most part, classify words not just according to their acoustic properties, which are measurable and concrete, but according to their meanings, which are more abstract. By contrast, our analysis of communication has thus far focused exclusively on how vocalizations *function* in vervet monkey society. As a result, despite the parallels we have drawn between vervet calls and human words, we have so far provided no evidence that monkeys understand the referential relations that exist between their calls and things in the external world or even that the animals can dissociate a call's meaning from its acoustic properties. In short, we have not yet considered the mental operations that might underlie the perception of vervet vocalizations and, as a result, any claim that the monkeys possess rudimentary "words" is premature.

The goal of this chapter, then, is to explore in greater detail what monkeys actually mean when they vocalize to one another and what listeners understand when they hear a vocalization. To do this, we approach the assessment of meaning from a number of different and complementary directions. First, we consider the relation between a signal and events in the minds of caller and recipient. As a guide to analyzing what may be the mental states of these individuals, we draw on the philosopher Daniel Dennett's scheme for assessing levels of intentionality in animal vocalizations. As we will discuss, the evidence suggests that the vocalizations of monkeys, birds, and many other species are not just involuntary reflexes. Some animals, for example, adjust their rate of calling when an audience is present. It is not clear, however, whether animal signals, like human language, involve the attribution of mental states to others. We do not know whether monkeys vocalize only with the intent to modify the *behavior* of others or whether they also call to modify what other animals *think*.

As a second approach to the study of call meaning, we examine the relation between a vocalization and the objects or events it denotes. We review a variety of evidence indicating that, among the other information they convey, primate vocalizations denote external referents. Although monkeys may not be aware of the relation between sign and referent, they nevertheless compare calls on the basis of their referents and not just on the basis of their acoustic properties. Furthermore, monkeys are sensitive to the breadth of referential specificity in different vocalizations; some calls are quite specific in the objects they denote, whereas others are more general. As is often the case with human language, the meaning of each call cannot be described in isolation but depends on its relation with other vocalizations in the animals' repertoire.

Finally, we adopt a developmental perspective and analyze the meaning of vervet vocalizations by comparing them with the earliest sounds of human infants. When human babies first begin to vocalize, their grunts, babbles, goos, and screams are clearly not words, nor do young infants understand the meaning of sounds they hear spoken by others around them. During the next 12 to 18 months, however, children begin to discriminate among other peoples' speech and to use words of their own. They behave as if they understand what a particular sound *means*. What standard of evidence is used to make this assessment, and how does the linguist or psychologist determine the content of word meaning in the mind of a 10- or 14-month-old child? What would our conclusions be if we applied the same standard of evidence to the vocalizations of vervet monkeys?

A Signaler's Intentions

The Problem

What goes on in a vervet monkey's mind when he gives a leopard alarm call, a grunt to another group, or any of the other vocalizations in his repertoire? The simplest explanation is that there is nothing "mental" at all about monkey vocalizations: they are just relatively inflexible responses to particular stimuli, like a cry of surprise given to someone who suddenly leaps out from a hiding place. Alternatively, a monkey might give an alarm call or grunt only after he has studied the situation carefully and taken into account a number of different factors. Calling might depend, for example, on whether a predator is hunting, whether there are other monkeys nearby, and whether these monkeys are kin, mates, or other close companions. Finally, calling might even be influenced by the states of mind that the signaler attributes to others. Conceivably, before giving an alarm call a monkey might ask himself, "Have other group members already seen the leopard? Have they responded appropriately? Wait a minute: do I *want* others to think that I've seen a leopard?"

The distinction between communication and simpler, more reflexive calls that can nevertheless convey information has been considered by many different philosophers. Grice (1957), for example, distinguished the "nonnatural" meaning of linguistic phenomena, in which a speaker intends to modify the beliefs or behavior of his audience, from the "natural" meaning of most other types of signals, in which, for example, thunder and lightning mean that it will soon rain (see also Bennett 1976; Tiles 1987). According to Grice's definition, communication does not occur unless both signaler and recipient take into account each others' states of mind. By this criterion, it is highly doubtful that *any* animal signals could ever be described as truly communicative.

Nevertheless, they might be. Keeping an open mind, how would we know if they were? Our first task is obviously to formulate hypotheses that address the mechanisms underlying communication in monkeys and other animals. Dennett (1971, 1978a, 1978b, 1983, 1987, 1988) argues that we can best understand the signals of other species if we adopt what he calls the *intentional stance* and assume, at least for purposes of analysis, that monkeys, birds, and other animals are capable of mental states like believing, wanting, and thinking. In our view, Dennett's analytical scheme is a useful one, but its description requires a brief digression.

In philosophical terms, intentional phenomena are largely restricted to mental states, such as beliefs, desires, and emotions (Dennett 1987). Intentional phenomena are always *about* some other thing, be it a physical stimulus or another mental state, and they probably constitute the major components of human thinking and language. Whenever an individual thinks, believes, wants, likes, or fears something, he is in an intentional state (from the Latin verb *intendo,* meaning *to point at*). Intentional statements exhibit a specific logical property called *referential opacity.* In these statements, unlike ordinary relational statements, there is no guarantee that "substituting equals for equals" preserves their truth value. Consider two words that are synonymous or even refer to the same thing; in an ordinary, nonintentional statement, one of these words can be substituted for the other without risk of damage to the whole statement. So, for example, since *vervet monkey* and *Cercopithecus aethiops* are words for the same creature, if the statement, "The vervet monkey was eaten by a leopard," is true, we can infer that the statement, "The *Cercopithecus aethiops* was eaten by a leopard," is also true. This is not necessarily the case for intentional statements. If I don't know that *Cercopithecus aethiops* is the Latin name for vervet monkey, I may *fear* that the vervet monkey was eaten by a leopard without manifesting the same anxiety for *Cercopithecus aethiops*. There have been centuries of debate about the proper analysis of intentionality; before we dissolve into referential opacity ourselves, we refer the reader to Quine (1960), Fodor (1975), Searle (1983), and Dennett (1987).

For our own empirical purposes, Dennett's intentional stance provides a useful method for investigating communication and the attribution of mental states in nonhuman species. We begin by assuming that a vervet monkey is an *intentional system*, capable of mental states like beliefs and desires. But what kind of beliefs and desires? Here Dennett's different "levels of intentionality" provide us with a number of alternative hypotheses.

First, we must entertain the possibility that vervets are *zero-order intentional systems*, with no beliefs or desires at all. A zero-order explanation holds that vervet monkeys give alarm calls because they are frightened. Vervets, moreover, experience a number of different kinds of fear, each associated with a different kind of predator. Each type of fear elicits a characteristic alarm call and a characteristic escape response.

Alternatively, vervets might be *first-order intentional systems*, with beliefs and desires but no beliefs *about* beliefs. At this level, vervet monkeys give leopard alarm calls, for example, because they believe that there is a leopard nearby or because they want others to run into trees. The caller does not need to have any conception of his audience's state of mind, nor need he recognize the distinction between his own and another animal's beliefs.

It is also possible that vervets are *second-, third-, or even higher-order intentional systems*, with some conception about both their own and other individuals' states of mind. A vervet monkey capable of second-order intentionality gives a leopard alarm call because he wants others to believe that there is a leopard nearby. At higher and increasingly baroque levels, both the signaler's and the audience's states of mind come into play. At the third level of intentionality, a vervet gives an alarm call because he wants others to believe that he wants them to run into trees. Linguistic communication, it has been argued, requires at least third-order intentionality on the part of both speaker and listener (Grice 1957, 1969; Bennett 1976).

Humans are intentional systems almost to a fault. Our legal systems revolve around Byzantine efforts to establish intent, and our elections reveal with depressing consistency that politicians worry more about what others think (and what others think they think) than about the actual course of events. All the same, it is unlikely that even humans regularly ascend (or stoop) to levels beyond third- or fourth-order intentionality in their daily lives. Whether animals are capable of second- or even first-order intentionality, however, is a hotly debated issue.

In Dennett's scheme, zero-order intentionality attributes no mentality at all to vervets. It therefore constitutes an essential null hypothesis that allows us to distinguish more complex communication from relatively inflexible responses. Explanations of communication in terms of zero-order intentionality are, for many animal species, perfectly plausible. In our discussion of referential, or semantic, communication, it is important to remember that an individual can in theory communicate "about" aspects of his environment—that is, in a functionally semantic manner—without any comprehension of the effect of his call on his audience.

By contrast, first-order intentionality implies that the signaler wants to modify the behavior of his audience. It suggests that signalers can "choose" to communicate or not, depending, for example, on which other animals are nearby. Note, however, that first-order intentionality still does not fulfill Grice's requirements for true communication. It does not require that the audience recognize the signaler's intention or even that the signaler has any conception about what his audience does or does not know. First-order intentionality demands only that the signaler recognize the effect of his call on the audience's *behavior;* it makes no claim concerning what a signaler knows about the effect of communication on his audience's *mind.* A vervet could, for instance, simply have learned that making a particular sound causes others to run into trees, and he could make this sound in order to elicit the predicted response (Dennett 1983).

A qualitative change in the nature of communication occurs at the level of second-order intentionality, the level at which a signaler attributes mental states to others and communicates in order to modify not just behavior but also these mental states. Higher, more recursive levels of intentionality are simply elaborations on the second-order system (Cargile 1970; Dennett 1983, 1987). All, like second-order intentionality, are fundamentally different from zero- and first-order explanations because they require that individuals be able to attribute mental states to others. Moreover, the transition from first- to second-order intentionality has important functional consequences for the animals involved. There will, for example, be far wider scope for deception and the manipulation of signals if animals not only can modify call production but also can attribute beliefs to others.

It is still unclear whether monkeys and apes are capable of attributing states of mind to themselves or others, and we reserve this question for more detailed discussion in chapter 8. Here we concentrate on the more basic question of whether vervets and other animals are capable of even first-order intentionality.

Evidence for Voluntary Signaling

Are animal signals ever voluntary? As we noted in chapter 4, a voluntary signal is traditionally defined as one whose delivery is not obligatory but can be varied to occur in predictable and appropriate (or adaptive) contexts. In an animal learning experiment, for example, a voluntary signal would be one that can be reinforced or extinguished through conditioning. A zero-order intentional system cannot modify signal production depending on contextual variables like the presence of an audience or the occurrence of a reinforcer; its signals are fixed, obligatory responses to particular sets of stimuli. Signals that are tied to some motivational state are by definition zero-order systems. The blue color of a vervet male's scrotum cannot be modified except under cases of severe physical or psychological stress. The same is true of some vocal signals, like screams of pain. A variety of evidence, however, suggests that many of the vocal signals given by vervet monkeys are under some voluntary control and that monkeys can vary their signaling rate according to changes in the social and physical environment.

Consider again the monkeys' alarm calls. The first point to emphasize is that there is no obligatory link between a particular alarm call and its associated response. Although eagle alarm calls cause vervets to look up into the air more often than do leopard alarm calls, vervets do not *always* respond to eagle alarm calls in this way. They may run into dense bushes or, if they are in a tree, run out of the tree. They may also completely ignore the call. Similarly, alarm calls do not always evoke other alarm calls. In our playback experiments, subjects who heard an alarm almost never gave alarm calls themselves, even though they consistently showed the appropriate escape responses (Seyfarth, Cheney, and Marler 1980b). Depending upon the circumstances, then, vervets can "choose" to give an alarm call without an escape response or to flee without giving a call.

We have also observed that solitary vervet monkeys (in every case a male, either traveling between groups or temporarily separated from others) do not give alarm calls when confronted with a predator. In the most dramatic example, Phyllis Lee, who studied the vervets in 1978 and 1979, once found adult male Rosebery feeding alone in two small acacia trees located on a small island in the middle of a swamp. The rest of Rosebery's group was about half a kilometer away. Apparently, Rosebery had not noticed that a leopard had stalked him into the swamp until it suddenly leapt into the tree with him. For almost an hour the leopard pursued Rosebery through the trees. Rosebery, however, was lighter and far more nimble than the leopard, and he was eventually able to jump out of the tree and make his escape through the swamp. What most struck Phyllis was the utter silence of the chase. In marked contrast to a leopard attack on a group of monkeys, when the air is filled with loud alarm calling by many individuals, it had all occurred without a sound.

As a more systematic test of whether monkeys can modify their alarm calling depending on the presence of an audience, we carried out experiments on four groups of captive vervet monkeys that live on the grounds of the Sepulveda Veterans Administration Hospital in Sepulveda, California (Cheney and Seyfarth 1985b). Here each group's cage consists of an indoor and an outdoor enclosure connected by a chute that allows some animals to be temporarily separated from the rest of the group. In our tests we locked most of the group indoors, isolating an adult female in the outdoor enclosure with either her offspring or a similarly aged, unrelated juvenile. The two animals were then approached by the captive monkey's version of a predator, a veterinarian wearing a surgical gown and mask and carrying a net. This nemesis (actually Marc Hauser, a graduate student at the time) walked around the cage for a predetermined time and then disappeared. Upon seeing this predator, all adult females gave alarm calls—calls that, to our ears at least, sounded similar to the ones vervets in Amboseli give to strange, potentially hostile humans like Maasai tribesmen. The females gave alarms, however, at significantly higher rates when they were with their offspring than when they were with unrelated juveniles. In another experiment, we locked subordinate adult males outside in the company of either a female or a dominant male. In all four cases, the males gave more alarm calls when they were with the female than when they were with the dominant male (Cheney and Seyfarth 1985b).

Vervets therefore seem able to modify their alarm calling rate depending on their audience. Whatever their precise motivational basis, the production of alarm calls is not obligatory but is influenced by the presence of kin, potential mates, and more dominant rivals. As we discuss further in chapters 7 and 8, withholding alarm calls may be an effective means of deceiving others, because this kind of "cheating" is difficult to detect. Indeed, deception through signal concealment appears to be widespread in monkeys and apes.

The ability to modify calling in the presence of social companions is not restricted to monkeys. In some species of ground squirrels, for example, females with kin give more alarm calls than those without kin (e.g., Dunford 1977; Sherman 1977). Similarly, downy woodpeckers (*Picoides pubescens*) that encounter a predator give no alarm calls if they are alone, if they are the only woodpecker foraging in a mixed-species flock, or if the only other woodpecker present is a member of the same sex. If a downy woodpecker of the opposite sex is nearby, however, the birds give alarm calls at high rates (Sullivan 1985). The vulnerability of kin also seems to influence alarm calling behavior. Patterson, Petrinovich, and James (1980) presented breeding white-crowned sparrows (*Zenotrichia leucophrys*) with either a snake, a hawk, a jay, or a junco. They found that although pairs of sparrows gave most alarm calls to the snake when they had nestlings, they gave most alarm calls to the hawk and jay when they had older fledglings. Juncos elicited little response at all stages of the breeding cycle. The pattern and level of response, therefore, was correlated with the vulnerability of offspring and the class of predator.

Alarm calls are not the only vocal signals that are subject to modification. House sparrows (*Passer domesticus*) modify the rate at which they utter food calls, apparently according to whether or not the food supply is divisible (Elgar 1986). At Gombe in Tanzania, solitary male chimpanzees who come upon a fruiting tree will sometimes utter loud pant hoots that attract others to the site (fig. 5.1). The probability that males will call is directly correlated with the abundance of fruit; males who encounter only a small amount of fruit are less likely to call than those who encounter enough to feed more than one animal (Wrangham 1975, 1977). Similarly, when captive male chimpanzees were given bunches of prunes of varying sizes, they gave the most rough grunts to the largest number of prunes, the least to the fewest, and an intermediate number to the intermediate amount (Hauser and Wrangham 1987).

In an explicit test of the effects of social context on call production, Marcel Gyger, Steve Karakashian, and Peter Marler (1986; see also Karakashian, Gyger, and Marler 1988; Marler, Karakashian, and Gyger 1990) presented male jungle fowl (*Gallus gallus*) with a silhouette of a hawk which "flew" over the birds' cage on a wire, replicating the classic experiments conducted by Tinbergen (1951). The roosters gave almost no alarm calls when they were alone and significantly more when in the presence of another male or a female. This "audience effect," moreover, was species, or perhaps

Figure 5.1. The loud pant hoots of male chimpanzees attract others to fruiting trees. Males seem to vary their calling rate depending upon the abundance of the resource. Photograph by Richard Wrangham.

size, specific: roosters paired with a female bobwhite quail gave no more alarm calls than when they were alone. Even for roosters, apparently, alarm calling was not simply a reflexive response but was modulated according to social context.

Were the roosters acting with an intent to inform their audience and to change their audience's *beliefs?* Not necessarily. Roosters alarm called at the same rate regardless of whether or not the hens could see the predator, suggesting that they failed to attribute ignorance or awareness of danger to their audience (Karakashian, Gyger, and Marler 1988; Marler, Karakashian, and Gyger 1990). Similarly, when a baffle was erected so that only the hens, and not the roosters, could see the predator, the males showed no alarm response even when hens took evasive action by fleeing and crouching (Gyger, Karakashian, and Marler 1986). Apparently, the roosters did not recognize that the hens' escape responses signaled danger; they only alarm called if they themselves could see the predator. In chapter 8 we discuss some similar experiments that investigate the extent to which monkeys adjust their alarm calls according to whether their audience is ignorant of danger.

There is evidence, therefore, that chickens, vervet monkeys, and many other animals modify their calls depending on their audience. The data prompt us to reject an explanation based on zero-order intentionality and to describe the animals as at least first-order intentional systems. Many species, in other words, appear to recognize the association between a particular sound and a specific escape behavior, and they make use of this knowledge to alter the behavior of others.

At the same time, none of the behavior that we have described requires the attribution of mental states found in second-, third-, and higher-order intentional explanations. None of the data demonstrate, for example, that a vervet male monitors what his audience thinks, as opposed to what it *does,* or that he adjusts his vocalizations on the basis of these attributions. Whether roosters recognize the distinction between their own knowledge and the knowledge of others, and can therefore be said to have an *intent* to communicate, seems doubtful. Whether monkeys or apes are better mind readers than chickens remains an open issue and one we discuss in chapter 8.

The most impressive examples of voluntary production and suppression of calls come, not surprisingly, from chimpanzees. Indeed, there is even some evidence that chimpanzees recognize how noise is produced and take active measures to control it.

Free-ranging chimpanzees at Gombe make periodic patrols in the peripheral parts of their own and neighboring communities' ranges. These patrols are potentially dangerous, because if males from a neighboring community discover a patrol, there may be an aggressive conflict in which animals can be injured or even killed. Describing these patrols, Jane Goodall comments: "Perhaps the most striking aspect of patrolling behavior is the silence of those taking part. They avoid treading on dry leaves and rustling the vegetation. On one occasion vocal silence was maintained for more than three hours. A male may perform a charging display during which he drums on a treetrunk, but he does not utter pant-hoots. Copulation calls are suppressed by females, and if a youngster inadvertently makes a sound, he or she may be reprimanded. By contrast, when patrolling chimpanzees return once more to familiar areas, there is often an outburst of loud calling, drumming displays, hurling of rocks, and even some chasing and mild aggression between individuals" (1986:490–91).

Goodall observes that juveniles and infants who vocalize or inadvertently hiccup during patrols may be hit or embraced until they become silent. Even a human follower who steps on a stick may be threatened. The chimpanzees behave as if they understand the need for silence and also even the causes of noise. During aggressive interactions within groups, too, chimpanzee mothers will occasionally silence their offspring by placing their hands over their mouths (de Waal 1986b).

What One Monkey's Call Tells Another Monkey's Mind

David Premack (1976, 1983b) used an artificial lexicon of plastic chips to study communication and intelligence in chimpanzees. The chips were differently shaped pieces of plastic that could be arranged on a magnetic board. A chip could represent an object, such as *apple,* or a concept such as *same, different,* or *name of.* It could also represent a more abstract descriptor such as *color.* To test whether subjects really understood the meaning of these symbols, the chimpanzee Sarah was first asked to describe the features of an actual apple. Was it red? Was it round? Did it have a stem? Then Sarah was asked the same questions about the symbol for apple, in this case a blue triangle. Similarly, Sarah was shown the symbol for a caramel candy and asked whether it was cube or disk shaped, white or brown, smooth or crumpled. In each case she used the same features to describe both the object and the sign that represented it. She described a blue triangle, for example, as being red and round.

Premack then reversed the question and asked Sarah to begin with an object and then describe the name for that object. Shown an apple, for example, Sarah correctly answered that the sign for this object was triangular not round, blue not green, and small not big (Premack 1976).

In a similar set of experiments, Sue Savage-Rumbaugh, Duane Rumbaugh, Steven Smith, and Janet Lawson (1980) first trained two chimpanzees, Sherman and Austin, to sort different objects into two groups: food and tools. The chimpanzees then learned to label each object by pressing an illuminated computer key that depicted the symbol for food or tool.

In the next experiment, Sherman and Austin learned to label photographs of objects by indicating the appropriate symbol. In the fourth and final experiment, they learned to label the symbols themselves. Shown, for example, the symbol that stood for *sweet potato,* a symbol they had never been tested on before, the chimpanzees correctly indicated *food* by pressing the computer key that depicted the symbol for food. Shown the symbol for *wrench,* the chimpanzees pressed the symbol for *tool.*

When does a piece of plastic (or a nonsense symbol on a computer keyboard or the noise made by another monkey) cease to become a piece of plastic and become a word (Premack 1976)? Premack argues that this transformation occurs when the properties ascribed to the symbol are not those of the piece of plastic but of the object that it denotes. There is nothing about a blue triangle that helps a chimpanzee to guess that it means *apple* or about a nonsense symbol on a computer keyboard that suggests it represents food or a tool (Savage-Rumbaugh et al. 1980).

By contrast, it is still unclear whether the noise that we have called a *leopard alarm call* is a word in the minds of vervet monkeys. True, there is nothing about the acoustic properties of a leopard alarm call (see fig. 4.4) that is obviously related to either a leopard or a particular escape strategy. Vervet leopard alarm calls, for example, do not sound like the noises made by leopards; the monkeys' vocalizations are not onomatopoeic. At the same time, the vervets' different responses to different alarm calls (or different grunts) do not, by themselves, prove that vervets understand referential relations like "leopard alarm calls stand for leopards." The monkeys have not, as in Premack's experiment, been presented with a call and asked to perform a feature analysis of its referent. Nor have they, as in Savage-Rumbaugh's experiment, been asked to classify signs according to the functional similarity of their referents.

In the absence of such data, it is sometimes suggested that the meaning of animal signals does not derive from any referential relation with a particular object or event but is instead a direct function of its acoustic properties. Morton (1977) notes that many birds and mammals give harsh, low-frequency calls when they are acting aggressively and higher-frequency, more tonal calls when they are frightened or behaving in a friendly manner. Morton argues that there are predictable "motivational-structural rules" linking the motivation of a caller and the physical structure of its vocalization and that animals use these rules to deduce the meaning of a call from its acoustic properties.

Arguing against this view, recall that the same predator can elicit vocalizations that are acoustically quite different from one another. For example, leopards elicit long, low-frequency barks from vervet males and short, high-pitched chirps from females (see fig. 4.4; chapter 4). Despite the calls' different acoustic properties, vervets respond similarly to the two vocalizations, suggesting that the monkeys classify calls according to their referents and not just their acoustic properties.

Similarly, although rhesus and pigtailed macaques both give acoustically distinct screams in different types of social interaction (chapter 4), the same type of interaction is correlated with acoustically different calls in the two species. In fights that involve actual physical contact, pigtail screams are tonal while rhesus screams are noisy (Gouzoules, Gouzoules, and Marler 1984; Gouzoules and Gouzoules 1989). Once again, there is no consistent relation between a caller's motivation and the physical structure of his vocalization.

Finally, different social circumstances can evoke acoustically similar calls. The vervets' four grunts, though physically very similar, are given in widely different circumstances and evoke measurably different, though subtle, responses (chapter 4). Apparently, the monkeys do not deduce a call's meaning solely from its physical structure.

One method for determining how animals assess the meaning of calls is to examine how they classify calls whose acoustic structures differ. If two calls are normally given in similar contexts, are they treated the same even if they sound very different?

To investigate how vervets classify their vocalizations, we designed a series of playback experiments that tested the monkeys' ability to transfer information about the reliability of a particular signaler from one call type to another. In these experiments, we repeatedly played recordings of one individual's intergroup call or alarm call in the absence of another group or a predator until subjects had habituated to the call. We then tested whether subjects transferred their habituation to acoustically different calls whose referents were either similar to or different from the call used as the habituating stimulus.

In the first series of experiments, we used two different calls given by female and juvenile vervets to the members of other groups: a chutter and a *wrr,* the long, loud trilling call described in chapter 3. Although the two calls are acoustically different (fig. 5.2), each occurs only in the presence of another group (Struhsaker 1967a; Cheney and Seyfarth 1982b). *Wrrs* are usually given when a neighboring group has first been spotted (fig. 5.3), and they seem to alert both the members of the caller's own group and the members of the other group that the other group has been seen. About 45% of all intergroup encounters involve only the exchange of *wrrs;* others, however, escalate into aggressive threats, chases, and even physical contact (fig. 5.4; chapter 2). When groups come together under these more aggressive conditions, females and juveniles often give the acoustically different chutter (Cheney and Seyfarth 1988). The chutter seems to be a call of greater intensity than the *wrr*. Although only 40% of all *wrrs* are given during aggressive interactions with the members of other groups, over 95% of all chutters are given in this context.

Although *wrrs* and chutters have broadly similar external referents, therefore, they do not always occur simultaneously or in precisely the same context. This raises a question: if Acton, for example, repeatedly hears Carlyle's intergroup *wrr* when there is no other group present, and Acton ceases to respond to that call, will she also cease responding to Carlyle's chutter? If the two calls have similar meanings, and if monkeys use meaning to judge the relationship between calls, habituation to *wrrs* should produce habituation to chutters. Alternatively, if monkeys use some other feature (like the calls' acoustic properties) to judge similarity or difference between calls, these features, and not the calls' referents, should determine whether Acton, having ceased responding to Carlyle's *wrr*, also ignores her chutter.

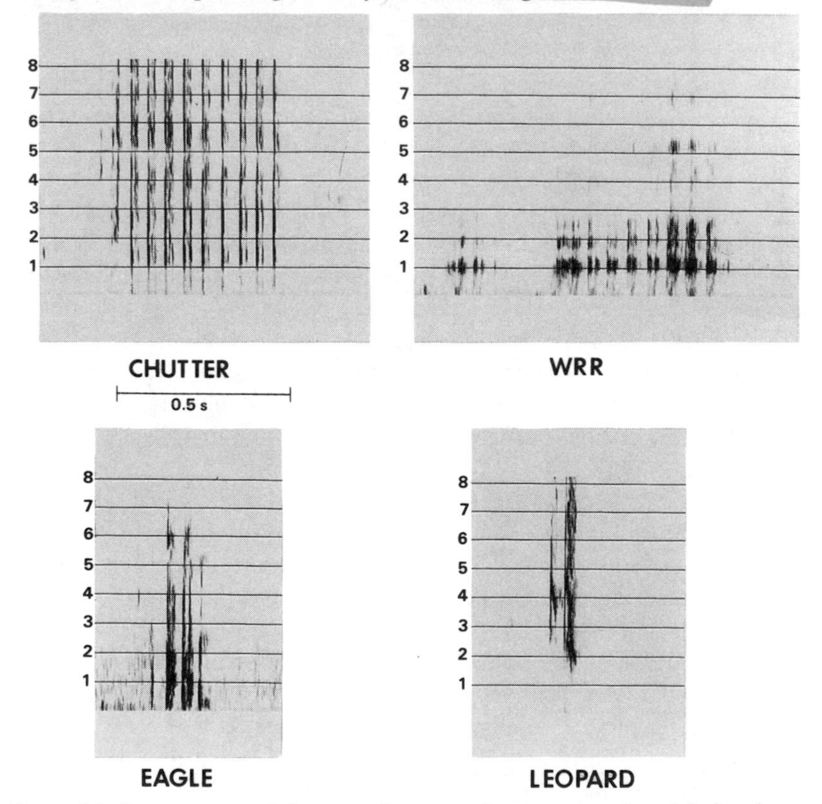

Figure 5.2. Spectrograms of chutter and wrr *vocalizations given by adult female Snickers and alarm calls to eagles and leopards given by adult female Amin. From Cheney and Seyfarth 1988, with permission.*

Figure 5.3. Adult females Carlyle and Acton coalesce with their offspring moments after Carlyle has given a wrr *upon spotting a neighboring group.*

Figure 5.4. The members of groups A and B threaten and chase each other during an aggressive intergroup encounter. During such interactions, females often give chutter vocalizations.

In conducting our experiments, we borrowed a method that has been used successfully in research on preverbal human infants (e.g., Eimas et al. 1971). On day 1, as a control, a subject was played a particular female's chutter in order to establish the baseline strength of the subject's response to this particular vocalization. Then, on day 2, the subject heard the same female's *wrr* repeated eight times at roughly 20-minute intervals. Because no other group was present at the time, we predicted that the subject would soon come to regard this female's *wrr* as an unreliable signal and would cease responding to it. Finally, roughly 20 minutes after the last playback in the habituation series, the subject heard the female's chutter again (the test condition). The magnitude of the decrement in response between control and test conditions measured the extent to which the subject judged the habituating and test stimuli to be the same. A large decrement indicated that the subject regarded the two calls as similar. So, if the subject had ceased responding to X's *wrr*, and treated *wrrs* and chutters as having roughly similar meanings, she should have responded much less strongly to X's chutter following habituation to X's *wrr* than she had on the previous day.

Since vervets clearly take note of the signaler's identity when attending to calls (see chapters 3 and 4), we also wanted to determine whether subjects would transfer habituation from one individual to another. Hence in a second series of experiments we varied the test procedure by playing two individuals' calls. On day 1, we established baseline data on the strength of a subject's response to Y's chutter. Then, on day 2, we played X's *wrr* to the same subject eight times. After the subject had habituated to X's *wrr*, she was then tested to see if she had also habituated to Y's chutter.

A third test examined whether vervets would also transfer habituation if the identity of the signaler remained the same but the call's *referent* was changed. We therefore repeated the procedure described for the first series but used as stimuli leopard and eagle alarm calls instead of *wrrs* and chutters. So, on day 1 we played X's eagle (or leopard) alarm call, to be followed on day 2 by eight exposures to X's leopard (or eagle) alarm call and then, once again, one exposure to X's eagle (or leopard) alarm call.

Finally, to determine whether subjects would habituate across both individuals and call types, we tested whether habituation to X's eagle (or leopard) alarm call would cause subjects to habituate to Y's leopard (or eagle) alarm call.

Results provided clear evidence that vervet monkeys compare different calls on the basis of their meaning and not just their acoustic properties. In all experiments, subjects rapidly habituated to repeated presentation of the same vocalization. And, when they were presented with the same individual's *wrrs* and chutters, two acoustically different calls with roughly the same referent, they transferred habituation across these different call types.

In other words, if a subject had habituated to X's intergroup *wrr,* she also ceased responding to X's intergroup chutter (fig. 5.5A).

By contrast, when subjects were played two calls whose referents were different, they did *not* transfer habituation across call types (fig. 5.5C). If a subject had learned that X was unreliable when signaling about leopards, she still responded at normal strength to X's eagle alarms.

Habituation was also not transferred when the calls had the same referent but were given by two different *individuals* (fig. 5.5B). Even if a subject had ceased responding to X's *wrr,* Y's chutter still elicited the same response as it did under normal conditions.

Finally, as might be expected, when the two calls both had different referents and came from different individuals, habituation was also not transferred (fig. 5.5D). Habituation to X's eagle alarm had no effect on the strength of response to Y's leopard alarm calls.

Compared with our earlier experiments on alarm calls and grunts, these tests address the question of meaning and reference more directly by asking animals to compare two vocalizations (that is, make a same/different judgment between them) and to reveal the criteria they use in making their comparison. Like humans (e.g., Yates and Tule 1979), vervet monkeys appear to process vocalizations according to an abstraction—their meaning—and not just according to their physical similarity. The fact that subjects did not transfer habituation when played the call of another animal suggests that they took into account both the signal's meaning and the signaler's identity when attending to a call.

It might be argued that vervets failed to transfer habituation from one alarm call type to another (that is, across calls with different referents) because alarm calls are simply too costly to ignore. If this were true, however, monkeys should have taken longer to habituate to repeated presentation of the same alarm call than to repeated presentation of the same intergroup call. This was not the case. Habituation to alarm and intergroup calls occurred at similar rates (compare the habituation curves in figs. 5.5A–D).

It might also be argued that our results do not reflect judgments about the referents of two calls but instead indicate a form of *sensory preconditioning,* a process by which two stimuli are treated as similar because of their prior temporal association (e.g., Brogden 1939; Jacobson and Premack 1970). By this explanation intergroup *wrrs* and chutters would have been judged to be similar not because they had similar referents but because they were usually heard together. If individuals often gave different intergroup calls in rapid sequence, subjects would have been particularly likely to habituate to trials involving the same signaler. In contrast, alarm calls to leopards and eagles would have been judged to be different because they rarely occurred in close temporal association.

Obviously, any calls (including human words) that are roughly synonymous will tend to be more closely associated in time than calls with different referents. Nevertheless, it should be emphasized that there was no consistent temporal link between *wrrs* and chutters. During the period we were conducting the playback experiments, *wrrs* and chutters occurred together in only 27% of all intergroup encounters. Moreover, they were usually given by different individuals; the two calls were given by the same individual in only 3% of all encounters. Given their rather loose temporal

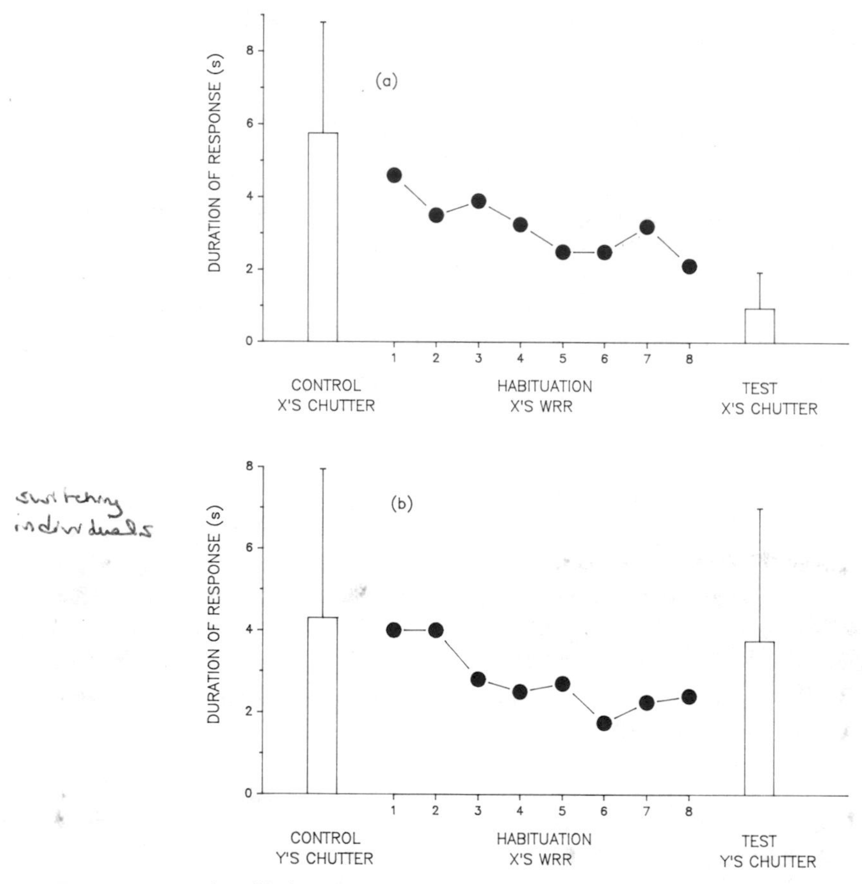

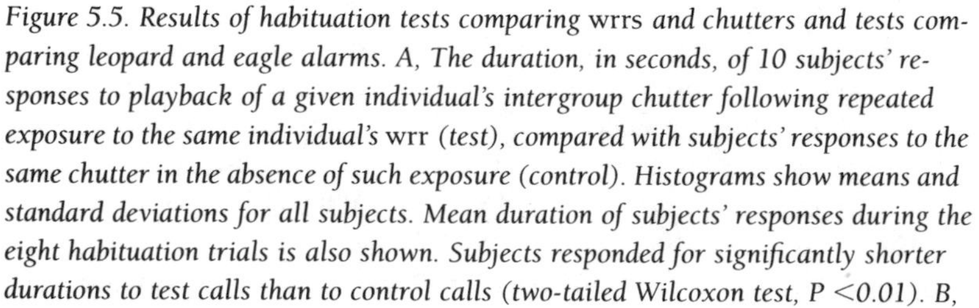

Figure 5.5. Results of habituation tests comparing wrrs *and chutters and tests comparing leopard and eagle alarms. A, The duration, in seconds, of 10 subjects' responses to playback of a given individual's intergroup chutter following repeated exposure to the same individual's* wrr *(test), compared with subjects' responses to the same chutter in the absence of such exposure (control). Histograms show means and standard deviations for all subjects. Mean duration of subjects' responses during the eight habituation trials is also shown. Subjects responded for significantly shorter durations to test calls than to control calls (two-tailed Wilcoxon test, $P < 0.01$). B,*

association, it seems unlikely that *wrrs* and chutters were judged as similar solely because of their temporal links. Furthermore, if sensory preconditioning had been occurring, we should have obtained results *opposite* to those we did in fact obtain. Since *wrrs* and chutters were more likely to be given by different individuals, subjects should have transferred habituation from one individual's *wrr* to another individual's chutter but not across *wrrs* and chutters given by the same individual.

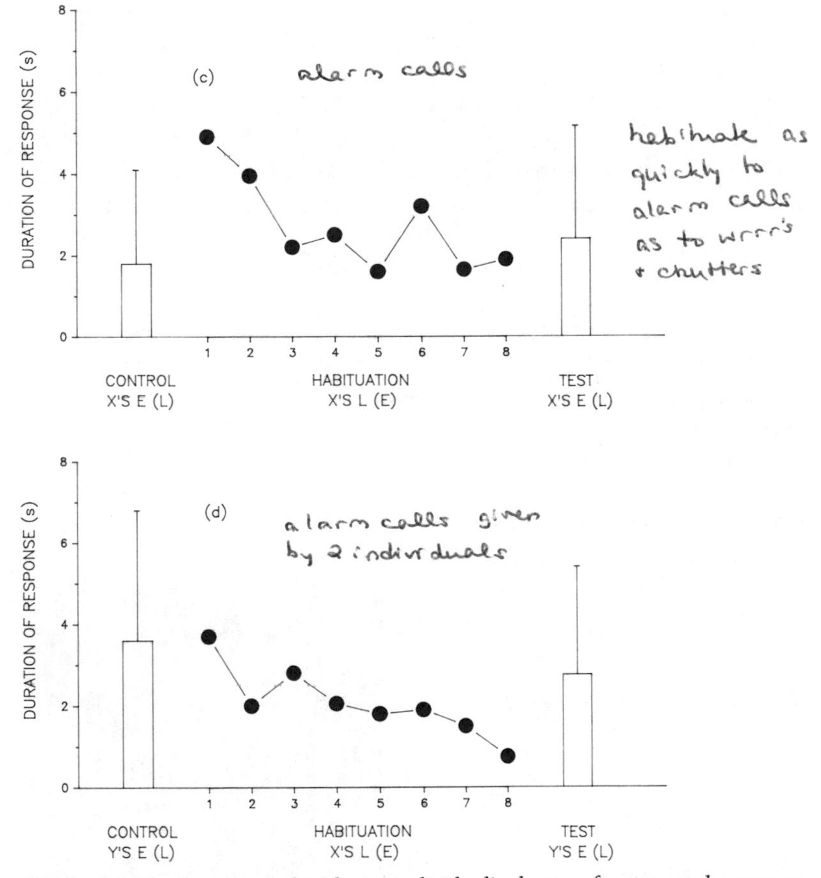

Results for 10 subjects tested with one individual's chutter after repeated exposure to another individual's wrr. C, Results for nine subjects tested with one individual's eagle (or leopard) alarm after repeated exposure to the same individual's leopard (or eagle) alarm. D, Results for nine subjects tested with one individual's eagle (or leopard) alarm after repeated exposure to another individual's leopard (or eagle) alarm. E = eagle alarm; L = leopard alarm. Redrawn from Cheney and Seyfarth 1988.

In sum, when one vervet hears another vocalize she appears to form a representation of what the call means. And if, shortly thereafter, she hears a second vocalization, the two calls are compared on the basis of their meaning rather than just according to their acoustic properties.

What Monkeys Know About the Alarm Calls of Other Species

In the experiments we have just described, vervet monkeys treated *wrrs* and chutters as if they were roughly synonymous, regardless of their different acoustic properties. If this interpretation is correct, it should be possible to find other examples of calls that are judged by the monkeys to be similar on the basis of shared meaning.

Perhaps more important, if vervet monkeys respond to vocalizations according to the objects and events they denote, we should expect the monkeys to be sensitive to the breadth of referential specificity exhibited by different calls. In our own language, for example, we use words with very specific meanings (or narrowly defined referents) like *praying mantis, hand calculator,* or *chocolate ice cream sundae*. We also use words like *insect, thing,* or *food* that refer to a much broader class of objects and have, as a result, meanings that are more general. In making judgments about substitutability we take these differences into account. Sometimes praying mantis and insect can be used interchangeably. More often they cannot, because the meaning of insect is too broad.

Vervet monkeys face similar problems when responding to the alarm calls of other species. In Amboseli, for example, vervets share their habitat with many other species of birds and mammals that also give different

Figure 5.6. Like many bird species, the superb starling has two different alarm calls for terrestrial and aerial predators.

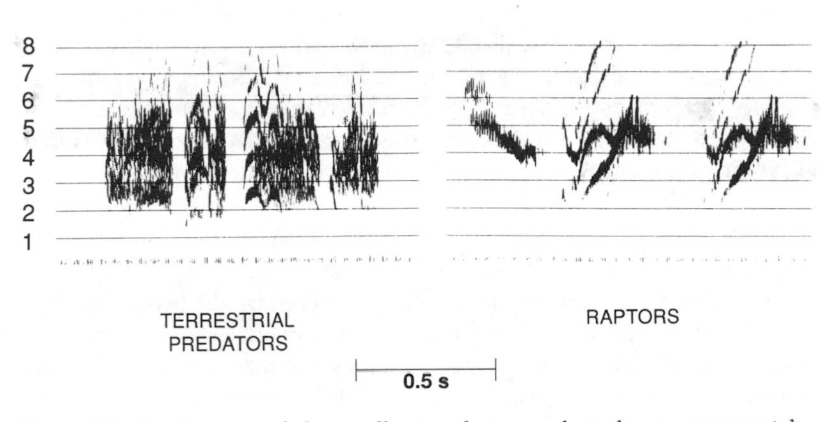

Figure 5.7. Spectrograms of alarm calls given by a superb starling to a terrestrial predator (left, *in this case, a slender mongoose) and a raptor* (right, *in this case, a pale chanting goshawk,* Melierax poliopterus).

alarm calls to different types of predators. One of these species, the superb starling (fig. 5.6), has at least two distinct alarm calls, neither of which bears any acoustic resemblance to the vervets' own alarms. The first, a harsh, noisy call (fig. 5.7), is given in response to various terrestrial predators (including vervets), all of which prey on starlings or their eggs but only some of which prey on vervets. The second, a clear rising or falling tone (fig. 5.7), is given in response to at least eight species of hawks and eagles, only one of which preys on vervets.

The association between a particular starling call and a particular predator is, therefore, complex. For a monkey to learn to respond appropriately to the starling's alarm calls, she must first learn which of the starling's vocalizations function as alarm signals. She must then learn to distinguish between the two types of alarm calls and recognize which predator species evoke the calls. This last problem requires that monkeys learn that although starlings most often alarm call at species that pose no danger to vervets, they also occasionally alarm call at species that *do* prey on vervets.

To determine whether vervets distinguish between the two different alarm calls of the superb starling, we conducted a series of playback experiments that followed the same protocol we had previously used in tests of the vervets' own alarm calls (chapter 4). First, we hid a loudspeaker near a group of one to five vervets. The monkeys were then filmed for 10 seconds to establish the probability that they would show a given response in the absence of any call. We then played one of the starlings' calls and continued to film the monkeys' responses for another 10 seconds. Three starling calls were used: their terrestrial predator alarm call, their raptor alarm call and, as a control, their song.

Just as vervets often responded to their own terrestrial predator alarm calls by running into trees, playback of the starling's terrestrial predator alarm call caused a significant number of monkeys to run toward trees. In contrast, though vervets almost never ran toward trees when played the starling's raptor alarm call, playback of this call caused a significant number of vervets to look up. The starling's song elicited little response (fig. 5.8; Cheney and Seyfarth 1985a).

Despite these rather extreme differences, however, the monkeys' responses to the starlings' alarm calls are not as clear-cut as these experiments suggest, perhaps because the information conveyed by the two alarm calls is not as precise as might first appear. In our original experiments, conducted in 1983, very few subjects looked up when they were played a starling terrestrial predator alarm call (fig. 5.8). By contrast, in later playbacks conducted in 1988, vervets looked up in 25% of the experiments involving starling terrestrial alarms, compared with 75% of all starling raptor alarms. We suspect that the more extreme difference in our original tests was due to the inclusion of juvenile subjects, who were more likely than adults to run to trees. The 1988 sample included only adult females and males, more than half of whom were already in a tree when the terrestrial predator

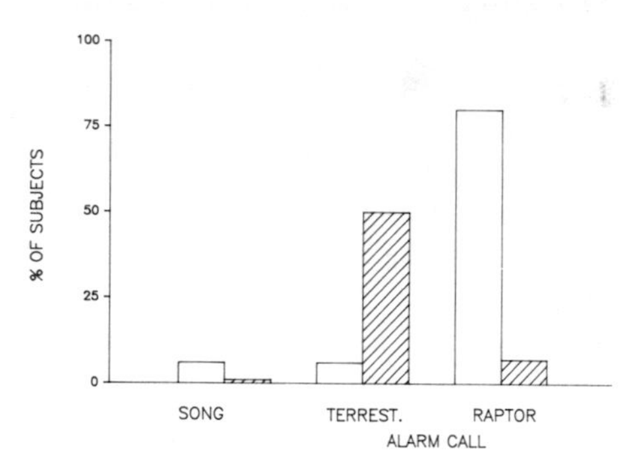

Figure 5.8. Responses of vervet monkeys to playback of three different starling calls. Open histograms show the proportion of subjects who looked up; shaded histograms show the proportion of subjects who stood bipedally or ran toward trees. The number of subjects for tests using song, terrestrial predator alarms, and raptor alarms were 17, 18, and 15, respectively. Subjects were significantly more likely to look up after hearing raptor alarms than after hearing either song or terrestrial predator alarms (chi-square test, $P < 0.01$ for both comparisons). Subjects were significantly more likely to run toward trees or stand bipedally after hearing terrestrial predator alarms than after hearing either song or raptor alarms (chi-square test, $P < 0.05$ for both comparisons). Redrawn from Cheney and Seyfarth 1985a.

alarm call was played. Adults in trees were more likely than juveniles on the ground to remain seated and simply scan the area around them.

The important point, however, concerns the relative imprecision of starling terrestrial predator alarm calls. Starlings give terrestrial predator alarm calls to an extremely broad array of animals. In addition to cats like leopards and servals, starlings give terrestrial alarm calls to at least three species of snake, slender and dwarf mongoose (*Herpestes sanguineus* and *Helogale parvula*), genets (*Genatta genatta*), and birds like fiscal shrikes (*Lanius collaris*) and lilac-breasted rollers (*Coracias caudata*), which hunt not by diving from above but by locating the starlings' nests and stalking them from a nearby branch. Starlings also give alarm calls to vervets, particularly when the birds are incubating eggs. The birds even give terrestrial predator alarm calls to elephants when the elephants reach up into trees to break off branches. From the vervets' perspective, therefore, starling terrestrial predator alarms are imprecise: when the monkeys hear this alarm call they know that something is nearby, but they have no idea whether it is harmful or not, or, if it is harmful, what escape strategy they should pursue (remember that monkeys respond differently to snakes and leopards).

By contrast, starling raptor alarms are given to a much narrower array of predators, all eagles or small hawks that attack from the air. Although most of these raptors pose no danger to vervet monkeys, the starling's raptor alarm call is occasionally given to martial eagles, a species that does prey on vervets. Starling raptor alarm calls, therefore, are much more precise and restricted in scope than starling terrestrial predator alarm calls. Although starling raptor alarm calls may be imprecise about the magnitude of danger, they do denote a specific type of predator, its location, and an appropriate escape strategy.

Given these differences, we designed a series of playback experiments to examine two related issues: first, whether vervet monkeys recognize the similarity between their own eagle and leopard alarms and the raptor and terrestrial predator alarms of starlings and, second, whether vervets are sensitive to the relative lack of precision in starling terrestrial predator alarms compared with starling raptor alarms (Seyfarth and Cheney 1990). If monkeys assess not only their own but also other species' alarm calls according to their meaning, habituation to the raptor (or terrestrial) alarm call of one species should produce habituation to the corresponding alarm call of the other. Moreover, if vervets are also sensitive to the specificity of each call's referent, they should transfer habituation more readily when the comparison involves raptor, rather than terrestrial, predator alarms.

As in our earlier experiments on *wrrs* and chutters, we began on day 1 by establishing a subject's baseline response to playback of a vervet (or starling) alarm call. Then, on day 2, the subject heard a series of eight starling (or vervet) alarms. Roughly 20 minutes after this series had ended, the subject was once again played the vervet (or starling) alarm call she had heard on day 1. As in our previous experiments, the magnitude of the decrement in response between control and test conditions measured the extent to which habituating and test stimuli were judged to be the same.

Results support the view that vervet monkeys treat starling raptor alarms

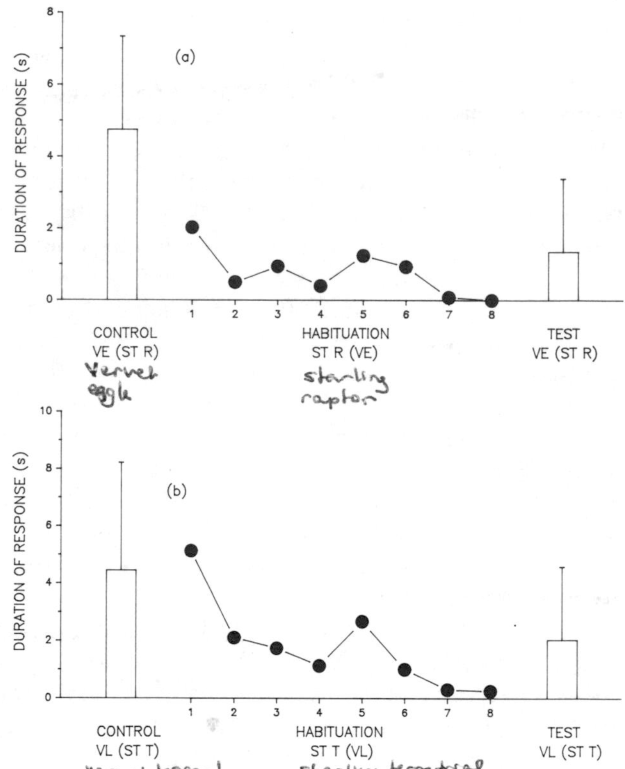

Figure 5.9. Results of habituation tests comparing vervet and starling alarm calls. A, Duration, in seconds, of eight subjects' responses to playback of a vervet eagle (or starling raptor) alarm following repeated exposure to a starling raptor (or vervet eagle) alarm (test), compared with subjects' responses to the same alarm call in the absence of such exposure (control). Histograms show means and standard deviations for all subjects. Mean duration of subjects' responses during the eight habituation trials is also shown. Subjects responded for significantly shorter durations to test calls than to control calls (one-tailed Wilcoxon test, P <0.01). B, Results for six subjects tested with a vervet leopard (or starling terrestrial predator) alarm after repeated exposure to a starling terrestrial predator (or vervet leopard) alarm.

as relatively precise signals, similar in meaning to vervet eagle alarms but different from vervet leopard alarms. In tests that compared starling raptor alarms with vervet eagle alarms, all eight subjects transferred habituation from one species' alarm call to the other's. As a result, there was a large decrement between control and test conditions (fig. 5.9A). The monkeys behaved as if, from their perspective, the two types of raptor alarm calls denoted similar referents despite having markedly different acoustic prop-

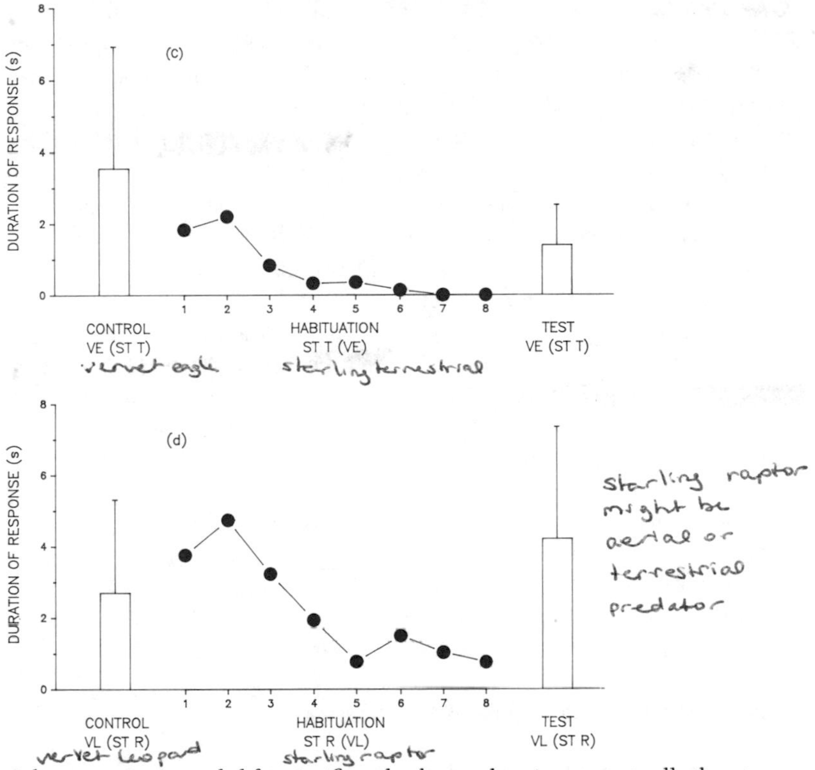

Subjects again responded for significantly shorter durations to test calls than to control calls (one-tailed Wilcoxon test, P <0.05). C, Results for seven subjects tested with a vervet eagle (or starling terrestrial predator) alarm after repeated exposure to a starling terrestrial predator (or vervet eagle) alarm. The difference between subjects' responses to test and control calls was barely significant (one-tailed Wilcoxon test, P = 0.055). D, Results for seven subjects tested with a vervet leopard (or starling raptor) alarm after repeated exposure to a starling raptor (or vervet leopard) alarm. VE = vervet eagle alarm; ST R = starling raptor alarm; VL = vervet leopard alarm; ST T = starling terrestrial predator alarm.

erties. Conversely, tests that compared starling raptor alarms with vervet leopard alarms revealed *no* transfer of habituation. In these tests, there was little or no decrement in response from control to test conditions (fig. 5.9D).

In contrast, the meaning of starling terrestrial predator alarms was apparently more ambiguous. In tests that compared starling terrestrial predator alarms with vervet leopard alarms, five of six subjects transferred habituation from one species' alarm call to the other's (fig. 5.9B). However, the decrement between control and test conditions, though significant, was less than we found in the comparison between starling raptor and vervet eagle alarms (compare fig. 5.9A and B). Apparently, although the monkeys treated starling terrestrial predator alarms and vervet leopard alarms as similar, the two calls were judged to be less synonymous than starling raptor alarms and vervet eagle alarms.

Moreover, six of seven subjects also transferred habituation between starling terrestrial predator alarms and vervet *eagle* alarms (fig. 5.9C). They behaved as if starling terrestrial predator alarms could have been directed at either a terrestrial predator or a raptor.

These results suggest that, when assessing the meaning of a vocalization or comparing the meaning of two calls, vervets are sensitive to the array of objects denoted by each call. Vervets treat the starling's raptor alarm as if it refers to a relatively specific class of predators, a class whose membership overlaps considerably with the class of predators denoted by the vervets' own eagle alarms. As a result, the two species' alarm calls are essentially interchangeable. By contrast, starling terrestrial alarms refer to a much broader array of predators. Consequently, they overlap to some extent with *both* vervet leopard and vervet eagle alarms. The monkeys behave as if they regard their own alarm calls as more precise and referentially distinct than those of the superb starling, as indeed they are.

Assessing the Meaning of Animal Vocalizations

Meaning and Mentality in Signaler and Recipient

We began this book by borrowing the philosopher Quine's (1960) notion of a linguist in a foreign land who tries to learn the meaning of other peoples' words and, in so doing, to understand how they think. While establishing the meaning of words is an obvious place to start, the meaning of *meaning* is in fact a complex issue that can be approached from a disconcerting number of directions. Does meaning derive primarily from the relationship between signs and the things for which they stand, or is it mainly concerned with the relationship between signs and the mental operations of those who use them? Can we discuss the meaning of single words (or signs, or animal vocalizations) in isolation, or is one word's meaning so embedded in its particular cultural context that it can only be defined in relation to other words? Here we consider animal vocalizations from each of these perspectives.

For some, like the philosopher John Locke (1690/1964:259), signals "stand for nothing but the ideas in the mind of him that uses them." If Locke ever wondered about the vocalizations of monkeys, he might well have concluded that their status as "words" is uncertain because it is impossible to specify the ideas, beliefs, or desires in the mind of a monkey when he gives or hears a vocalization. In the intervening 300 years, it seems we have made little progress, because this conclusion still rings true.

On the basis of recent work, however, we can rule out alternative explanations that, if true, would have eliminated the possibility that *any* ideas, beliefs, or desires underlie the vocalizations of monkeys and other animals. The simplest explanation, that animal signals are evoked by certain levels of arousal, can be ruled out for a number of reasons. The responses that vervet monkeys give to alarm calls are not based on a call's length or its amplitude, as might be predicted by the arousal hypothesis (chapter 4). Further, individuals can modify the production of calls depending on the presence and characteristics of an "audience." Animals as diverse as vervets, ground squirrels, roosters, and woodpeckers, for example, rarely give alarm calls when alone and call at higher rates in the presence of kin than when they are near other, unrelated group members. Third, call production and responses to calls have no obligatory link and are readily dissociated in nature. All of these data argue, in Dennett's (1983, 1987, 1988) terms, for at least first-order intentionality on the part of both communicator and recipient.

Our early observations and experiments on alarm calls and grunts tested each vocalization separately; in effect we were asking the monkeys to tell us, by their responses, what each call meant. By contrast, later experiments using *wrrs*, chutters, the alarm calls of vervets, and the alarm calls of other species probed further, asking the vervets to compare two calls and to reveal the factors used in making their comparisons. The results of these tests are difficult to explain without assuming that the monkeys have some representation of the objects and events denoted by different call types.

This is not to say that monkeys are necessarily aware of the distinction between signs and the objects they denote or aware of their ability to compare vocalizations on the basis of their referents. We cannot assume that an individual who can make same/different judgments about two vocalizations on an habituation test will be able to make conscious *use* of this distinction in his daily life. Indeed, there is evidence that infant chimpanzees who can perceive a relational distinction when tested with a habituation procedure are nevertheless unable to apply their apparent knowledge of this distinction in a match-to-sample test (Oden, Thompson, and Premack 1988; described in more detail in chapter 8). Habituation data alone, therefore, do not prove that monkeys understand the relation "*wrr* denotes another group," or "eagle alarm denotes eagle" in the same way that a chimpanzee understands the relation "blue triangle means apple" or "this symbol stands for the class of items called tools."

This state of affairs, in which animals use signals whose relations they do not, in fact, understand, is not as implausible as it first seems. Children, for instance, use words correctly long before they can articulate the semantic relations that underlie them. Indeed, young children's behavior conforms to the view espoused by many philosophers that relations between signs and the things for which they stand can hold independently of what the signers themselves think about them (Tiles 1987). In chapter 8 we consider further the rather complicated issue of what animals know about what they know.

In sum, although monkeys use vocalizations to denote features of their environment and make same/different judgments about vocalizations on the basis of their referents, we presently know little about the mental operations that underlie such communication in signaler and recipient. We can rule out the simplest explanations, which postulate no mentality on the part of signaler or recipient. On the other hand, as yet we have no evidence that monkeys are aware of the referential relations that characterize their communication or that they attribute thoughts or beliefs to others and vocalize in order to modify these mental states. In between these two extremes lies a variety of possible mechanisms; progress in our understanding of animal communication demands that we find a way to choose among them.

Meaning and External Referents

Taking a somewhat different perspective, we can examine the meaning of animal vocalizations not in terms of the mental states of signaler and recipient but in terms of the relationship between signs and the things for which they stand. Throughout this and the preceding chapter, for example, we have referred to leopard alarm calls, eagle alarm calls, grunts to another group, and other vocalizations as if we can specify with some precision the objects that these calls denote. Is this in fact the case? Consider first an alternative hypothesis.

As we noted earlier, for years scientists held the appropriately conservative view that no animal vocalizations had been shown to have external referents but instead had been shown to provide information only about the signaler's subsequent behavior. And, since the behavior of individuals responding to signals was known to be strongly influenced by context, the meaning of animal calls to nearby recipients was assumed to be highly context dependent. External referents were not ruled out in principle (see, for example, Smith 1977:73–74); there simply was no evidence for their widespread existence in the communication of nonhuman species.

Of course, given the close link in vervet monkeys between predator type, alarm call type, and escape response (or between social situation, type of grunt, and the response this grunt elicits), there will often be a close correlation between a call and the signaler's subsequent behavior. In itself, however, this does not rule out the possibility that vocalizations also serve a referential function. Human words—for example, *Boogie!*—offer many good examples of signals that both have external referents and are linked with specific behavioral responses.

At the same time, however, vervet monkey vocalizations often exhibit a very weak relationship between call type and the caller's subsequent behavior. As noted earlier, for example, alarm calls can be given by animals who do not themselves show any escape response. If monkeys spot a martial eagle soaring far in the distance, they often give alarm calls but continue to feed, taking no other evasive action.

Similarly, although infant vervets give alarm calls with the same acoustic features as those of adults, infants' responses to alarms vary widely. Very young infants run to their mother regardless of what sort of predator has been seen, and older infants can even respond inappropriately, in a manner that increases their vulnerability (chapter 4). Overall, then, the alarm calls of infants are not correlated with the same types of behavior on the part of the signaler as are the alarm calls of adults. Nevertheless, adults react to infant alarms with the same predator-specific escape responses they use when responding to the alarms of adults (Seyfarth and Cheney 1986).

Third, recall that vervet eagle alarm calls elicit at least four different responses. Animals on the ground may look up or run into a bush, whereas animals in a tree may run down out of the tree. In either circumstance, a listener can also do nothing. Moreover, vervets in a tree run down and out of the tree even if the caller himself is on the ground and is responding by looking up. In this case, the most parsimonious explanation would seem to be that calls denote a type or class of danger, not the caller's behavior, and that an individual's particular circumstances strongly influence the exact nature of his response.

Fourth, vervets respond to the raptor alarm calls of starlings by looking up. The escape responses of starlings obviously differ from those of vervets, since vervets do not fly. Here again, the vervets respond to calls as if the calls denote a particular type of danger, not the caller's behavior.

Fifth, vervet monkeys do not give eagle alarms in response, for example, to falling branches, where one can easily imagine that the appropriate response would be to run out of the tree. Again, the simplest explanation is that eagle alarm calls refer not to particular escape responses but to certain broad classes of danger.

Finally, the strongest evidence that vervet calls have referents external to the signaler, and that animals make judgments about the meaning of calls on the basis of these referents, comes not from alarm calls but from the habituation experiments that asked monkeys to compare intergroup *wrrs* and chutters. Although these calls both have the same broad referent—another vervet group—they are given in slightly different behavioral contexts. *Wrrs* usually occur when another group has first been spotted, whereas chutters are typically given during an actual fight. The two calls, associated with rather different behavioral responses, are nevertheless treated by the vervets as similar. Having habituated to one individual's *wrr*, the monkeys transfer their skepticism and ignore the same individual's chutter. It is hard to explain these results without assuming that *wrrs* and chutters have the same general referent and that this referent, together with the caller's identity, is a major determinant of call meaning.

It would obviously be as incorrect to claim that monkey vocalizations convey information *only* about external referents as it is to argue that they convey information *only* about the signaler's motivational state or subsequent behavior. Clearly, as Smith (1986, 1990) has argued, whenever one animal vocalizes to another, a variety of information is made available—information about the identity and physical characteristics of the signaler, about what the signaler is likely to do next, and about events in the environment. From the listener's point of view, the sort of information that is most important will vary from one situation to the next. In human language, where the existence of external referents is not disputed, words and sentences convey information about a speaker's identity, mood, and subsequent behavior in addition to information about particular external referents (e.g., Johnson-Laird 1987). Language, in this respect, is both expressive and denotative: it conveys the emotions, thoughts, and beliefs of the speaker while simultaneously referring to objects or events in the external world. In emphasizing the external referents of primate vocalizations, our aim is not to minimize the importance of emotion, contextual cues, or the caller's subsequent behavior. Instead, we hope to show that the communication of monkeys and apes—long known to be highly expressive (e.g., Jolly 1985)—can be denotative as well.

But what, precisely, do their calls denote? We have been describing leopard alarm calls, eagle alarm calls, chutters, *wrrs*, and grunts as if each referred to a quite specific set of objects or events. Is this in fact the case? Can we specify with precision the meaning of leopard alarm call to vervets or accurately state what they are thinking about when they hear an intergroup *wrr*?

The philosopher Stephen Stich (1983) argues that when we see a dog barking at the foot of a tree and say that the dog does so because it thinks that a squirrel is up the tree, we demonstrate the weakness of applying our human-based folk psychological concepts to animals. Can we be sure that the dog is *really* thinking of a squirrel and only a squirrel? Would it behave differently toward a cat? A mechanical squirrel? A squirrel-sized Australian marsupial that it had never seen before but that behaved just like a squirrel? We have no way of knowing. Since there is an infinite number of logically possible objects that the dog might treat indistinguishably from a squirrel, Stich maintains that we cannot really say that the dog is actually thinking thoughts about a squirrel. In fact, to Stich, the whole exercise demonstrates the impossibility of applying terms like *thinking* or *meaning* to animals other than ourselves (see also Allen 1989).

But is this objection really valid? Surely there are many cases in human interactions where there exists, for example, an infinite number of logically possible objects that are indistinguishable from squirrels; yet this does not cause many problems for the specification of the contents of human beliefs about squirrels (Allen 1989). We make distinctions when it is important for us to do so—important in the sense of necessary to achieve our objectives in interactions with others.

Similarly, when we try to state what a vervet monkey is referring to when he gives a leopard alarm, we assume that the meaning of leopard alarm is, from the monkey's point of view, only as precise as it needs to be. In Amboseli, where leopards hunt vervets but lions and cheetahs do not, leopard alarm could mean "big spotted cat that isn't a cheetah," or "big spotted cat with the shorter legs," or however you want to describe it. In other areas of Africa, where cheetahs *do* hunt vervets, leopard alarm could mean "leopard *or* cheetah."

This point is nicely illustrated by the behavior of vervet monkeys in the Cameroon. Vervets on the Cameroon savanna are sometimes attacked by feral dogs. When they see a dog they respond much as Amboseli vervets respond to a leopard; they give loud alarm calls and run into trees. Elsewhere in the Cameroon, however, vervets live in forests where they are hunted by armed humans who track them down with the aid of dogs. In these circumstances, where loud alarm calls and conspicuous flight into trees would only increase the monkeys' likelihood of being shot, the vervets' alarm calls to dogs are short, quiet, and cause others to flee silently into dense bush where humans cannot follow (Kavanaugh 1980). For Cameroonian vervets in the savanna, then, dogs are included within the term leopard alarm, creating a call whose meaning appears to be "terrestrial predators from which you can escape by running into trees." As long as new predators fall within this category the same alarm call will presumably be used. For vervets in the forest, however, referring to dogs with a leopard alarm call would be dangerously inaccurate, and therefore another call is used.

The data from the Cameroon suggest that leopard alarms in Amboseli denote a particular hunting strategy (It's that thing that chases and grabs us on the ground!). Once again, however, the explanation is not this simple. Recall that, in addition to leopard alarms, vervets in Amboseli have a separate, acoustically distinct alarm call for what Struhsaker (1967a) called "minor mammalian predators": animals like lions, cheetahs, hyenas, and jackals that could pose a danger but, as far as we can tell, rarely hunt vervets (chapter 4). When the monkeys see one of these species, they give minor mammalian predator alarms, become vigilant, and may move toward trees. Two different alarm calls, therefore, are given to predators that have the same hunting strategy. Leopards differ from lions, cheetahs, jackals, and hyenas only in terms of the magnitude of danger involved. If one of these species began hunting vervets regularly, we imagine that the vervets would move it from membership in their minor mammalian predator category to membership in the category leopard.

Function, within a particular ecological context, thus determines the precision of word meanings and the specificity of external referents. The meaning of a call cannot be defined in absolute, context-free terms, as in "leopard alarm denotes the African leopard, *Panthera pardus*." Instead, each call is best defined relative to the meaning of somewhat similar calls in slightly different contexts. Leopard alarm therefore denotes a predator or class of predators that is sufficiently different from those denoted by minor mammalian predator alarm, eagle alarm, or snake alarm to warrant a different vocalization. The criteria that determine whether two, three, or more alarm calls will evolve—and hence the criteria that determine each call's referential specificity—seem to be the predator's hunting strategy, the likelihood that it will attack, and the monkey's mode of escape. If two predators are similar according to these criteria, they will elicit the same alarm call. If they are different, they will not.

By contrast, the overlap in meaning between *wrrs* and chutters is more complicated. On the one hand, at least by one test, the monkeys treat the calls as broadly similar, since habituation to one produces habituation to the other. At the same time, however, observations suggest that the calls are associated with subtly different activities—different enough, apparently, to warrant the evolution of two vocalizations. Hence we conclude that chutters and *wrrs* from the same individual are not exactly the same, but that they refer to objects or events that are more like each other than either is like the objects or events denoted by, for example, an eagle alarm.

Similarly, calls can differ in the breadth of their referential specificity. From a vervet's point of view, raptor alarm calls by starlings denote a class of referents with well-defined parameters. The predators denoted by starling raptor alarms are similar to those denoted by vervet eagle alarms and different from those denoted by vervet leopard alarms. In contrast, the meaning of starling terrestrial predator alarms is less specific, since these calls overlap in the species they denote both with the vervets' leopard alarm and, to a lesser extent, with the vervets' eagle alarm.

Hence the external referents of primate vocalizations are best defined in relation to the external referents of other calls within the animals' repertoire. A complete definition of the vervets' leopard alarm call requires that we specify both the objects that it denotes (leopards, other small cats, and rarely even a stooping martial eagle [chapter 4]) and the objects that are *excluded* from its denotation, either because such objects are denoted by another call (predatory eagles, jackals, snakes, or other groups of vervets, for example) or because they elicit no call at all (for example, hippos, zebras, or elephants). We need, in short, to specify both positive and negative instances of the vervets' class *leopard*. And the only way we can learn which referents are excluded from leopard alarm (the negative instances) is by knowing the referents that *are* contained in eagle alarm, snake alarm, *wrr*, chutter, and so on, together with the stimuli that elicit no vocalizations at all. Thus our dictionary of vervet words builds up slowly, with each term being defined in relation to the others. Under these conditions, as Quine's imaginary linguist would no doubt have discovered, the more words you have, the more precise your definitions can be. For us, in the land of vervet monkeys, words are relatively few and as a result their meanings are ill-defined.

This position may seem too much like a strategic retreat: after arguing that the vocalizations of vervet monkeys can refer to objects and events in the environment, we are now not really able to state precisely *what* such calls denote. There are, however, good reasons for such imprecision; reasons that become even more apparent when we consider a parallel case, namely the earliest speech of human infants.

From their first day of life babies emit an extraordinary variety of grunts, babbles, and goos. Despite an understandable tendency for excited parents to interpret *nuklgfsst* as *nuclear physics*, most observers believe that in its earliest stage babbling has no real meaning for the infant (e.g., Brown 1973; Gleitman and Wanner 1982). The infant's sounds are given at no particular time, and babies themselves give no indication that their noises are related in any way to specific objects or behavior.

During the next few months, however, life becomes more complicated. Infants are selective in the timing of their vocalizations (babbling, for example, is more likely after an adult has spoken), and they begin to accompany their sounds with gestures, like looking and pointing, that direct the parent's attention toward a particular object (e.g., de Villiers and de Villiers 1978). Still later, nonsense sounds are replaced by more precisely articulated noises, like *dadoo* when pointing to an other person or *kitty* when both parent and child have been looking at a cat. Moreover, the child begins to use her vocalizations only in particular social circumstances. She says "bye-bye," for example, only when someone is leaving and only when the audience is appropriate. Children say bye-bye less readily to strangers than to grandmothers. Such observations suggest that the child is using words that have a specific meaning to her.

Even so, it is difficult to state precisely *what* the child means at this one-word stage of development. For a brief period, human children—like young vervet monkeys—may overgeneralize or undergeneralize the meaning of a word. This behavior helps us to understand what they have in mind and demonstrates that meaning is not always the same for infant and adult. The baby's use of dadoo to refer to any person, or ball when pointing to any round object, tells us that her understanding of what these words mean is slightly different from our own. The problem facing developmental psychologists is to understand what criteria the child uses to assign word meaning—in other words, how she classifies objects and events in the world around her. At present there is evidence that children may classify objects according to their function (Nelson 1973), their physical features (Clark 1973), or a "prototype" that serves as a basis of comparison for newly encountered stimuli (Rosch 1973; see also chapter 4). These hypotheses, however, are not mutually exclusive, and it seems likely that there is some truth to all of them (e.g., Gleitman 1986; Gardner 1987).

There is a further complicating factor. Children at the one-word or two-word "telegraphic" stage may know much more about their system of communication than is revealed in their own speech. In one study, for example, children at the two-word stage were significantly more likely to respond to a grammatically well-formed command than to an ungrammatical command. Interestingly, the very utterance types that they themselves did *not* use were more effective in eliciting a response than were utterance types they *did* use (Shipley, Smith, and Gleitman 1969). This suggests a child's competence—in grammar or in the assignment of meaning to words—will be underestimated if it is simply inferred from spontaneous speech.

Finally, there is the possibility that a child's one-word utterances are not single words but rudimentary attempts at a sentence. Consider the following example. As our colleague Lila Gleitman fed her infant daughter one evening, the child repeatedly refused to eat and gestured instead toward the refrigerator behind her, loudly saying "Zert!" Unlike other mothers, Lila's immediate reaction was not mild irritation but curiosity as to whether this was a one-word statement (Dessert!) or a rudimentary sentence, expressing the child's attitude toward a particular item or idea (I want to eat dessert! or Give me dessert now!). To test between these two hypotheses, Lila repeated the word dessert in a friendly, conversational tone. Understandably, this attempt at casual conversation was immediately rejected by the outraged baby, as was holding up some dessert while saying the words, "Ah, yes, this indeed is a dessert." The only response deemed satisfactory from the baby's point of view was to provide a dessert and allow her to eat it. Lila concluded that at this stage, the utterance "Zert!" was not a word but a proposition. Its meaning included both reference to an object and a disposition to behave toward that object in a particular way.

This brief excursion into the earliest speech of children is instructive for two reasons. First, the parallels between children and vervet monkeys should by now be obvious. Although children quickly go on to much greater things, for a few brief months communication by child and monkey is similar in the following respects. Both use sounds in a limited set of circumstances, *as if* they are denoting particular features of their environment or "commenting" on the situation in which they find themselves. For the child, such designations may be made explicit by pointing or other gestures; for the monkey they are not. For both species, however, communication involves no obligatory links with subsequent behavior. Both child and young monkey overgeneralize in a way that is not entirely random, and for both child and monkey production lags behind comprehension.

Second, the comparison we have drawn indicates that observers of children and monkeys, presented with similar data, have similar difficulty in specifying the exact meaning of words. Unable to specify the meaning of vervet calls precisely, we find some consolation in the fact that knowledge of our own species is also imprecise.

Summary

In chapter 4 we presented evidence that vervet monkeys and other nonhuman primates use vocalizations in a manner that effectively denotes objects and events in their environment. We also noted parallels between the way vervet monkeys learn calls and the way human infants learn words. Parallels between monkey calls and human language must be drawn with care, however. Here we have taken a closer look at the "meaning" of vocalizations from the monkeys' perspective.

One approach examines meaning in terms of the relationship between vocalizations and the mental states of those who use them. Despite some recent progress, we still know relatively little about the cognitive mechanisms that might underlie vocal signals. At one extreme, simple explanations that posit no mentality at all (for example, Monkeys give alarm calls because they are excited) can be ruled out for a variety of reasons. On the other hand, there is as yet no evidence that monkeys attribute mental states to others. Taking an intermediate position, we conclude that vervet communication is most consistent with Dennett's (1983, 1987) first-order intentionality: monkeys give leopard alarms because they want others to run into trees, not necessarily because they want others to think that there is a leopard nearby.

A second approach examines meaning in terms of the relation between calls and the things they denote. Although vocalizations obviously convey a wide variety of information to those nearby, there is good evidence that some (and perhaps much) of this information concerns referents external to the signaler. For two reasons, however, it is difficult to state precisely what each call denotes. First, call meaning cannot be described in absolute terms (for example, *snake alarm* means the African python, *Python sebae*). Instead, the meaning of a call can only be stated relative to the meaning of other calls in a species' repertoire. Among vervets, for example, snake alarm denotes something that is both different from the objects that elicit eagle and leopard alarms and different from objects that elicit no alarm at all, like harmless snakes or lizards. Hence the meaning of each call is relative and depends on the meaning of others. Function seems to determine the number of calls within a species' repertoire and, as a result, the referential specificity of each vocalization.

Our assessment of call meaning is also imprecise because, as yet, we cannot tell whether a vervet's call should be glossed as a word (simply, snake) or as a proposition (Snake! Let's approach and mob it!). Hence we make no absolute distinction between a call that provides information about an external referent and a call that combines referential information with information about the caller's attitude or disposition toward that referent.

Finally, we note that our inability to state the precise meaning of vervet vocalizations is not peculiar to work on nonhuman species but parallels similar difficulties in the assessment of word meaning among very young children.

CHAPTER SIX

SUMMARIZING THE MENTAL REPRESENTATION OF VOCALIZATIONS AND SOCIAL RELATIONSHIPS

At this point, we want to pause briefly to reconsider the problem of how vervets and other monkeys represent their world. Thus far, we have referred twice to the "representation" of knowledge in the minds of nonhuman primates. In chapter 3 we argued that monkeys and apes make good primatologists. On the basis of their observations, they not only recognize the relationships that exist among others but also compare *types* of social relationships and make same/different judgments about them. To do this the animals must have some way of representing the properties of social relationships. This representation, we noted, is not made explicit: we have no evidence, for example, that the monkeys have labels that describe mothers and offspring or that distinguish closely bonded individuals from others in their group. Nonetheless, we can hypothesize that the social complexity of nonhuman primate groups is based, at least in part, on processes that go beyond the formation of associations between particular individuals. Monkeys observe one another, note who associates with whom, and then infer properties of social relationships in a way that allows relationships to be compared regardless of the particular individuals involved.

Similarly, in chapters 4 and 5 we argued that vervet monkey vocalizations denote objects or events in the environment and that monkeys compare vocalizations on the basis of these referents. Once again, to make such comparisons the monkeys must be able to make representations of the objects denoted by a vocalization, even if the objects themselves are not present.

Although in each of these instances there seems to be a strong case for a mental representation, or concept, it is important to emphasize again how little we know about what actually exists in the minds of our subjects. At this stage, for example, it is unclear precisely how the monkeys' representations might differ from associations formed through classical conditioning, associations which can themselves be extremely complex (e.g., Dickinson

1980; Rescorla 1988). Similarly, we cannot specify how much information is contained within a representation, how the information is structured, or how it is coded in the nervous system. We can, however, consider what representations are good for and how under natural conditions monkeys might benefit from having them. Perhaps, as Fodor and Pylyshyn (1981) suggest, the content of representations can be elucidated, at least in part, from their function (see also Herrnstein 1990).

Why Monkeys Need Mental Representations

Throughout this book we have adopted a functional, evolutionary approach to the study of primate intelligence. If representations of certain aspects of the world exist in the minds of monkeys, we assume that they do so because they confer a selective advantage on those who make use of them. We also assume that what is represented, as well as the structure of information contained within a representation, will be determined by the relative utility of one sort of mental operation as opposed to another.

The notion that monkeys might *need* representations of social relationships is buttressed by the experience of those who study them. Primatologists have long recognized that, in order to explain and predict the behavior of their subjects, they cannot simply describe or list who does what to whom and how often. Instead, they must step back from the minutiae of social behavior and identify, at a more abstract level, social relationships and the general principles that underlie them. Hinde (1976a, 1976b, 1983a), for example, defines a relationship "in terms of the content, quality and patterning of interactions" between two individuals over time. By this definition, a relationship cannot be described by any single interaction, nor is it enough simply to list what two individuals did with one another during a particular period (for example, that they groomed three times, hugged each other once, fought once, and spent 23% of the observation period together). What matters—and what defines a relationship—is not simply the behaviors themselves but also the temporal relations among behaviors and the way each activity is carried out. Some pairs of animals groom whenever they are together, others groom only briefly; some separate after a fight, others reconcile; for some a hug is prefunctory, while for others it is a lengthy embrace. The point is that if either we (as observers) or the monkeys (as participants) want to explain or predict social behavior, we must change our focus of analysis from a set of interactions that is simple and concrete to a relationship that is more complex and abstract.

We hypothesize that the monkeys' ability to represent social relationships has evolved because it offers the most accurate means of predicting the behavior of others (see also Humphrey 1976, 1983; Whiten and Byrne 1988b). But there are also other advantages. Because relationships conceived in this way are abstractions, they can be more parsimonious and simpler than absolute judgments, which require learning the characteristics of every interaction (Kummer 1982; Premack 1983a; Dasser 1985; Allen 1989). If a monkey can assess the relationships of others—rather than having to remember or observe all of their interactions—he may be able to predict what opponents will do next even when he has seen them interact only once or twice. In other words, a monkey would be considerably better off if he had some representation of a social relationship.

A similar argument can be made for the representation of meaning in vocal communication. Assume, for the moment, that we are dealing with cases in which a signaler gains by providing truthful information to her audience—a female communicating with her offspring or another close relative. (Of course, this will not always be the case. Since the signaler's interests will not always coincide with the recipient's, there will be some occasions when the signaler does better to provide unreliable information. We discuss the evolution of deceptive signals further in chapter 7.) When communicating reliably, individuals can clearly benefit from the use of vocalizations that denote specific aspects of their environment. If the best escape from a leopard is to run into a tree and the best escape from an eagle is to run into a bush, the adaptive value of acoustically distinct calls for different predators is obvious.

Monkeys, moreover, often vocalize when out of sight of each other. This favors the evolution of calls whose meaning can be derived from acoustic properties and does not depend crucially on contextual cues.

Further, in many species the appropriate response to a vocalization may differ markedly from one individual to the next. Upon hearing an eagle alarm, vervet monkeys on the ground should look up, but those in a tree should run down and out of the tree; upon hearing a grunt to another group, infants should run to their mothers, but adults should direct their attention toward the border of their range. This favors the evolution of calls whose meaning does not depend on the caller's subsequent behavior. That is, it favors callers who can communicate about events in ways that are relatively independent of their own behavior or emotional state and listeners who can interpret a call's meaning in a manner that is relatively independent of what the caller himself is likely to do next.

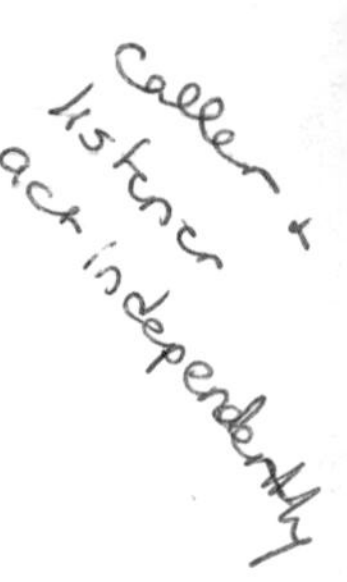

We hypothesize, therefore, that the monkeys' ability to represent the meaning of vocalizations has evolved because of the advantages that accrue to individuals who can interpret sounds without relying on contextual cues or on the behavior of those who vocalize. In more complex systems of communication, the ability to link one or more vocalizations with a common referent (for example, giving both *wrrs* and chutters in response to the approach of another group) and to compare calls on the basis of the things they represent, allows individuals to develop a rich semantic system in which some calls, like eagle and snake alarms, are markedly different, whereas others, like *wrrs* and chutters, show more subtle differences from each other and can be used to represent shades of meaning within a general class.

The Content of Representations and Their Limitations

When we talk of monkeys recognizing a close association between two other animals, it is important to distinguish between *association* as referring strictly to an observable fact (that two animals are often together) and *association* as referring to a more structured and differentiated representation of a social relationship. Monkeys are undoubtedly capable of recognizing that certain other individuals interact at high rates, and their comparisons of different relationships are probably often based on differences in rates of interaction. However, an association that is based exclusively on interaction rates cannot incorporate any other qualities of a relationship. If a monkey learns to associate other animals solely on the basis of the rate at which they interact, he will be unable to distinguish between two different types of relationship that both involve similar rates of interaction. There will be no way for him to distinguish, for example, a female's relationship with a juvenile male (her son) and the same female's relationship with an adult male (her "friend" or long-term mate).

As we discussed in chapter 3, however, it seems likely that monkeys are sensitive to more than just interaction rates when assessing other animals' social relationships. Apparently, they also attend to subtler distinctions, including the types and quality of interactions, the age and sex of participants, their dominance ranks, their past history of behavior, and so on (see Hinde 1983a, 1983b). Recall, for example, that Dasser's (1988a) long-tailed macaques correctly identified numerous mother-offspring pairs despite marked variation in the ages, sex, and interaction rates of the individuals involved. Similarly, vervet monkeys reconciled primarily with their opponents' *kin* following fights with unrelated animals but with their opponents *themselves* following fights with members of their own matriline. These data suggest that vervets distinguish their own close associates from the close associates of others despite similarly high rates of interaction within all matrilineal kin groups. To give one final example, numerous studies of baboons, macaques, and vervets have shown that high-ranking females are more attractive grooming and alliance partners than low-ranking females, regardless of the rate at which they reciprocate (see chapters 2 and 3; see also Walters and Seyfarth 1987 for a review). This observation suggests that females assess the benefits of social relationships not just in terms of interaction rates, but also according to the potential benefits that different individuals can offer.

We may hypothesize, therefore, that the primate mind is predisposed to organize data on social behavior according to both the individuals involved and the content, quality, and pattern of their interactions. The resulting representation has an abstract component because it is more than the sum of its parts. A social relationship cannot be described simply in terms of the participants' physical resemblances, individual identities, or any single measure of activity like time spent grooming or the proportion of fights followed by a reconciliation. Instead, it must incorporate information about all of these features.

If future investigations support the hypothesis that monkeys' representations of social relationships are not based solely on association rates, two further, related issues will deserve particular attention. First, how many kinds of relationships are recognized? Is *mother* different from *sister*; is a *friend* of the opposite sex different from a *friend* of the same sex? At the moment, we cannot address this issue because few studies have examined the extent to which monkeys discriminate among relationships in which interaction rates are similar. Second, what are the consequences of having different representations for different types of relationships? How might they give one individual a selective advantage over others?

Like representations of social relationships, representations of call meaning must be sufficiently rich to incorporate information about individual identity, previous interactions (for example, whether the individual in question has used a signal reliably in the past), breadth of external referent, and the immediate social and ecological context. In addition, representations of call meaning must be stored in a way that allows different calls to be compared. Social bonds must be compared according to the type of relationship they represent; vocalizations must be compared according to the things for which they stand.

Beyond this, however, our understanding of monkeys' representations of social relationships and call meaning is not only ill-defined but also likely to remain fuzzy and imprecise even as more data accumulate. Given the difficulty of stating precisely what is contained within a monkey's mind, we turn instead to what may not be.

First, even if monkeys do distinguish among different types of social relationships, their ability to compare relationships may be relatively inflexible and limited to circumstances in which the individuals involved are familiar. In all of the studies described to date, subjects have of necessity been tested only with the familiar social companions that make up their group. As a result, we cannot state conclusively that a monkey confronted with an entirely new set of individuals—a young male transferring into a new group, for example—would be predisposed to look for close bonds among matrilineal kin, linear dominance relations, and so on. More to the point, how long would it take for a vervet or baboon to learn that not all primate species have the same patterns of social interaction? If a vervet male transferred into a gorilla group, where females are seldom closely related (Stewart and Harcourt 1987), how long would it take for the male to *cease* expecting the females to interact at high rates? Would he ever?

There is no doubt that monkeys can learn to adjust to novel patterns of behavior, as Kummer, Goetz, and Angst (1970) demonstrated when they experimentally transferred females between groups of hamadryas and savanna baboons in Ethiopia. Unlike savanna baboons, who live in large, multimale groups, hamadryas baboons form small, relatively stable one-male units (see chapter 3). The spatial integrity of these units is strictly enforced by the males who lead them, and male unit leaders herd and threaten their females whenever the females stray from their units (Kummer 1968). When Kummer and his colleagues artificially introduced female savanna baboons into hamadryas groups, the females learned within an hour to follow the specific males who had chosen them as their own. In particular, the females learned to approach males who threatened them, rather than to flee from them as they normally would have done in a savanna baboon group. Similarly, female hamadryas baboons who were introduced into a savanna baboon group soon learned to cease following males and formed no particular attachments with any individuals. Interestingly, males who were transferred from one species to another failed to modify their behavior. Savanna male baboons who were introduced into hamadryas groups, for example, never learned to herd females as hamadryas males did.

Did the females' ability to adjust to their adopted groups involve any hypothesis about the nature of the social structure and relationships in these groups? Was their rapid learning due entirely to the experience of being attacked, or did it also involve observation and deduction? We simply do not know the basis of a monkey's understanding of its social environment. While a monkey's conception of social relationships may be abstract and independent of the particular individuals involved, it may also be relatively stimulus bound and limited to the general types of bonds to which the monkey has been exposed (see also D'Amato, Salmon, and Colombo 1985). It remains possible, in short, that monkeys are primatologists who have spent too much time studying a single species or living in the same group.

Second, although monkeys may be able to represent social relationships, their ability to make use of such representations in reasoning or computation may be limited. Consider, for instance, the different ways in which human primatologists on the one hand and monkeys on the other deal with the simultaneous existence of close bonds among kin and the attractiveness of high rank. Humans can readily see that these two principles will be additive for high-ranking families and counteractive for low-ranking families.

We therefore deduce that high-ranking families will be more cohesive than low-ranking families, a prediction that is borne out by data (see chapter 2). At present, however, we have no evidence that the monkeys themselves recognize this difference: no evidence, for example, that a middle-ranking female distinguishes the relations that exist in high-ranking matrilines from the relations that exist in low-ranking matrilines. More important, even if such data were to emerge, it would be essential to distinguish between information that the middle-ranking female had acquired through observation and experience (high-ranking mothers, for instance, support their offspring in alliances at higher rates than do low-ranking mothers) and information that the middle-ranking female had acquired through deduction. Indeed, with the exception of data on the recognition of other animals' dominance ranks, we presently have no evidence that computation plays a major role in the monkeys' representations of social relationships or in their representations of word meaning. In this respect, representations of social phenomena may differ fundamentally from the representations of rate, time, and space used by birds and other animals when computing and comparing feeding returns at alternative food patches (see chapter 9; see also Gallistel 1989a, 1989b).

Third, as we noted in chapter 3, we have no evidence that monkeys can label social relationships or give names to the criteria they use in classifying them. Although certain primate vocalizations do function in a manner that effectively labels different predators, like leopards and eagles, or different classes of conspecifics, like dominant or subordinate, monkeys apparently have no calls referring to *close partners, friends,* or *enemies* that could be used to classify relationships. Whether they could learn such terms under the appropriate conditions remains an open question: none of the ape language studies has ever asked subjects about each others' relationships. Among adult humans, accurate use of a word like *friend* implies that we recognize the necessary and sufficient characteristics for membership in this category, and hence that we can apply the category's label correctly in novel situations. If the presence or absence of a label is some measure of an individual's awareness of classes and of relations between classes, then the ability of monkeys to compare relationships and generalize to novel social situations may be severely limited.

Similarly, although monkeys use vocalizations to signal about things and compare calls on the basis of their referents, we have no evidence that monkeys explicitly recognize the referential relations that hold between their calls and objects in the world around them. In contrast, chimpanzees can be taught not only to identify the relation between a sign and its referent but also to group signs within a larger superordinate class (e.g., Savage-Rumbaugh et al. 1980; Premack and Premack 1982). Premack (1976), for example, taught the chimpanzee Sarah not only words for specific colors but also the word *color.*

The vervets' apparent lack of any vocalization for a superordinate class, like *danger* or *family,* provides indirect support for the hypothesis that monkeys can use sounds to represent things but are not explicitly aware of the referential relations that exist between sounds and the objects they denote. Vervets, for example, do not transfer skepticism about one individual's leopard alarm calls to her eagle alarm calls, suggesting that they treat the two types of calls as denoting very different classes. Conceivably, humans in the same circumstances would behave differently, and on the basis of unreliable leopard alarms by a particular individual would begin to doubt that individual's reliability in situations involving any predator.

Of course, the monkeys' disinclination to lump predator calls into one broad category does have functional advantages. Given the different hunting strategies of, for example, leopards and eagles, it would clearly be maladaptive to respond to these two types of predators with a single call or a single mode of escape. Nevertheless, human words (and the systems of classification that underlie them) derive considerable communicative power from the ability to be either specific *or* general. We are able to use superordinate terms while still maintaining, in our minds, the distinction among the elements that comprise a superordinate category. We can, for example, differentiate among different kinds of hazard (like a fire and a rapidly approaching car) and nevertheless still refer to both as dangerous.

The monkeys' apparent lack of vocalizations to label superordinate classes may be symptomatic of a larger problem: the monkeys are unaware of their own knowledge. In Paul Rozin's (1976) terms, a monkey's knowledge of social relationships or word meaning may be *inaccessible*. While the monkey can classify familiar relationships into types and even compare social relationships involving different individuals, he may not be able to examine his own knowledge, label it, apply it to new stimuli, or use it to deduce new knowledge. In addition, perhaps because the monkey cannot reflect on what he knows about others, he may be unable to attribute motives and hence understand *why* some relationships are alike and others are quite different.

We have argued that monkeys, in order to succeed socially, must be able to predict the behavior of others. To do this well, they cannot rely on memorizing single interactions but must instead deal in abstractions, deducing the relationships that exist among others. For humans, the quest to predict behavior prompts us to search still further for the factors that cause some relationships to differ from others. A monkey who can compare social relationships is better able to predict the behavior of others than one who simply memorizes all the interactions he has observed. Vastly more powerful abilities to interpret other animals' behavior accrue to the individual who can attribute motives to others and classify relationships on the basis of these motives (Humphrey 1983; Whiten and Byrne 1988a, 1988b).

In the end, then, we confront the relation between an individual's behavior and his recognition of the mental states of others. Given that monkeys recognize relationships, do they understand the motives that underlie them? Given that monkeys can communicate selectively about things, do they use these calls simply to affect each others' *behavior,* or do they ever attempt to change each others' *minds?* These questions strike at the heart of how monkeys see the world and lead us directly into the next three chapters on deception, attribution, and the limits of primate intelligence.

CHAPTER SEVEN | DECEPTION

Leslie, a high-ranking vervet female, had just chased Escoffier away from a grooming bout with Leslie's mother, Borgia. After grooming Borgia briefly, Leslie approached Escoffier, who cowered. Leslie lipsmacked—a sign of appeasement—and began to groom Escoffier. After a few minutes, Escoffier relaxed visibly and stretched out to allow Leslie to groom her on the back. At this point, Leslie picked up Escoffier's tail and bit it, holding it in her teeth while Escoffier screamed.

What should the observer make of this? Did Leslie actively lie to Escoffier, plotting her revenge until the moment that it would be maximally effective (fig. 7.1)? Or did her grooming of Escoffier simply remind Leslie that she was angry with her? Just as juries are admonished to consider intent in criminal trials, our judgment about Leslie's behavior depends on our perception of her motives. Yet before we can begin to analyze Leslie's motives we must first establish whether she even has the *ability* to signal falsely. Is a vervet monkey in fact able to withhold signals in some contexts or use false signals in others? How reliable, in general, *are* animal signals?

Do monkeys have the ability to deceive?

In chapters 4 and 5 we concentrated on the information conveyed by primate vocalizations. To what use, though, are these vocalizations put? Humans, after all, use speech not just to convey information about matters of immediate concern but also to cajole, convince, entreat, manipulate, and lie. The moment that a signaling system becomes even rudimentarily referential, its power to convey both accurate and inaccurate information increases tremendously; information can be withheld and even specifically falsified. And yet it is far from clear whether primates ever do use their vocal signals (or indeed any other communicative gestures) to convey false information.

Whether animal signals are considered "truthful" or "deceptive" depends to some extent on the sort of data one decides to emphasize. In most cases, signals do seem to provide accurate and reliable information to those nearby. In other cases, however, they clearly do not. As one of the many examples of truthful signaling, consider the courtship display of male smooth newts (*Triturus vulgaris*). After copulation, male newts require a number of hours to replenish their supply of spermatophores (Halliday 1976). When a male meets a female, therefore, he may have a great deal of sperm available or very little. If a male encounters a female, he gives a courtship display that includes a variety of fanning tail movements. Since the rate at which he performs this display is strongly correlated with how much sperm he eventually deposits, fanning provides the female with accurate information about the male's suitability as a mate (Halliday 1983).

There are many other examples of signals that provide accurate information about an individual's resources, its intentions, or some aspect of its environment. Male mockingbirds (*Mimus polyglottos*) with large song repertoires have the best-quality territories (Howard 1974), the "play face" of many mammals (fig. 7.2) accurately signals that aggression is unlikely (e.g., Fagen 1981), and the alarm calls of many birds (e.g., Marler 1956b), rodents (e.g., Sherman 1977; Dunford 1977; Leger and Owings 1978; Hoogland 1983), and monkeys (chapter 4) provide accurate information about not only the presence of a predator but also whether the predator in question is an eagle, a snake, or a carnivore.

At the same time, however, there is evidence that at least some animal signals provide *inaccurate* information and in this sense apparently function to deceive others. Here we use an operational definition: a deceptive signal is one that provides others with false information. Whether any animal ever

Figure 7.1. Adult female Newton grooms another female's tail. Females sometimes seem to "deceive" their grooming partners by lulling them into a relaxed position and then biting or attacking them.

attributes false beliefs to others, or is conscious of its own attempts to deceive, remains an open question to which we will return in chapter 8.

In many instances animal signals seemingly *function* to mislead opponents or to present opponents with ambiguous information about subsequent behavior. To cite one of the most common examples of ambiguous signaling, the threat displays of many species of birds and mammals are often relatively poor predictors of subsequent behavior and cannot reliably be used by recipients to gauge the probability of attack (Caryl 1979; but see Hinde 1981). In other cases, unreliable signals involve active falsification. A male scorpionfly (*Hylobittacus apicalis*), for example, can only copulate with a female if he first provides her with a nuptial gift of a dead insect. Some males catch insects on their own, while others steal insects by approaching males who already have them and adopting the posture and behavior of females (Thornhill 1979).

Similarly, polygyny in the pied flycatcher (*Ficedula hypoleuca*) has been interpreted as an example of successful deception of females by males (Alatalo et al. 1981; but see Stenmark, Slagsvold, and Lifjeld 1988). Male pied flycatchers form territories, and females fly from one territory to another, eventually choosing a mate. A female's preference is affected by the quality of a male's territory (Alatalo, Lundberg, and Stahlbrandt 1984) and by whether or not the male is already mated. Secondary females have lower reproductive success than monogamously mated females, and females generally avoid males who already have a mate. Males, however, often hold two territories that can be separated by as many as 3.5 km. One explanation for these widely dispersed territories is that they function to deceive, since they prevent females from determining if a male is already mated. Indeed, a fe-

Figure 7.2. Two juvenile males, Macaulay and Trollope, play. The two males' "play faces" are typical and seem to be reliable indicators that aggression is unlikely.

male often unwittingly becomes a male's secondary mate. By the time she has laid her eggs and he has returned to his first female, it is too late in the season for the female to start another clutch.

The study of deception in animal communication is a young one, and many unresolved issues remain. For example, only the most preliminary analyses have been conducted on the rate at which animals signal truthfully as opposed to unreliably or on the effects of context on supposedly "deceptive" signals. More important, there is still little consensus on whether deceptive or unreliable signals occur at all or how flexible animal signals actually are. Do animals really have the *ability* to modify their signals, and, if they do, how flexible is signal production (chapter 5)?

Here we review some of the evidence for deceptive communication and the detection of unreliable signals in primates and other animals. We describe the form that deceptive signals appear to take and discuss their possible function. We concentrate in particular on signals that convey information about the environment or the signaler's probable behavior, since these signals offer the widest scope for modification. Our primary purpose is to provide a framework, however rickety, within which we can consider the far more speculative and controversial issue of what deception might tell us about other animals' minds. We reserve this topic for chapter 8.

Theoretical Background

In some cases, signals cannot easily be falsified. In particular, signals that reflect or are tied to some physiological state—such as body size or reproductive condition—are usually reliable simply because they are difficult to modify. But this is not necessarily true of signals that convey information about the environment or about the signaler's probable behavior. In fact, there are good reasons for predicting that animals should *seldom* communicate precisely about their probable behavior, except in cooperative interactions.

Consider, for example, the threat displays used by many birds and mammals to signal the probability of attack or flight. Maynard Smith and Price (1973) argue that natural selection will always favor those individuals who "cheat" and threaten others at the highest level of intensity (falsely signaling imminent attack), regardless of their actual intentions (see also Maynard Smith 1979, 1984; Caryl 1979; Krebs and Dawkins 1984). Building on this argument, Andersson (1980) offers a scenario that attempts to explain how deceptive signals might evolve, why there is often a poor correlation between signals and subsequent behavior, and why animal species have so many different threat displays. Originally, Andersson argues, a movement or posture (such as the lunge of a monkey) is an effective threat display because it reliably predicts subsequent attack. Over time, however, the display gains increasing use as "bluff," presumably because signalers find that the display alone is sufficient to deter opponents, even if it is not followed by an actual attack. As the frequency of bluffing increases, skeptical recipients note that the display is no longer an accurate predictor of attack, and the display loses its effectiveness. Signalers, in turn, respond by introducing a new display that predicts attack more reliably than the old one. However, this signal, too, will eventually be used as bluff. Continuing competition between deceptive signalers and skeptical recipients ultimately produces a proliferation of displays, each of which loses its effectiveness the more often it is used as bluff (Andersson 1980; see also Paton 1986). Although Andersson's scenario is no doubt oversimplified, the general point that displays about intention also function as signals of assessment, and therefore seldom signal precise courses of action, is an important one.

Given theoretical arguments that deceptive signaling should be evolutionarily successful, and empirical evidence that animals sometimes, but not always, signal false information, we need to specify the conditions under which deceptive signaling can succeed and be beneficial to the signaler. Conversely, it will be necessary to identify those factors that constrain deception and give animals no choice other than to signal accurate information.

Constraints on the Use of Deceptive Signals

Despite the apparent advantages of unreliable (or at least unpredictable) signals, a number of factors limit the ability of animals to deceive one another. First, as we have already mentioned, deception will be constrained by a signaler's ability to control and invent new signals. Some signals simply cannot be faked because they reflect and depend on some physiological attribute, such as size or age. Male toads (*Bufo bufo*), for example, fight for access to females, and larger males typically win. Since fights usually take place in murky pond water or at night, visual cues are absent and males use auditory signals to assess the size of their opponents. Larger males have larger vocal cords and hence give calls with a lower fundamental frequency, or pitch. When a male hears a rival calling, he is more likely to continue fighting if the rival's pitch is higher and less likely to continue fighting if the rival's pitch is lower than his own (Davies and Halliday 1978, 1979). Because their acoustic features are so closely linked to anatomical structures that cannot be altered, signals like the toads' calls will always be truthful (Maynard Smith 1984, 1986).

Even displays that are tightly linked to some physiological attribute, however, cannot be entirely divorced from motivation. Many additional factors, such as possession of a territory or mate, can influence the willingness of an animal to fight (Parker 1974; Bachmann and Kummer 1980). As a result, even in relatively simple cases where information about each opponent's physiological state is evident, fights can escalate slowly and involve numerous displays that permit the evaluation of less accessible motivational information (Markl 1985).

Signals that are not dependent on physiological attributes but instead function as cues to probable courses of action, the possession of a resource, or some aspect of the environment are more open to deceit. Even these signals, though, will obviously be constrained by the signaler's ability to assess its opponent and to create novel signals. In the case of threat displays, for example, there will probably be few behavior patterns directly associated with attack and fewer still that can be separated from attack to serve an independent signal function (Moynihan 1970; Andersson 1980).

A second important constraint arises from a species' social structure. Animals that live in stable social groups face special problems in any attempt at deceptive communication. If individuals recognize one another and remember past interactions, a deceptive signal will be far easier to detect than in species like scorpionflies, where individuals encounter each other rarely. Among socially living animals, deceptive signals will probably have to be more subtle and occur at lower frequencies if they are to go undetected. Equally important, if animals live in social groups in which some degree of cooperation is essential for survival, the need for cooperation can reduce the rate at which unreliable signals are given. Indeed, it has been argued that bluffing about one's intentions cannot evolve under these conditions (van Rhijn and Vodegal 1980). Since social animals act as both signalers and recipients, mutual cooperation, even if based on skepticism, may be more evolutionarily stable than mutual exploitation (Markl 1985; see also Axelrod and Hamilton 1981). Constraints imposed by social structure therefore suggest that many of the theoretical arguments about deception, which assume no individual recognition and little memory of previous interactions (e.g., Maynard Smith 1974, 1982; Krebs and Dawkins 1984), may not be applicable to a variety of group-living animals.

Unfortunately for our purposes, both the risk of detection and the need for cooperation are likely to make deception in social groups rare and difficult to study. Observational studies will almost by necessity be anecdotal, simply because most forms of deception will occur at rates too low to permit systematic analysis.

Finally, deception will be constrained by the skill with which receivers can assess the meaning of signals and can incorporate this information into what they have learned about the signaler from past interactions. Although the assessment of meaning has received little attention, we consider it in some detail below.

Deception Through Silence

One of the most effective means to mislead an opponent is through the withholding or concealment of information. Although signal concealment involves no active falsification of information and cannot distract attention from one event to another, it can certainly *function* to deceive. More important, because deception through concealment involves no signal production, it is almost impossible to detect; as a result, it may occur at higher rates than more active signal falsification.

Alarm calls would seem to be ideally suited for deceptive signaling. Consider once again the alarm calls of vervet monkeys described in chapter 4. Different alarm calls signal the presence of different classes of predator with different hunting strategies, and they evoke different escape responses. Given this vocal repertoire, the adaptive significance of false alarm calls would seem to be obvious. A vervet monkey could spot a leopard in a bush, wait until a rival approached the bush, then give an eagle alarm and watch the rival run into the bush and be eaten. Curiously, however, vervet monkeys never seem to do this. Perhaps such deceptive alarm calling would simply be too transparent. Although a false alarm might dispatch a rival once, other group members might quickly detect the mismatch between alarm call type and predator. They might then cease paying attention to the signaler or, worse, punish or ostracize him.

There is a more subtle strategy, though, that is almost as effective as outright deception and much less easy to detect. The vervet monkey could spot a predator and simply remain silent, giving an alarm call when only she or her kin were in imminent danger. There is some evidence that this sort of deception occurs regularly in vervets and that it may also be widespread in other species.

Among the Amboseli vervet monkeys, high-ranking males and females were significantly more likely to give alarm calls than were low-ranking animals. Apparently, this result was not due to any greater ability on the part of high-ranking animals to detect predators, since films of the animals gave no indication that high-ranking individuals had better vantage points or spent more time scanning for predators. Alarm calling frequencies were also not correlated with the presence or absence of kin, since high-ranking females in our study population did not have more offspring than low-ranking females. Instead, it seemed possible that low-ranking males and females might have spotted predators equally often but simply failed to warn others in their group (Cheney and Seyfarth 1985b).

Similarly, among prairie dogs and a number of different squirrel species, females with kin are more likely than those without kin to give alarm calls (e.g., Dunford 1977; Sherman 1977; Hoogland 1983; see chapter 5). Since we have no reason to believe that animals with kin are more likely to spot predators or spend more time scanning than animals without kin, these observations provide additional indirect evidence that individuals may sometimes see predators but fail to inform others.

Numerous studies of nonhuman primates have reported instances of apparent withholding or concealment of information (see examples in Byrne and Whiten 1988c; Whiten and Byrne 1988c; Ristau 1990; Jolly 1990). Most examples are anecdotal, which is perhaps not surprising given the expectation that deception in social animals will be relatively infrequent.

Hans Kummer (1982) reports observing a female hamadryas baboon who spent 20 minutes gradually shifting her way in a seated position toward a rock where she began to groom a subadult male—an act that would not normally be tolerated by the dominant adult male. From his resting position, the dominant adult male could see the back and head of the female, but not her arms. The subadult male sat in a bent position and was also invisible to the adult male. What made Kummer doubt that this arrangement was accidental was the exceptionally slow, inch-by-inch shifting of the female toward the rock (see also Whiten and Byrne 1988c).

The chimpanzees and rhesus macaques observed by Frans de Waal (1982, 1986b) also provide examples of apparent deception through information concealment. In both studies, dominant males often attacked subordinate males if the subordinates failed to show submissive behavior when threatened. Dominant animals were rarely aggressive, though, if it appeared that the subordinate male had not noticed the original threat. Under these conditions, subordinates frequently ignored threats (by sitting very still, looking down at the ground or up in the air) that, in de Waal's opinion, they had almost certainly seen.

Other forms of feigned ignorance seem to be common among chimpanzees. De Waal (1982, 1986b), for example, recounts a case in which a young male, Dandy, walked over a place where experimenters had hidden some grapefruit underneath the ground. Because Dandy did not react in any way, the experimenters assumed that he had not noticed the fruit. Over 3 hours later, however, when the other chimpanzees were asleep, Dandy walked straight to the spot and dug up and ate the grapefruit (for further examples of feigned ignorance see Goodall 1986). De Waal also describes a number of instances in which male chimpanzees concealed their penises behind their hands when dominant males interrupted their courtship.

There is also some suggestion that chimpanzees occasionally withhold food calls and remain silent if the food they have found is insufficient to share (Wrangham 1977; see also chapters 5 and 8). Goodall (1986) describes an instance in which a 9-year-old male chimpanzee, Figan, gave loud food calls upon being handed a bunch of bananas. The whole group immediately converged on him, and he lost most of his prize. The next day he was again given a bunch of bananas. This time, although Goodall could hear faint choking sounds in his throat, he remained silent and was able to eat his bananas undisturbed.

Finally, chimpanzee females typically utter loud copulation calls when mating. The calls attract other males to the copulating pair, and dominant males seem to use these calls to monitor and disrupt the mating attempts of more subordinate males. Both de Waal (1982) and Goodall (1986) note that females sometimes suppress copulation calls when they are mating with subordinate males, even though the same females will continue to give the calls when mating with dominant males (see chapter 8 for an example of monkeys suppressing copulation calls).

What do these anecdotes tell us? They suggest, but clearly do not prove, that nonhuman primates do not simply monitor physical aspects of their world, such as the location of a food item or another individual, but also monitor and predict the mental states of other animals and the consequences of their own behavior on the behavior of others. What is presently lacking is a method for systematically observing the frequency and consequences of such apparent attempts at deception. More important, we need some way of discriminating between explanations that posit that animals have the ability to monitor the thoughts of others and simpler interpretations that do not rely on the recognition of such mental processes. In the case of the female hamadryas baboon, for example, explanations based on learned behavioral contingencies are as plausible as those based on mental attributions; the female could have groomed the subadult male behind a rock simply because she had learned from past experience that she could avoid attack by grooming other males out of sight of the dominant male. Choosing between simpler and more complex explanations becomes particularly difficult when we attempt to compare apparent acts of deception across species. Although house sparrows, for example, also modify the rate at which they utter food calls depending on the size of the food source (see chapter 5), we are intuitively inclined to believe that chimpanzee food calls are governed by different mechanisms than those of house sparrows. In the absence of any systematic information about the flexibility and modifiability of calls in each of these species, however, we are left simply with two very similar patterns of behavior.

Those taking an exclusively functional or evolutionary perspective (e.g., Krebs and Dawkins 1984) might argue that the mechanisms underlying the food calls of chimpanzees and house sparrows are irrelevant as long as the calls *function* to manipulate others. Mechanisms become more important, though, if we wish to use deception as a means to study the mental states and capacities of animals. Moreover, some knowledge of the proximate mechanisms underlying deceptive behavior is essential if we are ever to understand the constraints within which communication operates and the different forms that manipulation can take.

Other hints that animals occasionally manipulate each other by withholding information can be found in the experiments on vervet monkeys and roosters described in chapter 5 (Cheney and Seyfarth 1985b; Gyger, Marler, and Karakashian 1986; Karakashian, Gyger, and Marler 1988; Marler, Karakashian, and Gyger 1990). Recall that both vervet monkeys and roosters modified their alarm calls depending on their audience and apparently withheld alarm calls from rivals (in the case of vervets) or another species (in the case of chickens). Here too, however, it is unclear whether call concealment constitutes deception, because a number of crucial questions remain unanswered. In the vervet experiment, for example, the predator was presented to signaler and audience simultaneously. As a result, we could not consider whether signalers ever take into account their audience's current state of knowledge before giving alarm calls. In the case of roosters, it is even unclear whether the birds took into account their audience's *behavior,* since males failed to alarm call when they could observe females taking evasive action but could not see the predator themselves (Karakashian, Gyger, and Marler 1988). In each case, the experiments demonstrate that animals can change their alarm calling rate depending on social context but not that they are capable of conscious decisions to give or withhold information. It even remains possible that audience-dependent alarm calls reflect different levels of excitement or fear rather than any systematic ability to manipulate call production.

Although withholding information clearly has the potential of misleading others to the signaler's personal gain, it could be argued that signal concealment cannot, strictly speaking, be called deception unless the signaler has an intent to communicate or conceal information and unless others have some expectation of being informed. The intent to communicate and the expectation of being informed clearly demand that both the signaler and the audience have some ability to attribute motives or intentions to others (chapter 5). To date, however, there have been few systematic attempts to address this question, at least in part because of the methodological difficulties surrounding any investigation of attribution in social groups. Nevertheless, the problem does not seem entirely intractable. In brown capuchins, for example, the dominant adult male is the primary individual in each group to inform others of avian predators and to call others to palm nuts, which he alone is strong enough to open (C. van Schaik, pers. comm.). It might be possible to test females' expectations of information by presenting them with predators or palm nuts out of the sight of the dominant male and then determining whether the male's apparent failure to inform them affects their subsequent interactions with him. Does a male who fails to inform females about the presence of an eagle receive some form of retribution?

Finally, a study by Woodruff and Premack (1979) illustrates how withholding information can appear as an early, rudimentary form of deception among animals that eventually practiced more explicit forms of deceit. In these experiments a chimpanzee was first shown two containers, one with food hidden inside. The chimpanzee was then introduced to two trainers, neither of whom knew the location of the food. One was a "cooperative" trainer: if the chimpanzee signaled which container held the food, the trainer collected the food and shared it. The second trainer was "uncooperative"; when shown the location of the food he ate it himself. Over time, chimpanzees were tested with each trainer in trials where chimpanzee and trainer alternately served as sender and recipient of information. When interacting with the cooperative trainer, chimpanzees from the very beginning were able to produce and to comprehend accurate cues about the location of food. When interacting with the uncooperative trainer and playing the role of sender, the chimpanzees after many trials began to "deceive" this trainer. They first did so by withholding information: turning their backs and sitting motionless so that the trainer was given no clue as to where the food was hidden. Only later, after considerably more trials, did some of the chimpanzees signal falsely to the trainer by gesturing or pointing to the wrong container (Woodruff and Premack 1979).

To summarize, in social species where individual recognition, the memory of past interactions, and presumably the detection of false signals are well developed, withholding information provides an effective means by which animals can deceive one another without being detected. Unfortunately, however, just as silence may be difficult for conspecifics to detect, it may also be difficult for observers to detect. The frequency of deception through silence may therefore easily be underestimated.

Signaling False Information

Although signal concealment provides a functionally effective means of deception, more explicit deceit occurs when one individual actively falsifies the information he conveys to another. The falsification of signals potentially provides a much wider scope for deception than concealment, since it can actually change an opponent's knowledge or beliefs. It therefore provides an ideal tool to examine whether animals ever attribute beliefs, knowledge, or motives to others (chapter 8).

Explicit falsification can occur when an individual grossly distorts information, as in the case of the male scorpionfly that mimics a female in order to deprive another male of his nuptial gift. It can also occur when a stimulus is present and an individual signals false information about it. De Waal (1986), for example, describes how one male chimpanzee, injured in a fight with his rival, limped for a week afterward, but only when his rival could see him. Such active falsification is clearly more complex than simple concealment, because it demands that the signaler not only withhold something from his rival but also actively distract his rival from one feature of the environment to an entirely different one (Byrne and Whiten 1988c; Whiten and Byrne 1988c).

Similarly, on six occasions de Waal (personal communication) observed chimpanzees making false overtures for reconciliation. They invited their opponent toward them with a friendly gesture (fig. 7.3), only to turn aggressive at the last second, when the other had approached to within arm's reach. As we have seen, similar false reconciliation occurs in vervets and other monkeys. There are also numerous anecdotal accounts of chimpanzees and gorillas actively leading rivals away from a hidden food. On one occasion at Gombe, for example, the adolescent male Figan was unable to obtain bananas at a provisioning site because of competition from others. Suddenly, Figan stood up and strode out of the provisioning area in a manner that caused all others nearby to follow him. Shortly thereafter, Figan abandoned his companions and circled back to eat the bananas (Goodall 1971, 1986; for more examples of signal falsification in monkeys and apes, see Menzel 1971; de Waal 1982, 1986b; Byrne and Whiten 1988c; Whiten and Byrne 1988c).

Even more interesting are the apparent cases in which animals attempt to hide some physiological manifestation of anxiety from a rival. To give one example, Luit and Nikki, two adult males in the captive group of chimpanzees studied by de Waal (1982), were engaged in a prolonged struggle for dominance. During one fight, Luit chased Nikki into a tree and then took up a position of vigilance at its base. As he sat, Luit began to *fear grin*

Figure 7.3. A juvenile chimpanzee extends his hand to another for reassurance and appeasement. Occasionally, an individual who has invited an opponent to reconcile in this manner will attack when the opponent gets to within arm's reach. Photograph by Frans de Waal.

nervously. Quickly, Luit turned away from Nikki, put his hand over his mouth, and pressed his lips together, apparently to hide this sign of submission. Only after three attempts, when he had succeeded in wiping the fear grin from his face, did Luit turn to face Nikki again. Luit's actions suggested that he was aware of his nervousness, of the external manifestation of his fear, and of the need to hide this sign from his rival. Luit appeared to be attempting to manipulate Nikki's beliefs.

Monkeys and apes also occasionally falsify vocal signals by giving calls in inappropriate contexts. There is anecdotal evidence, for example, that adult male vervet monkeys sometimes give false alarm calls during intergroup encounters or when an immigrant male is approaching their group (discussed in chapter 8). Such calls are at least temporarily effective, since they invariably cause others to flee. Given the apparent ability of vervets to give contextually false alarm calls, it is even more puzzling that they never seem to "mislabel" predators with the wrong alarm call.

Finally, there are numerous suggestions that monkeys may use *each other* as social "tools" to gain an advantage over their rivals. Indeed, in their extensive review of deception in nonhuman primates, Dick Byrne and Andy Whiten (1988c) conclude that the use of others may constitute one of the most common forms of social manipulation by monkeys. To cite just one example, Kathy Rasmussen (quoted in Whiten and Byrne 1988c) describes a situation in which one male baboon recruited three other males in a fight over access to a rival's female (fig. 7.4). The male approached his rival and then screamed, as if he had been attacked. Three other males converged on the rival until the rival "cracked" and chased them, leaving his female. At this point the first male (who had *not* joined the coalition) quickly ran up to the female and chased her in the opposite direction.

Like redirected aggression and reconciliation, these triadic interactions certainly require that monkeys recognize and manipulate *relationships* among other group members. But do they really constitute deception rather than simply cases of ally recruitment? It seems possible, for example, that the males who were seemingly duped into aiding the supplicant were individuals who regularly cooperated with him and who would have come to his aid even if they had seen the original fight (see chapter 2). To demonstrate that this form of recruitment really constitutes manipulative "use" of alliance partners, it would be necessary to compare the responses of recruits when they could see the original aggressive interaction, and therefore could not be fooled, with their responses when they could not.

Even if careful observations were eventually to demonstrate that the use of third parties as social tools is a form of deception, we would still be left with some vexing questions. Why, for example, does this type of deception apparently occur so often in monkeys, and why do recruits continue to be

Figure 7.4. A male baboon helps his alliance partner attack a rival. Do males sometimes "dupe" potential allies into thinking that they have been threatened? Photograph by R. S. O. Harding.

fooled? If recruits *are* dupes rather than willing allies, it suggests rather poor powers of attribution on the part of everyone involved. In their analysis of apparently deceptive acts in monkeys and apes, Whiten and Byrne (1988c) note with some puzzlement that the use of third parties as social tools appears to be relatively rare among chimpanzees. One possible explanation is that chimpanzees have some ability to attribute motives to others and therefore recognize that they are unlikely to be able to recruit allies on a regular basis by feigning insult or injury (see chapter 8).

We must, in addition, entertain the possibility that the paucity of reports of social tool use in chimpanzees is an error of omission rather than cognition. Perhaps such interactions have been recorded less often in chimpanzees because of the sampling regimes of the observers, because chimpanzees have been observed for far fewer total hours than baboons or macaques, or because field observers did not interpret triadic interactions as deceptive. De Waal's (1982) account of shifting alliance patterns among male chimpanzees at the Arnhem Zoo certainly suggests that triadic interactions of the sort reported for baboons are also common in chimpanzees, although perhaps in a more subtle and flexible form.

To suggest that the study of deception in animals is still in its infancy is not to trivialize the extensive and provocative examples that have recently been compiled (e.g., Byrne and Whiten 1988c; Whiten and Byrne 1988c). These examples must be taken seriously, if only because they are not the casual remarks of naive observers but have been solicited from investigators who have spent literally thousands of hours studying monkeys and apes. At the same time, anecdotes are limited by the fact that, as observers, we cannot help but notice "smart" behavior in our subjects more than we notice "stupid" or even "normal" behavior. Because anecdotes can by definition never be systematically recorded, it is also almost impossible to make cross-species comparisons of their relative frequency of occurrence.

As is true also for examples of signal concealment, examples of signal falsification are potentially subject to overinterpretation. For example, if we were to consider examples of false reconciliation in isolation, we could easily argue that these gestures have nothing to do with deceit but result simply from the conflicting tendencies to show aggression and appeasement. However, the literature now contains numerous examples of apparent deceit, which taken together illustrate a wide variety of techniques for concealment or falsification. The examples are particularly rich in the case of chimpanzees. The variety of gestures used to deceive or conceal information is crucial, since it argues against the hypothesis that the animals' behavior simply reflects a displacement activity or a ritualized display. Through their number and variety, anecdotes gain in persuasive power and suggest at least the possibility of some degree of intentional falsification of signals.

At the same time, we must be careful not to exaggerate the abilities of nonhuman primates. Before we conclude that signal falsification in monkeys and apes indicates an ability to impute mental states to others, we should remind ourselves that the ability to modify and falsify signals is not restricted to nonhuman primates. In fact, there is some suggestion that signal falsification may be quite common in birds.

Munn (1986a, 1986b) studied mixed species flocks of birds in the Amazon basin and found that within each flock the members of one of two species, either the ant-shrike (*Thamnomanes schistogynus*) or the shrike-tananger (*Lanio versicolor*), led flock progressions and were the first to give alarm calls to predators. These "sentinel" species, however, also frequently gave alarm calls when no predator was present. Such false alarms were especially common when a member of the sentinel species and a member of another species were chasing an insect. Typically, false alarms caused the other individual to hesitate briefly, whereupon the sentinel species grabbed the prey. Similarly, Moller (1988) found that 63% of all alarm calls given by great tits (*Parus major*) in winter foraging flocks were false and were emitted when no predator was present. False alarm calls seemed to be given deliberately to drive away more dominant individuals from concentrated food sources, and they allowed callers to gain access to feeding perches from which they would otherwise have been excluded. Subordinate birds gave false alarm calls in the presence of both dominant and subordinate individuals. In contrast, dominant birds gave false alarm calls to other dominant individuals but not to subordinate individuals, whom they could easily supplant with threat displays.

Analyzed from a functional perspective these alarm calls are clearly deceptive, since they falsely manipulate the recipient's behavior to the signaler's benefit. What information do they reveal, however, about the signalers' ability to assess the knowledge and probable responses of others? No doubt any analysis of such calls will be influenced by the fact that they are given by birds, rather than apes, simply because we are usually less inclined to attribute complex cognitive abilities to birds than to primates.

Part of the problem of investigating deception in animals is therefore methodological; it is exceedingly difficult to study what animals intend to communicate solely by observing the effect that signals have on those nearby. A first step, however, would be to examine the proximate mechanisms underlying apparently false alarm calls. In the case of Munn's birds, for example, do sentinel species ever use *other* signals to manipulate flock members? How often can false alarm calls be given before the recipients cease to respond? In other words, how flexible is the behavior of signalers and recipients? How easily can signalers modify their calls to falsify information? Under what circumstances do recipients detect deceptive signals?

To give another example, in England cuckoos (*Cuculus canorus*) parasitize reed warblers (*Acrocephalus scirpaceus*) by replacing one of the warbler's eggs with an egg of their own (Davies and Brooke 1988). The cuckoo's egg is of the same cryptic color and size as the warbler's, cuckoos parasitize eggs primarily in the afternoon, and the cuckoo removes only one of the warbler's eggs. Each of the cuckoo's actions seems especially adapted to fool warblers: warblers reject foreign-looking eggs, they abandon nests that have had more than one egg removed from them, and they are more likely to detect an egg replacement if it has occurred in the morning than in the afternoon. Clearly, the cuckoo's behavior functions to deceive the warbler. How flexible, though, is it? Would the same individual cuckoo be less careful in removing a single egg and timing her egg laying if she were parasitizing a species less sensitive to parasitism?

Detection of Deception and Assessment of Signal Meaning

If one animal provides another with false information, how does this affect the recipient's subsequent behavior? The degree to which animals can deceive one another depends crucially on how recipients assess the meaning of signals. The success or failure of a deceptive signal will depend on at least three factors: the ability of the recipient to discover that a given individual's signal is false; whether, once the recipient has identified that individual's signal as false, he continues to be skeptical about the signaler and his call; and whether the recipient's skepticism then expands, and he begins to doubt the signaler's credibility in other, quite different, spheres of interaction. A fourth and equally crucial criterion for the detection of false signals is that the recipient restrict his skepticism only to the unreliable signaler and not generalize his skepticism to other individuals.

To take a commonplace example, suppose a friend offers you advice about the stock market. You follow his suggestions, and you quickly lose a substantial amount of money. Fairly rapidly you will begin to doubt your friend's expertise, and you will cease paying attention to his suggestions. Now suppose your friend offers you advice about which banks offer the best interest rates on their savings accounts. Whether or not you follow his new advice will depend to a great extent on how you assess the meaning of his remarks and how you classify what we might call *spheres of meaning* in the world around you.

In this example, because dealing in stocks and dealing with banks concern fairly similar issues (or, we might say, fall within the same sphere of meaning), you may well transfer skepticism about your friend's knowledge of the stock market to skepticism about his knowledge of banks. Because his information was false in one context, you assume it will also be false in other, closely related contexts. On the other hand, despite past events you may still be willing to consider your friend's advice about restaurants, since it concerns an entirely different sphere of activity.

This discursion illustrates an important point: the spread of deception in any population depends considerably on how animals assess the meaning of signals and how signals are classified on the basis of their meaning. If an individual successfully uses bluff in a threat display, can he also deceive others in courtship? Will he also be able to falsify information about predators?

Conversely, if a recipient learns that a particular individual is deceptive in its threat displays, how does this affect the recipient's subsequent behavior? Will he begin to doubt everyone's threat displays, or will his skepticism be limited to one individual? Will the recipient be skeptical only in the domain of aggression, or will he also begin to doubt the signaler when he signals about food or alarm calls? The limits to deception will be set, to a considerable extent, by the skill with which recipients can take information gained in one sphere of meaning and transfer it to other spheres. Deception therefore depends crucially on how animals classify signals, and the study of deception can reveal how animals categorize events in the world around them.

The experiments described in chapter 5, in which vervets were asked to compare the meaning of *wrrs*, chutters, leopard alarm calls, and eagle alarm calls, can now be reexamined from a somewhat different perspective. These, after all, were classic "cry wolf" experiments: one signaler was made to appear unreliable by her (apparent) repeated warnings of things that were not there. Other group members rapidly habituated to this individual and did so not just when she was using the same call over and over again, but also when she switched to a different-sounding call that had the same referent (Cheney and Seyfarth 1988; chapter 5). Subjects did not, however, transfer their skepticism to a *different* individual's call, suggesting that doubts about one individual's reliability did not generalize to include other group members. Finally, skepticism was also not extended to calls whose referents were very different. Even when subjects had ceased responding to a given individual's leopard alarm call, for example, they nevertheless still responded to her eagle alarm call.

These experiments suggest that the detection of unreliable signals is influenced by the ways in which animals assess the meaning of signals. Vervets and probably other species as well seem able to transfer information gained in one sphere of meaning to another closely related one. Of course, the experiments do not prove that under natural conditions vervets *do* deceive each other through signal falsification. Rather, they investigate the potential scope for deception by testing the skepticism of recipients and the ability of recipients to transfer their doubts to other contexts. Classification of signals according to their meaning suggests that an individual seeking to deceive others consistently will have to alternate among signal types if she is to succeed. If a vervet, for example, tries to deceive others by falsely signaling the presence of another group with a chutter, fellow group members will soon recognize that this chutter is no longer reliable and will transfer their skepticism to other intergroup calls. The signaler will then no longer be able to deceive with *any* of her intergroup calls, even though their acoustic properties might be quite different. On the other hand, the signaler *could* deceive others by giving false alarm calls, because, to vervets at least, an individual who is unreliable when signaling about one event is not automatically regarded as unreliable when she is signaling about a completely different event. Vervet monkeys seem to view leopard and eagle alarm calls as so different in meaning that experience gained in one sphere is not transferred to another. Had the boy who cried wolf been dealing with vervets, he might still have caused panic by switching his shouts of alarm to "Rustlers!"

At least two additional factors might affect the responses of recipients to "deceptive" signals. First, the relative costs of responding or failing to respond to a potentially false signal will almost certainly influence whether or not a recipient continues to attend to unreliable signals. For example, although recipients who respond to false alarm calls may incur costs in the form of wasted energy or decreased feeding opportunities, the cost of failing to respond to an alarm call is potentially so high that animals might continue to attend to alarm calls even after they become skeptical of the signaler's reliability. In contrast, since the cost of failing to respond to a food call is far less, recipients might become skeptical of false food calls at a much faster rate. As a result, the opportunities for deception will be greater in some circumstances than in others, simply because recipients cannot always afford to become skeptical.

It is also possible that the failure of recipients to transfer skepticism across very different spheres of meaning is related less to their ability to generalize across widely disparate contexts than to the ways in which they assess the *motives* of the signaler. To return to the example of the unreliable friend and his advice about stocks, suppose that you and your friend are in a building together when he suddenly shouts, "Fire!" Whether you heed his advice and run for the nearest emergency exit will depend to a large extent on how you perceive your friend's motives. If you perceive your friend as merely incompetent in financial matters, you will probably take his advice, because ineptitude in one sphere does not necessarily imply ineptitude in another. If, however, you are beginning to suspect that your friend is maliciously trying to drive you to financial ruin or worse, you may well decide that his warning is simply another deceptive ruse and ignore his advice.

In the case of vervet monkeys, as with most social animals, some degree of cooperation among group members is essential for individual survival. It therefore seems unlikely that an individual will come to be regarded as intentionally unreliable unless many other aspects of his behavior are unreliable. In our habituation experiments an individual was made to be unreliable in a single context for only a few hours. Assuming for the moment that vervet monkeys are capable of assessing each others' motives, an assumption for which there is at present no convincing evidence, it seems more likely that the experiments caused the signaler to be regarded as mistaken rather than deceitful. A signaler who is regarded as mistaken might be attended to for longer, and in a wider variety of different contexts, than a signaler who is regarded as intentionally deceitful.

Clearly, whether or not we can talk in terms of intentions and attribution in any nonhuman species is a controversial issue. However, assumptions about intention and attribution are often implicit in much of the debate about the function of deception in animal signals. We should be aware of these assumptions and tread lightly around them, while simultaneously hoping that we can eventually use communication to gain a window on animals' minds. In the next chapter, we move on to this elusive issue.

Summary

The manipulation of others through false or unreliable signals can take a number of forms. Among group-living animals, one of the most effective means of deceiving others is through silence, by withholding information that might be beneficial to others. This is the method of deceptive communication least likely to be detected. More direct manipulation can occur through the active falsification of signals. This strategy will be most effective if it occurs at low rates and if the circumstances surrounding successive acts of deception are varied; for example, if a false food call is subsequently followed by a false alarm call rather than simply by another false food call. Variation in the context of deception, at least among vervet monkeys, potentially allows individuals to maintain the highest rates of deceptive signaling without producing permanent skepticism among others in their group.

At the moment, we have no evidence that any animal species regularly varies the rate and context of false signals. Through more systematic observations and experiments it should eventually become possible to determine whether the intriguing anecdotes reported in the literature represent, at least in some cases, intentional signal falsification, as well as to specify more precisely the constraints under which deceptive communication is practiced.

CHAPTER EIGHT

ATTRIBUTION

When a new male transfers into one of our study groups from a distant, unfamiliar group, he rapidly learns to tolerate our proximity, even though he would have fled from humans in his previous group only a few days earlier. His rapid habituation to us is essential if he is to integrate himself into the group because, if he were to continue running away whenever we were around, his opportunities to approach females or to challenge resident males would be limited. The new male behaves as if he recognizes that other animals have no fear of us, and this diminishes his own fear.

Curiously, however, the reverse is not the case. We have often attempted to follow a completely habituated male when he transfers into a group that we have never observed before and whose members are unused to our presence. While the male still allows us to approach to within a few feet, the members of his new group flee from us in utter panic. Despite his habituation, their nervousness can persist for weeks, months, or even years. Indeed, the bewildered male often spends the first weeks in his new group standing bipedally, running into trees, and looking nervously around in search of invisible predators. He is, apparently, unaware that the other group members consider *us* to be the source of danger. His tolerance of us remains unswayed, even when his new companions give *strange human* alarm calls as we approach.

The contrast between these two examples is striking, all the more so because similar observations have been reported for many primate species, including gorillas (A. Harcourt, pers. comm.). In the first example, the strange male seems to learn rapidly that his new companions' tolerance of us is proof that we are not as threatening as we might appear. In the second example, the unhabituated group members never seem to learn that we are harmless, even though they are presented with the same information, albeit from fewer individuals. Perhaps it is largely a matter of how many animals flee, since the likelihood that a judgment is aberrant or mistaken presumably decreases as the number of individuals sharing the judgment increases. Nevertheless, in the first case the male seems to be attributing knowledge to other group members; in the second, the other group members fail to attribute knowledge to the male.

Do monkeys know as much about each others' beliefs, emotions and intentions as they do about each others' behavior? If Lockheed, a female vervet, threatens Maginot after Maginot's son, Trollope, has fought with Lockheed's son, Wordsworth, what does this tell us about Lockheed's knowledge of Maginot? We might conclude that Lockheed recognizes Maginot's association with the members of a particular matriline or that Lockheed can accurately place Maginot within the adult female rank order (see chapter 3). We have no evidence, however, that Lockheed knows anything about Maginot's *mind*—her beliefs, intentions, or emotions. Lockheed might be able to recognize that Maginot and Trollope associate closely without also understanding that Maginot *likes* Trollope or that Maginot's feelings toward Trollope are the same as her own feelings toward Wordsworth. There is, in fact, no evidence that Maginot herself recognizes that she likes Trollope. To recognize that Maginot has emotions and beliefs, and that these emotions and beliefs may be different from her own, Lockheed must be capable of attributing states of mind to others, an ability for which there is as yet patchy and puzzling evidence in monkeys and apes.

To attribute beliefs, knowledge, and emotions to both oneself and others is to have what Premack and Woodruff (1978) term a *theory of mind*. A theory of mind is a theory because, unlike behavior, mental states are not directly observable, although they can be used to make predictions about behavior. Monkeys are clearly adept at recognizing the similarities and differences between their own and other individuals' social relationships. What is not known is whether they are equally adept at recognizing the similarities and differences between their own and other individuals' states of mind. "The test of a first-rate intelligence," wrote F. Scott Fitzgerald in *The Crack-up*, "is the ability to hold two opposed ideas in the mind and still retain the ability to function." This remark did not always apply to the company Fitzgerald kept, and it may be equally inapplicable to other primates.

Why is it of interest to determine whether any species other than our own is capable of attributing mental states to others? After all, even without a theory of mind, monkeys are skilled social strategists. It is not essential to attribute thoughts to others to recognize that other animals have social relationships or to predict what other individuals will do and with whom they will do it. Moreover, it is clearly possible to deceive, inform, and convey information to others without attributing mental states to them. The alarm calls of vervet monkeys, roosters, and ground squirrels (chapter 5), for example, *function* to inform others of danger even if the signalers fail to attribute ignorance to those nearby. Similarly, the inappropriate alarm calls given by great tits at feeding perches (chapter 7) *function* to deceive other animals regardless of whether the birds actually attribute false beliefs to their audience. Even among humans, many complex skills are acquired through observational learning without requiring explicit pedagogy (e.g., Boyd and Richerson 1985).

However, the moment that an individual becomes capable of recognizing that her companions have beliefs, and that these beliefs may be different from her own, she becomes capable of immensely more flexible and adaptive behavior. To cite just a few examples, the individual who understands that her behavior can influence not just her companions' actions but also their beliefs can manipulate others in a far wider variety of contexts than an individual who simply recognizes a contingency between a given behavior and the response it evokes in others. By judging others according to their motives rather than just their prior behavior, she is also better able to generalize her past experiences from one circumstance to another and to predict whether an individual who has cheated her in one context is likely to cheat her in another. Furthermore, if she can identify ignorance in others, she can selectively reveal or withhold information from uninformed companions. Similarly, she can overcome the necessity of transmitting novel information through the relatively slow process of observational learning by actively teaching and instructing those who are ignorant.

Our aim in this chapter is to review the evidence for a theory of mind in nonhuman primates and to consider (or at least circle around) the following questions: Are monkeys and apes as skilled at monitoring each other's states of minds as they are at monitoring each other's behavior? Do they recognize the distinction between mental states and behavior, either in themselves or in others? Do they know what they know?

Lest we give the impression that these are issues buttressed by respectable scientific evidence, we should state at the outset that there has been almost no systematic research on the attribution of mental states in social animals. Although many anecdotes from studies of nonhuman primates can be interpreted as evidence for the attribution of mental states, most of these examples can also be explained more parsimoniously in terms of learned behavioral contingencies, without recourse to theories of mind. This does not mean, however, that the most parsimonious explanation is necessarily the correct one. Similarly, while many examples suggest that monkeys and apes do *not* attribute mental states to others, such negative results are also inconclusive. In many cases, animals may be able to behave adaptively simply by attending to the behavior of others; they may not *need* also to attend to their mental states. If this chapter raises more questions than it answers, therefore, it has addressed the issue honestly.

The Problem

Attribution is perhaps best considered in terms of Dennett's (1983, 1987, 1988) intentional stance (chapter 5). Recall that Dennett describes a number of different levels at which we could interpret, for example, a vervet monkey's alarm call. Is a vervet's leopard alarm call obligatory, in the sense that it cannot be suppressed or modified (zero-order intentionality)? If zero-order explanations can be ruled out, can we think of an experiment that would allow us to distinguish between a vervet who calls because he wants others to run into trees (first-order intentionality) and one who calls because he wants others to think that trees are a safe place to run to (second-order intentionality)? As we discussed in chapters 5 and 7, we now have observational and experimental evidence from species as diverse as chickens and chimpanzees that animals are capable of using calls selectively, to inform some but not all others of danger, the presence and quantity of food, and so on. The data are consistent with a first- but not necessarily a second-order intentional explanation.

Most of the controversy surrounding animal communication thus centers on second- and third-order intentionality—whether or not animals are capable of acting as if they want others to believe that they know or believe something. It is at this level that the most intriguing anecdotes surface, if only to be offered as sacrificial lambs (or bees, or chimpanzees) to Occam's razor, and it is at this level that experiments must eventually focus.

Higher-order intentionality implies the ability to attribute knowledge, beliefs, and emotions to others. Attribution, in turn, demands some ability to represent simultaneously two different states of mind. To do this an individual must recognize that he has knowledge, that others have knowledge, and that there can be a discrepancy between his own knowledge and theirs. In humans, this ability emerges only gradually and remains incomplete during early childhood.

Even very young children are able to impute simple mental states to others. By around 20 months of age, children begin to express explicit verbal knowledge of their own and others' intentions, moods, and actions, and they clearly differentiate between themselves and others (Bretherton, McNew, and Beeghley-Smith 1981; Bretherton and Beeghley 1982). Very young children also engage in pretend play, which suggests that they recognize the distinction between appearance and reality and that they can entertain multiple representations of objects or behavior. Nevertheless, before about 4 years of age, children seem to have difficulty in recognizing that other individuals' beliefs or thoughts might be different from their own.

To cite perhaps the best-known example of this failure to attribute false beliefs, Wimmer and Perner (1983) presented 3- to 9-year-old subjects with scenarios in which they had to describe the knowledge of others. In one case, the children watched a puppet show in which a boy, Maxi, puts a piece of chocolate into a blue cupboard. Maxi then leaves the room, and in his absence his mother removes the chocolate from the blue cupboard and places it in a green one. The children were then asked where Maxi would look for the chocolate. Children under 4 years of age consistently indicated the green cupboard, the cupboard in which they themselves knew the chocolate to be located. In contrast, about half of the 4- to 6-year-old children, and over 80% of the 6- to 9-year-old children, correctly pointed out that Maxi would still think that the chocolate was in the blue cupboard. The younger children's errors were not due to a failure of memory, because most of the children who gave an incorrect answer to the question nevertheless gave a correct answer when asked if they remembered where Maxi had put the chocolate. Rather, it seems that children's ability to represent two incompatible beliefs does not become established until around the ages of 4 to 6 years (see also Shultz and Cloghesy 1981; Dennett 1978b; Hogrefe, Wimmer, and Perner 1986; Leslie 1987; Perner, Leekam, and Wimmer 1987; Sodian and Wimmer 1987; Wellman and Bartsch 1988; Rakowitz 1990; and especially the chapters in Astington, Harris, and Olson 1988).

There is some suggestion that it is easier for young children to attribute ignorance to others than to attribute false beliefs. The attribution of ignorance does not require any precise identification of another individuals' knowledge, nor does it require the child to entertain two simultaneously contradictory mental representations. Instead, it simply demands some recognition of the fact that, although other people have beliefs, they may not always be privy to the same information as oneself. The recognition of ignorance in others seems to emerge at relatively younger ages, before children can differentiate clearly between their own knowledge and the knowledge of others. When 3-year-old children, for example, were read stories in which a character was excluded from some crucial bit of information, they correctly identified the character as being ignorant (Hogrefe, Wimmer, and Perner 1986).

Nevertheless, even in these relatively simple cases, in which the task is only to assess whether someone lacks or possesses information, children under the age of 4 years often find it difficult to regard problems from another person's perspective. Taylor (1988) showed drawings of animals (say, a giraffe) to children and then covered the drawing so that only a small, unidentifiable portion of the giraffe was still showing. She then asked children whether a puppet would be able to identify the drawing correctly. Most 4-year-old children attributed complete knowledge to the puppet and answered that the puppet would identify the giraffe. In contrast, most 6-year-old children correctly attributed ignorance to the puppet (see also Flavell 1988; and Wellman and Bartsch 1988 for another view).

The apparent failure of young children to recognize the potential discrepancy between what they believe to be true and what others, perhaps falsely, believe to be true is clearly not due to an inability to represent mental states or to understand that other people have thoughts. Instead, young children seem not to understand that mental states are causal agents that can influence behavior. They may have difficulty in regarding the mind as an "interpreting, executing, mediating entity" that *causes* behavior in the external world (Wellman 1988:88; Leslie 1988; Perner 1988; Wimmer, Hogrefe, and Sodian 1988).

The studies of young children's theories of mind suggest that attribution can be analyzed roughly at three levels, each demanding greater awareness of the distinction between one's own beliefs and the beliefs of others. At the most basic level, an individual imputes mental states to others but nonetheless fails to recognize that others' mental states can differ from his own. At an intermediate level, an individual recognizes that others' mental states can be false but can only describe these mistakes in terms of ignorance or the absence of information. As a result, behavior is somewhat inconsistent: the individual can specify what it is that another individual does not know, but he cannot describe what the other individual's false belief actually *is*. This is the simplest level at which Grice's nonnatural language operates, focusing as it does on the intent to inform. As we discussed in chapter 5, it is far from clear whether any nonhuman primates ever communicate with the intent to inform in the sense that they recognize that they have information that others do not possess. Finally, at the most complex level, an individual recognizes that others have mental states, appreciates that these mental states may be different from his own, and is able to specify what these mental states are. At this level of attribution, the individual communicates not just to inform but also to persuade others to give up a view that he himself believes to be false.

No animal studies have distinguished explicitly among these alternatives, largely because there is as yet little evidence of any higher-order intentionality among nonhuman species. The data we discuss below argue that apes, and perhaps also monkeys, do attribute states of mind to others, but that they do not always recognize the discrepancy between their own minds and the minds of other individuals.

Tests of Attribution in Captive Chimpanzees

In an experiment designed explicitly to test whether or not chimpanzees impute mental states to others, Premack and Woodruff (1978) showed Sarah, an adult chimpanzee, videotapes of human trainers struggling to solve a variety of problems. The trainer, for example, was shown trying to operate a record player whose cord was not plugged into the wall socket. After each videotape, Sarah was given several photographs, one of which depicted the correct solution to the problem. In the most subtle tests, the various solutions to the record player problem might include a cord that was plugged into the wall, a cord that was not plugged in, and a cord that had been cut. Sarah consistently chose the correct photograph. Interestingly, when tested with videotapes of a favorite and less favored trainer, Sarah chose correct solutions for the favorite trainer but incorrect ones for the trainer she did not like.

Premack and Woodruff interpreted Sarah's choice of the correct alternatives as evidence that she recognized the videotape as representing a problem and inferred purpose to the human trainer. The fact that Sarah picked out incorrect solutions for the unfavored trainer suggests that she was not simply choosing actions that would constitute the best solution for herself. Premack and Woodruff argued that even if Sarah's choices *had* simply reflected appropriate solutions for herself, some form of attribution would have to be invoked, since it is hard to imagine how she could recognize the correct solution without interpreting what the human was trying to do.

Although Sarah appeared to have little difficulty imputing purpose to others, she seemed less adept at attributing false beliefs. In a later experiment (Premack 1988), Sarah was taught to push a button that controlled the lock to a cabinet mounted on a wall just outside her cage. One side of the cabinet contained pastries that a favorite trainer would share with Sarah when she released the lock. The other side of the cabinet contained unambiguously repulsive items like rotting rubber and feces. Sarah quickly learned to push the button to release the lock on the "good" cabinet, and she did so with very little delay whenever the trainer visited her. Then, one day a "villain" wearing a mask entered Sarah's room and, while Sarah watched, removed all the delicacies from the good cabinet and replaced them with the repulsive items from the bad cabinet. Sarah responded aggressively to this villain and hurled things at him from her cage. Yet when the good trainer entered the room a few minutes later, Sarah showed no change from her normal behavior. She never made any obvious attempt to warn the trainer or even to hesitate before pushing the button. Sarah acted as if she did not recognize the discrepancy between her own knowledge and the trainer's knowledge. Although a number of other interpretations of this experiment are possible, the results at least suggest an inability to attribute false beliefs to others. They also to some extent contradict Premack's earlier experiments, in which Sarah learned to point to the wrong container when confronted with the "bad" trainer (Woodruff and Premack 1979). Recall, however, that these false gestures only appeared after a considerable period of time when the chimpanzee simply withheld information from the trainer (Dennett 1988; Premack 1988). As we will see below, chimpanzees often show a similar lack of differentiation between their own and others' states of minds in their interactions with each other.

Individuals who cooperate to solve a problem can also be said to impute beliefs and intent to others, if only because they must recognize each other's aims and purposes in order to arrive at a common goal. Savage-Rumbaugh, Rumbaugh, and Boysen (1978) trained two chimpanzees, Sherman and Austin, to cooperate in the acquisition of food. Before the experiment began, both chimpanzees had learned to label a variety of items (including food and tools) using lexigrams on an illuminated keyboard. The chimps were then placed in different rooms connected by an open window, and one chimp saw a site baited with food. He could also see which of five or six tools was needed to reach the food. Using a lexigram, he then "requested" the appropriate tool from his partner, who had access to the tools but could not see the food. Sherman and Austin quickly became proficient in acquiring food, and both performed equally well in either role. Apparently, each chimpanzee had learned not only what he himself needed to do but also what the other animal needed to do in order to solve the problem.

Even in the absence of any human training, chimpanzees occasionally cooperate to solve an apparently mutually recognized problem. Free-ranging chimpanzees, for example, frequently help each other into trees (Goodall 1986). Similarly, in both Menzel's (1973) and de Waal's (1982) studies of socially living captive chimpanzees, males cooperated to build and hold ladders for each other in order to climb forbidden trees or to escape from the compound. In each case, the animals acted as if they recognized each other's motives and purposes, and they worked together to achieve a common goal. De Waal (1989) also describes a spontaneous game played by captive bonobos at the San Diego Zoo in which individuals would deliberately strand each other in a dry moat by pulling up a chain rope that led down into the moat. Other animals would then "rescue' their companions by dropping the chain rope back down to them. The game suggested an ability to recognize both the purpose of the chain and the needs of the trapped animals.

Although direct comparisons of the behavior of monkeys and apes on logically similar problems are difficult both to find and to interpret, another study, conducted more than 15 years before that of Savage-Rumbaugh and her colleagues, offers some intriguing hints of an ape-monkey difference in the ability of animals to understand other individuals' roles and knowledge. In 1962, Mason and Hollis trained pairs of rhesus macaques to cooperate with one another in obtaining food. In each test, two individuals (usually cage mates) sat facing each other. Food was then hidden behind one of several boxes that could be opened by pulling a handle. One monkey, designated the *informant,* knew where the food was located but had no access to the handles. The other, designated the *operator,* had access to the handles but did not know the food's location. The informant had to indicate to the operator which handle to pull by altering his behavior in some way. Once the animals had become familiar with the apparatus, they were able to communicate with each other to obtain food. When the roles were reversed, however, performance returned to chance levels. Apparently, each animal had learned what was required in his own specific role, but neither had learned what the other animal needed to know. Unlike the chimpanzees, the macaques gave no indication of recognizing their partner's role in the task.

There are, of course, many ways in which we might interpret this difference in the performance of macaques and chimpanzees. Chimpanzees, for example, could simply be better observers than macaques and be better at noting the steps necessary to acquire food. This explanation implies that chimpanzees differ from macaques in their ability to observe crucial behavioral contingencies. It is also possible that chimpanzees are better able to understand the problem at hand and the steps necessary to solve it, even if they themselves are not active participants in all stages of the task. As we will see below, numerous other experiments and observations demonstrate that chimpanzees can easily learn to solve tasks that require a number of discrete steps for their solution. Finally, unlike the macaques, the chimpanzees might have realized that their partner's behavior was the result of certain knowledge and motives, and they might have taken note of this knowledge even though it was not relevant to their own behavior at the time. This explanation implies a qualitative difference between chimpanzees and macaques in the ability to impute purposes and goals to their partners.

Deception as a Measure of Attribution

In his review of the lessons of the ape language projects, Premack (1986) concluded that chimpanzees, like young children, do show some ability to attribute beliefs to others, but only if those beliefs are the same as their own. In other words, chimpanzees seem incapable of recognizing the potential discrepancy between their own and other individuals' states of mind. Given the paucity of systematic research on attribution in nonhuman primates, it is impossible to refute this conclusion. Nevertheless, a number of anecdotal accounts of apparent acts of deception provide tantalizing, if untested, hints of higher-order intentionality in nonhuman primates.

Since humanlike deception requires that a signaler create or support a false belief in another (Rakowitz 1990), it potentially provides evidence for a theory of mind. This is not to say, of course, that all deceptive acts require a theory of mind. Indeed, most examples of signal falsification in animals can probably be explained without recourse to higher-order intentionality, at least partially because deceptive signals in most species are relatively inflexible and occur in only a narrow range of contexts. Hence, for our present purposes, it becomes crucial to distinguish between the function of deception and the mechanisms underlying it. A scorpionfly may deceive another male by mimicking the behavior of a female, but it is not only phylogenetic chauvinism that prevents us from concluding that he attributes beliefs to his rival. Scorpionflies, after all, apparently never attempt to deceive each other in any other context or with any other pattern of behavior. What revisions to our thinking would be required, however, if we were to find that scorpionflies also occasionally give false alarm signals to drive their rivals from females? And what would we conclude if we learned that a particular individual scorpionfly, having deceived a rival with a false alarm call, could no longer successfully deceive him by mimicking the courtship behavior of a female?

Recently, Byrne and Whiten (Byrne and Whiten 1988c; Whiten and Byrne 1988c) devised an index for considering apparent acts of tactical deception in nonhuman primates. Their analysis suggests that nonhuman primates sometimes falsify signals in ways that are consistent with higher-order intentionality. Monkeys and apes appear capable of falsifying or concealing vocal signals, facial expressions, and body gestures in contexts as diverse as courtship, aggressive interactions, and feeding competition (chapter 7). Many of these examples are difficult to interpret except by assuming some attribution of thoughts and beliefs to others.

At the same time, however, careful scrutiny of the animals' behavior just prior or subsequent to their apparently deceptive acts often reveals curious gaps and lapses in their theories of mind, suggesting that they do not distinguish clearly between their own knowledge and the knowledge of others. To cite an example that strikes close to home, Dennett (1987) repeats an anecdote of ours in which two vervet groups were involved in an aggressive skirmish on the boundary of their territories. A male member of the losing group suddenly ran up into a nearby tree and began giving leopard alarm calls, thus causing all participants to race for the trees. This effectively ended the skirmish, restored the boundary to its former position, and avoided a defeat for the male's group. These "false" alarm calls, moreover, did not occur in an unreasonable context. On three other occasions we observed a real carnivore attempt an attack on vervets during an aggressive intergroup encounter, when the monkeys' attention was entirely absorbed in their interactions with each other.

During our study we observed a handful of similar instances, far outweighed by the frequency of "true" or reliable alarm calls, in which a male gave apparently false leopard alarm calls when a new male was attempting to transfer into his group (fig. 8.1). During our two 8-month study periods in 1983 and 1985–86, for example, males gave leopard alarm calls in four (2%) of the 264 intergroup encounters we observed; these "false" alarm calls occurred on 13% of the 32 days on which at least one male gave a leopard alarm call. False alarm calls did not appear to function as threat displays, because they were not accompanied by any other threat gestures or vocalizations, nor were they ever given during aggressive interactions involving resident adult males. The perpetrator of all but one of these false alarms was a perpetually low-ranking male, Kitui, who could reasonably have expected to become subordinate to the interloper if he transferred successfully. In years past Kitui had been the second ranking of two males, third ranking of three males, and so on throughout his ignoble career. Kitui's false alarms were effective, because they caused the new male to remain in his tree and delayed his approach toward the group.

So far, so good. The alarm calls appeared deceitful because they signaled danger that Kitui, but not the interloper, knew to be false, and they kept the interloper temporarily at bay. Kitui acted as if he knew that leopard alarm calls caused others to run to safety and as if he wanted the new male to believe that he, Kitui, had seen a leopard. It is, of course, possible to explain Kitui's behavior without reference to any sort of theory of mind; Kitui might simply have learned to associate alarm calls and other animals' flight.

Figure 8.1. Adult male vervets sometimes give apparently "false" alarm calls when a migrant male is attempting to transfer into their group. Photograph by Marc Hauser.

Nevertheless, we are tempted to describe this behavior as a rare example of higher-order intentionality in primates.

But did Kitui really attribute a false belief to the interloper? At this point, we should probably move on to another anecdote, because Kitui's subsequent behavior was often rather deflating. As if to convince his rival of the full import of his calls, Kitui twice left his own tree, walked across the open plain, and entered a tree adjacent to the interloper's, alarm calling all the while. Kitui acted as if he had got only half the story right; he knew that his alarm calls caused others to believe that there was a leopard nearby, but he did not seem to realize that other aspects of his behavior should be consistent with his calls. To leave his own tree and to walk toward the other male simply betrayed his own lack of belief in the leopard.

Many similar examples abound for the imperfectly rational (that is, deceitful) minds of children. A child of 3 denies having been to the cookie jar when there are crumbs on his face; only in older children is deception sufficiently refined that the telltale crumbs are removed. Just as Kitui's theory of mind is incomplete, so is the child not fully able to discriminate between his own knowledge and the knowledge of others.

Other anecdotes are equally provocative and equally insubstantial. Subordinate female baboons frequently raise their tails when approaching or interacting with more dominant individuals. Raised tails seem to reflect anxiety and are often accompanied by other signs of subordination, such as fear grimaces or presenting of the hindquarters. Frequently, the dominant animal reaches out to touch the subordinate one (Smuts 1985). In an earlier study of baboons, we observed an apparent attempt by a female to suppress this sign of subordination from the most dominant male in her group. The female, the Lady from Philadelphia, was negotiating a path on a narrow rock ledge that led directly past the resting place of the male, Rocky. Ordinarily, the Lady from Philadelphia would not have approached Rocky that closely except to interact with him. In this case, though, she was following the rest of the group on a rapid move, and was running to catch up with her daughter. As she approached Rocky on the ledge, her tail began to rise. Looking back at her tail, she pressed it down, holding it down until she had passed him. She acted as though she recognized that her raised tail was a sign of anxiety that she preferred to conceal from Rocky because, at that moment, she did not wish to interact with him.

It is impossible to know what to make of this anecdote in the absence of other, more systematic evidence that baboons attempt to hide their emotions and anxiety from others. A parsimonious explanation could easily argue, for example, that the Lady from Philadelphia's tail began to itch just as she was passing Rocky. Alternatively, she might simply have learned that a raised tail required a prolonged interaction with Rocky that at the moment she wished to avoid. But how could she monitor the behavior of her own body without, in some sense, being aware of its actions? How could she recognize the effects of a raised tail on her interactions with Rocky without some understanding of this signal's function? The anecdote certainly implies that the Lady from Philadelphia recognized the distinction between her actual state of mind and the state of mind she wished to convey to Rocky.

Even if her actions represent a true deceptive tactic, however, the Lady from Philadelphia's behavior remains puzzlingly incomplete. Why, for example, did she push down her tail in Rocky's full view, where her actions were less likely to have fooled him? In this respect, the Lady from Philadelphia's theory of mind seems as poorly developed as Kitui's. On the other hand, both individuals succeeded: Kitui kept his rival at bay and the Lady from Philadelphia avoided interacting with Rocky. Perhaps, if other individuals' theory of mind is as incomplete as your own, it really doesn't matter if the story is only half right.

Once again, we are left with observations that can be interpreted at a number of different levels. In addition to hinting at higher-order intentionality, each of the examples we have described can easily be explained in terms of acutely observed behavioral contingencies derived from past experience: if I do X, he will do Y. Kitui could simply have learned that a leopard alarm caused other animals to run away, and the Lady from Philadelphia could simply have learned that a lowered tail allowed her more easily to pass by Rocky. In each case, attribution might have occurred only in the mind of the human beholder.

Given the current lack of experimental evidence, even the most compelling observations are only consistent with higher-order intentionality and cannot be used to confirm it. In a study of captive stumptailed macaques, de Waal (1989) observed several occasions when females who were copulating with subordinate males attempted to suppress their partners' copulation grunts before they attracted the attention of the dominant male. Most of these attempts took the form of threats, but on one occasion a female placed her hand over her partner's mouth. Especially in the latter case, an explanation based on higher-order intentionality is tempting and perhaps even appropriate. Nevertheless, it remains possible that the females had learned to associate copulation grunts with aggression from the dominant male, and that they acted on the basis of this contingency. Although such reasoning is certainly complex, it does not demonstrate that monkeys are skilled at changing other animals' beliefs—only that they are skilled at changing other animals' behavior. We still lack examples that cannot be explained *except* in terms of a theory of mind.

Examples of deception in chimpanzees are replete with similar gaps and lapses, but they also offer more convincing hints of a true theory of mind.

Recall, for example, de Waal's (1982) accounts of the courting male who covered his penis when approached by a more dominant male or of the male who rearranged his facial expression before confronting a rival (chapter 7). Each of these anecdotes suggests, though obviously does not prove, that chimpanzees recognize the effect of their own actions on the knowledge of others. In the first example, it might be possible to argue that the chimpanzee had learned to associate the effects of a particular stimulus (in this case, his overly alert genitalia) on the aggressive behavior of others; by concealing this stimulus, he attempted to avoid aggression. In the second example, too, the chimpanzee might simply have learned a behavioral contingency: that a particular facial expression caused aggression in others. Nevertheless, it seems at least possible that the masking and manipulation of a facial expression requires some ability to recognize the effects of one's own anxiety on others and to distinguish between other animals' current and desired states of minds. That is, Luit acted as if he recognized not just that his fear grin might influence Nikki's behavior, but that a rearrangement of his facial expression might have a completely different effect on Nikki's mind. A similar interpretation is possible for the example of suppressed calls and noise on chimpanzee border patrols (see chapter 5).

Similarly, when a dominant male chimpanzee (or even a rhesus macaque) refrains from attacking a subordinate who pretends not to see him (chapter 7), does he attribute ignorance to the subordinate? Here too, the most parsimonious explanation is not terribly convincing. It is unlikely, for example, that the dominant male's attack is triggered only by the sight of the subordinate male's facial expression, because monkeys do occasionally bite or attack unsuspecting animals. Instead, it seems possible that dominant males monitor whether subordinate males have had the opportunity to observe and respond to their approaches. This second interpretation suggests that dominant males attribute knowledge about their behavior to others and expect subordinate animals to retreat only when it appears that this knowledge is available to them.

Some forms of spite also seem to require a theory of mind. Could vindictive behavior be possible without attributing purpose to others? Once again, we are restricted to a few intriguing anecdotes. In *The Mentality of Apes*, Wolfgang Kohler (1925) reports that many chimpanzees in his captive colony on the Canary Islands learned to stack boxes to obtain bananas that were otherwise out of reach above them. A few dominant chimpanzees, however, routinely supplanted subordinate individuals from their towers just after they had been built, thereby gaining access to the bananas without having to exert any effort themselves. While retreating from approaching dominants, subordinate animals often seemed to destroy their towers deliberately by sliding roughly off them. Both parties apparently regarded this destruction as willful, since the subordinate animal would then flee, pursued by the outraged dominant.

Drawing on their analysis of tactical deception, Byrne and Whiten (1988c, 1990) conclude that chimpanzees, but not monkeys or perhaps even other apes, show positive evidence for higher-order intentionality. We agree that the examples are certainly consistent with this interpretation, but we also note that other aspects of chimpanzee behavior suggest the opposite conclusion. As we discuss below, chimpanzees fail to behave as higher-order intentional systems in a number of contexts when such behavior might strongly be predicted. For example, chimpanzees and other apes seem not to instruct those who are ignorant, and they fail to show empathy for those who are grieving. Perhaps, like small children, chimpanzees can identify the belief that they want another individual to have and also recognize that their behavior can influence that belief. Having created the desired state of mind in another, however, chimpanzees may be unable to recognize that the other individual now has a belief that is incompatible with their own. They may be unable to accommodate simultaneously both a true and a false belief.

Informing as a Measure of Attribution

An individual who deliberately informs or teaches someone else must attribute knowledge (or, in this case, ignorance) to others; he conveys information that he knows others do not have. As we have mentioned, children may find it easier to attribute ignorance to others than to attribute false beliefs. The former demands only an understanding that others do not recognize the true state of affairs; the latter requires the individual to represent and keep separate two incompatible states of mind (Hogrefe, Wimmer, and Perner 1986). If the attribution of ignorance does indeed place fewer cognitive demands on individuals, we might expect to find more evidence in nonhuman species for informing and teaching than for signal falsification and deception. Surprisingly, however, this is not the case.

As we discussed in chapter 5, numerous species of birds and mammals give more alarm calls in the presence of kin than in their absence. This "audience effect" suggests that animals are acutely sensitive to social context. What is not clear, however, is if they are also sensitive to whether or not their audience is ignorant or already aware of the predator. Any call that is given to some individuals and withheld from others certainly *functions* to provide or conceal information. It is extremely difficult, however, to demonstrate that a signaler *intends* to communicate information, much less that he recognizes a discrepancy between his own knowledge and the knowledge of others.

A reconsideration of alarm calls reveals somewhat contradictory evidence for the ability to attribute ignorance to others. On the one hand, as we have seen, alarm calls in many species are not obligatory but depend on social context. Individuals do not give alarm calls when it would make no sense for them to do so (for instance, when they are alone), and they alarm call less when their kin are absent and when there is no functional advantage to be gained by alerting others (chapter 5).

On the other hand, when animals do give alarm calls, they seem to call regardless of whether their audience is already aware of danger. Vervet monkeys, for example, will continue to give alarm calls long after everyone in the group has seen the predator (Cheney and Seyfarth 1981, 1985b). In some cases, when the monkeys have seen a snake, for example, continuous calling may serve a mobbing or protective function. Vervet monkeys, however, give prolonged alarm calls even to species such as leopards, from which they simply flee. To confuse the issue even further, we also have evidence that vervets give most alarm calls to those species to which they *themselves* are vulnerable and not necessarily to those predators to which their *audience* is vulnerable (Cheney and Seyfarth 1981). For example, because of their size infant vervets are far more vulnerable than adult vervets to attack by baboons. Nevertheless, adults alarm call at far lower rates in response to baboons than in response to leopards, even when calling would carry no cost to themselves and presumably offer great potential benefit to their offspring. All of these observations suggest that signalers cannot differentiate easily between their own and their audience's knowledge or vulnerability.

It might be argued that there is no selective advantage in being able to inform or to communicate in Grice's (1957) sense (see chapter 5). There may simply be no *need* for vervet monkeys to recognize whether their audience is ignorant or knowledgeable before uttering an alarm call; as long as the call functions to inform others of danger, the audience's state of mind is irrelevant. In at least some cases, however, animals who give alarm calls put themselves at greater risk than those who remain silent, because their alarm calls attract the attention of predators (see e.g., Sherman 1977, 1985 for ground squirrels). Under these conditions, an individual would clearly be at an advantage if he could determine whether an alarm was necessary before uttering the call.

The same could be said for food calls. It is easy to imagine situations in which it might be advantageous to inform only some of the individuals within calling range of the presence of food. An individual who came upon a fruiting tree that could only be exploited by a small number of animals, for example, would do better to alert kin than nonkin and to refrain from giving food calls if his relatives were already aware that the tree had fruit. One experiment conducted in captivity has suggested that chimpanzees recognize that an individual who has witnessed an event possesses different knowledge than one who has not (Povinelli et al. 1990). It is therefore possible that chimpanzees are capable of distinguishing between ignorant and knowledgeable audiences. Under natural conditions, however, it is unclear whether chimpanzees ever alter their calling behavior depending upon their audience's state of mind. Although chimpanzees' loud calls appear to convey quite precise information about the presence of food (chapter 5), calling rate seems to be influenced primarily by the relative abundance of the resource rather than by what the audience knows (Wrangham 1975; Hauser and Wrangham 1987). Indeed, chimpanzee loud calls may not even *function* to alert others of food, since it seems quite likely that most community members already know which trees in their range contain fruit (Wrangham 1977). Perhaps one way to tackle this issue would be to investigate whether chimpanzees, or any other species, fail to call when there could be no doubt that their audience already knows about the availability of food. Similarly, it might be possible to test whether listeners had an *expectation* of being informed (see chapter 7). Any evidence that animals in some way punished knowledgeable individuals who failed to inform them of food or danger would be indicative of intent and also of attribution.

To examine the issue of informing in more detail, we designed an experiment that investigated the extent to which macaques take into account their audience's state of mind when uttering alarm or food calls. In captivity, Japanese and rhesus macaques often give alarm calls when they spot technicians with nets, and they also utter a *coo*like food call when they are fed preferred foods like fruit (pers. obs.; Green 1975; Owren, in prep.). Our experiments tested whether mothers would give more food or alarm calls when their offspring were ignorant about the presence of food or danger than when they were not.

The trials were conducted on two groups of rhesus macaques and two groups of Japanese macaques, housed at the California Primate Research Center in Davis, California (Cheney and Seyfarth 1990). Each group's enclosure was divided into two outdoor arenas, connected by a 5-m-long chute that could be used to capture and temporarily separate one or more animals from the rest of the group. It was possible, therefore, to provide one individual with information that only she possessed and to test whether this "informant" ever selectively altered her behavior according to whether her audience was ignorant or knowledgeable.

We began each trial by locking all but two members of a given group in one of the arenas. The remaining animals, a mother and her juvenile offspring, were locked in the chute that connected the two arenas. In the "knowledgeable" condition, the mother and offspring were seated next to each other. Each could see the other and both could see the empty arena. In one set of trials, the animals then watched a human place a highly preferred food (apple slices) in a food bin in the empty arena. In another set of trials, the animals saw a "predator" in the form of a menacing technician, who brandished a net and then hid himself behind a partition next to the arena. After each of these exposures, the offspring, but not the mother, was released into the empty arena.

In the "ignorant" condition, mother and offspring were also locked in the chute, but the offspring was seated behind the mother and separated from her by a steel partition. Now, only the mother was able to observe the food or the predator. After the food had been placed in the bin or the technician had hidden himself, the offspring, but not the mother, was once again released into the arena.

If monkeys are sensitive to the mental states of others—that is, if they take their audience's knowledge into account when giving food or alarm calls—mothers should have uttered more calls (or in some other way altered their behavior) when their offspring were ignorant than when they were already informed. On the other hand, if informants are unaffected by their audience's mental states, the mothers' behavior should have been similar regardless of whether or not their offspring had also seen the food or danger.

In fact, the mothers' behavior was apparently unaffected by their offspring's knowledge. In the food trials, the seven mothers whose offspring were ignorant showed no difference in their behavior or calling rate compared with the seven mothers whose offspring were knowledgeable. Mothers and offspring did exchange vocalizations at low rates, but there was no difference in calling rate between the mothers of knowledgeable and ignorant offspring (Cheney and Seyfarth 1990).

The mothers' apparent failure to communicate information about food to their ignorant offspring had direct, functional consequences; the mean latency for finding and eating food was significantly shorter for knowledgeable offspring than for ignorant ones (fig. 8.2). The primary factor deter-

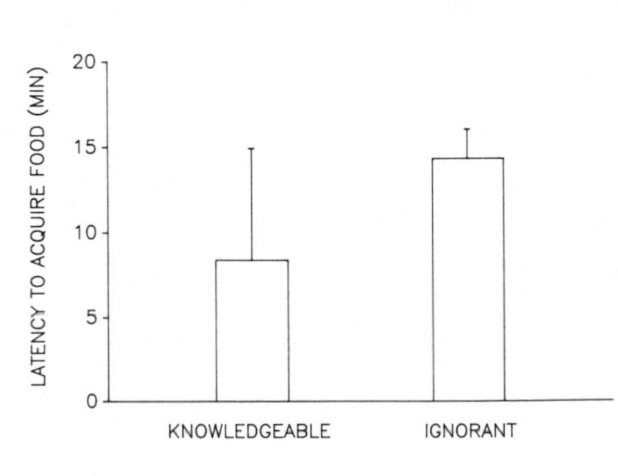

Figure 8.2. The latency, in number of minutes, with which knowledgeable and ignorant subjects acquired food. Histograms show means and standard deviations for seven knowledgeable and seven ignorant juveniles. Knowledgeable subjects acquired food significantly faster than ignorant ones (Mann-Whitney U test, $P < 0.05$).

mining whether an offspring acquired food, therefore, was the *offspring's* knowledge and not the mother's.

All mothers in the "predator" trials showed distress at the sight of the technician and struggled to free themselves from the chute; none, however, gave alarm calls. The lack of alarm calls clearly made it more difficult for us to measure any transmission of information from informant to audience. Nevertheless, several measures suggested that mothers did not behave differently when their offspring were ignorant than when they were knowledgeable. For example, mothers of ignorant offspring did not orient themselves toward or look at their offspring more than mothers of knowledgeable offspring.

Despite the similarity in the mothers' behavior under the two conditions, however, the behavior of ignorant and knowledgeable offspring differed significantly. Upon seeing the technician, knowledgeable offspring showed distress by crouching and sitting next to their mothers at the entrance of the chute. After being released, knowledgeable offspring spent significantly more time than ignorant ones sitting within arm's reach of their mothers (fig. 8.3). Paralleling results in the food trials, therefore, the determining factor in the amount of anxiety shown by offspring was their *own* knowledge and not their mothers'.

In summary, therefore, mothers made no special attempt to inform their ignorant offspring of food or danger. Although it is possible that mothers gave their offspring some subtle behavioral cues that were not apparent to us, the juveniles' behavior suggested that any such cues were not also apparent to them. They obtained food primarily when they had seen the food themselves, and they showed anxiety primarily when they had seen the technician themselves.

Might the mothers have failed to alert their offspring because they did not know that their offspring were ignorant? Although we cannot answer this question definitively, the mothers certainly had ample *opportunity* to determine whether their offspring were ignorant. Under the ignorant con-

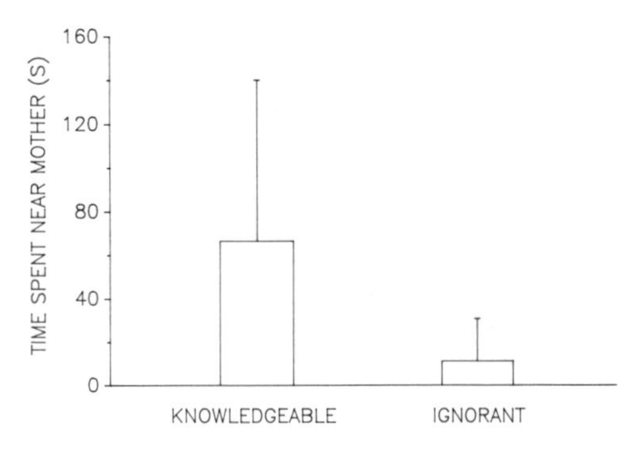

Figure 8.3. The number of seconds spent by knowledgeable and ignorant subjects within arm's reach of their mothers. (See Figure 8.2.) Knowledgeable subjects spent significantly more time near their mothers than ignorant ones (Mann-Whitney U test, $P < 0.05$).

dition, no other animals were in sight when the mothers were exposed to the food or the frightening technician, and their offspring were released from behind an opaque barrier. Most important, the offspring's own behavior was markedly different when they were ignorant than when they were knowledgeable. The question may, however, be moot. If the mothers were incapable of attributing ignorance to others, they could not, by definition, infer whether or not their offspring had seen the food or the predator.

Clearly, these negative results do not allow us to distinguish between the inability to attribute states of mind to others and the failure to make use of such an ability. It is certainly possible that monkeys *do* recognize the difference between their own knowledge and the knowledge of others, but that this recognition simply has no effect on their behavior. Whenever another species' knowledge is defined operationally, through behavior, there is a danger of concluding that an ability is absent when it is simply not manifested. Even if nonhuman primates are capable of distinguishing ignorance and false beliefs in others, however, their apparent failure to act on this knowledge is striking.

Teaching as a Measure of Attribution

Even more than informing, teaching would seem to demand some ability to attribute states of mind to others, at least in the form of ignorance, because it demands that the instructor recognize the distinction between his own knowledge and the knowledge of his pupils. Knowledge and novel skills can, of course, be acquired and transmitted without active pedagogy. In all human societies individuals learn a huge amount through observational learning and imitation, including especially many aspects of human language. Nevertheless, while much can be learned through observation alone, it is also the case that active pedagogy allows for much greater flexibility in the transmission of information. In particular, it permits individuals to inform each other about objects, events, skills, and ideas that are removed in time and space.

As is true also of deception, however, we cannot simply take evidence of pedagogy as evidence of a theory of mind. While an individual who possesses a theory of mind might be able to teach, it does not necessarily follow that all individuals who teach have a theory of mind. Take, for example, the way in which female domestic cats teach their kittens to hunt. No one who has ever owned a cat can easily escape the impression that cats actively instruct their kittens in the art of hunting. The mother brings partially crippled prey to her kittens and then gives them the opportunity to pounce on and bat at it. If the unfortunate mouse or bird manages to run a short distance away, the mother will retrieve the animal and return it to her kittens. Only after her kittens have played with the prey for an extended period of time will the mother finally kill and consume it. The impression of active pedagogy is supported by observations indicating that mother cats adjust their own behavior to accommodate their kittens' hunting skills, using more deliberate, simple movements for naive kittens than for more experienced ones (T. Caro, pers. comm.).

Like deception by scorpionflies and great tits, however, instruction by mother cats seems to be relatively stereotypic. Cats do not appear to teach their kittens in any contexts other than hunting and, across cats, there is little evidence of deliberate instruction. The mechanisms underlying the mother cat's teaching of her kittens seem fundamentally different from the mechanisms underlying human pedagogy, which occurs in a variety of different contexts and is highly modifiable depending on the specific deficits of the pupil.

What is the evidence for pedagogy in nonhuman primates? Once again, the anecdotes are provocative and potentially subject to both over- and underinterpretation. For example, Kohler (1925) reports an incident in which the male chimpanzee Sultan was forced to watch a number of other chimpanzees attempt to stack boxes to obtain bananas that had been placed out of reach. This was a trick that Sultan had long ago mastered on his own. Sultan watched with increasing agitation as the other, less insightful chimpanzees failed to solve the problem. Finally he ran into the room, stacked the boxes, and quickly ran out of the room again without attempting to obtain the bananas for himself. Kohler did not interpret Sultan's behavior as altruistic, because the chimpanzees in his colony were more likely to obstruct each others' efforts than to support them. Rather, Sultan seemed motivated primarily by the urge to do *something* to solve the problem that baffled the other chimpanzees. Kohler concluded that Sultan could view the problem from the standpoint of the other animals.

Perhaps even more provocative, Fouts, Hirsch, and Fouts (1982) on three occasions observed the language-trained chimpanzee Washoe mold her adopted son's hands into the appropriate sign for an item. By the age of 4, her son had acquired 39 signs. While the majority of his signs seemed to have been learned through imitation rather than active tutelage, direct instruction from Washoe may well have played an important role in his development (Fouts, Fouts, and Schoenfeld 1984).

Given these suggestive examples from captive chimpanzees, it is sobering to discover that the most common approximation to teaching in free-ranging monkeys and apes occurs in the form of punishment for some social transgression. Mothers aggressively interfere in rough play between their offspring and other juveniles, retrieve their infants from females who are handling them roughly, push infants from their nipples during the weaning period, and so on. These corrective actions, however, occur primarily as threats, and they seem to derive less from pedagogical intent than from an attempt by others to remedy what is aversive or unpleasant to *themselves*. There is no evidence that animals who threaten or restrain others from behaving in a particular way recognize any discrepancy between their own knowledge and the knowledge of their pupils. To the extent that they correct others, they seem to be attempting to make others behave in a way that is beneficial to themselves.

Outside the social domain, examples of teaching are even rarer and are limited to a handful of instances in which animals attempted to prevent others from approaching a novel item that they themselves were avoiding. So, for example, macaque mothers have been observed to pull their offspring away from a strange object presented to them by a human (Kawamura 1959; Menzel 1966; see also Fletemeyer 1978).

Free-ranging monkeys and apes are surrounded by dangerous predators and poisonous foods, and infants rapidly learn which foods and predators to avoid. Even in the most well-documented cases, however, active instruction by adults seems to be absent. Recall, for example, that infant vervet monkeys often make "mistakes," giving alarm calls to species that pose no danger to them (chapter 4). Adults nonetheless respond to infant alarm calls, albeit in some cases quite briefly. If, for example, an infant gives an eagle alarm in response to a stork, adults will look up and then quickly go back to what they were doing. By contrast, if an infant is the first member of his group to give an eagle alarm in response to a martial eagle, adults will look up and then give alarm calls themselves. At first glance these "second alarms" by adults seem to be instructive, since they reinforce the infant's correct behavior. Adults, however, are no more likely to give second alarms after correct alarms by infants than they are after correct alarms by adults. Even though infants make many more errors than adults, adults make no special effort to reward them when they are correct. We would expect such special efforts if adults attributed ignorance to infants but not to adults.

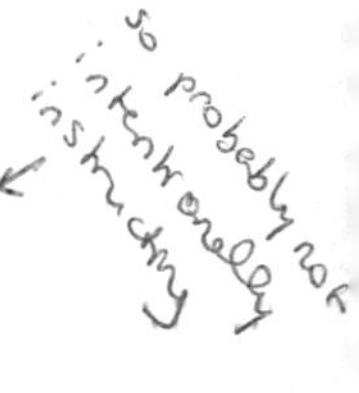

A similar picture emerges when we consider infants' responses to alarm calls. Here again, young infants make many mistakes. In some cases they even increase their vulnerability to predation by, for example, looking up when they hear a snake alarm or running into a bush when they hear a leopard alarm (chapter 4). In analyzing films of behavior by infants and mothers, we looked carefully to see whether an infant's behavior influenced what his mother did—whether, in this respect, mothers ever corrected their infants' errors. We found no such evidence.

Further investigation of maternal responses clouds the issue even more. Female vervets will often leave their infants behind in trees when the group moves off at the approach of baboons, or when it moves on to another feeding grove. A mother will watch impassively as her infant screams and struggles to descend a tree, often falling and even injuring himself. Does this mean that vervets are totally incapable of inferring need in others or of recognizing the difference between their own capabilities and their infants'? Not necessarily, because the same female who abandons her infant in a tree will race to retrieve her infant and carry him to safety at the sound of an alarm call. Furthermore, although the cries of an abandoned infant might fall on his mother's deaf ears, they usually mobilize a phalanx of older sisters to climb the tree and carry the infant to safety. These observations offer frustrating, inconsistent evidence that vervets might, after all, recognize that infants have special needs and a knowledge of danger that is different from that of adults.

A similarly inconclusive picture emerges from anecdotes that describe food avoidance. Monkey mothers are occasionally reported to knock poisonous fruits from their infants' hands, but the humans who observe such events are often uncertain about what prompted the mothers' actions. Did she realize the fruit was poisonous or was she just reacting to the item's novelty? Could she simply have bumped against her infant by chance? What is most striking about these anecdotes is how rarely they are reported.

Once again, chimpanzees may be different. Nishida (1983, 1987) reports that mothers and other group members occasionally take food from infants if the food is one that is not normally part of their diet (see also Goodall 1973). Nishida emphasizes, however, that the primary way that infants seem to learn about which foods to eat is by picking up scraps that the mother has dropped or by taking food directly from her mouth. In no study have chimpanzees been reported to *teach* each other new skills. Learning is passive and involves little active intervention.

It is surprising that no monkey or ape appears to have a call that conveys a negative command. Many nonhuman primate vocalizations, as we have discussed, seem to have rudimentary referential properties that denote objects or events in the external world. On the face of it, it would seem as easy to say "No!" as to say "Leopard!" There would also seem to be great selective advantage to the ability to shout "No!" loudly at an infant about to put poisonous fruit in his mouth or at a rival male approaching an estrous female. In fact, many of the methodological difficulties surrounding investigations of primate intelligence would be considerably facilitated if we could ask animals about their behavior and elicit simple yes or no responses. The issue speaks directly to the problems faced by Quine's (1960) imaginary linguist, whose translation task would be made impossible without yes and no responses.

The closest approximations to negative commands in monkeys and apes are threat vocalizations, which are sometimes given with the apparent intent of preventing an animal from doing something. Vervet males, for example, will often give threat-grunts at rivals who attempt to copulate with an estrous female, and females will give similar calls when interfering in disputes on behalf of their offspring. These calls, however, are usually accompanied by other threatening gestures or facial expressions, and they do not occur in nonaggressive contexts. To give another example, silverback male gorillas sometimes utter threatening cough-grunts to juveniles who play too boisterously near them or who come too close to the human observer (A. Harcourt, pers. comm.). But do these cough-grunts serve as negative commands or as threats? Even if vocalizations do sometimes serve to warn and to admonish, it is puzzling that they occur only in the form of threat vocalizations and not in the form of a vocalization with a more specific function. Moreover, the apparent lack of *positive* commands remains difficult to explain.

Perhaps the problem with commands like *yes* and *no* is that they demand some ability to separate one's own beliefs and purposes from the beliefs and purposes of others. A silverback gorilla who threatens a juvenile from a potentially dangerous human might attribute purpose to the juvenile but still be unable to distinguish clearly between what is potentially dangerous or inadvisable to the juvenile and what is dangerous or inadvisable to himself. If, for example, the human was only dangerous to the silverback, but not to the juvenile, would the silverback still attempt to prevent the juvenile from approaching too closely? A human father can admonish his daughter about the danger of matches even as he adds logs to the fire, because he can distinguish his own knowledge and experience from his daughter's. We do not know if the same is true of apes. It is possible, therefore, that the apparent lack of *yes* and *no* commands in nonhuman primates may in itself suggest only limited powers of attribution. Threat vocalizations may be given with the intent of modifying behavior, but there is little indication that they are given with the intent of modifying states of mind.

Imitation as a Measure of Attribution

In contrast to pedagogy and even informing, imitation would seem to require only a very limited theory of mind, since it demands simply that an individual be able to act with foresight to copy the form and function of a model (Visalberghi and Fragaszy, in press). A close examination of the process by which monkeys acquire novel tool-using skills, however, reveals little evidence that they actively imitate others or recognize the goals and purposes of other animals' behavior. Although monkeys sometimes acquire novel patterns of behavior from other group members, imitation usually seems to play only a minor role in the learning process (see Westergaard and Fragaszy 1987; Galef 1988; Visalberghi and Fragaszy 1990; and Chevalier-Skolnikoff 1989 for reviews and discussion).

More than any other monkey species, capuchins are renowned for using sticks as rudimentary "tools" (see chapter 9 for a discussion). Placed in the same room with food that is out of reach and a number of sticks that could be used for poking or prodding, most capuchins readily learn to use the sticks to acquire the food (see e.g., Visalberghi and Trinca 1989). Monkeys are attracted to the tools and often begin handling and manipulating the tools after observing other monkeys do so, suggesting that social companions enhance and facilitate tool use. Several factors, however, argue against purposeful copying by imitation. Individuals typically require extensive practice through trial and error before they acquire the skill; as a result, different individuals adopt different idiosyncratic styles, and dissemination of the skill is usually very slow. Furthermore, some individuals never even attempt the novel behavior.

The same seems to be true of other monkey species. Even in the famous case of potato washing by Japanese macaques, only 11 out of 25 monkeys adopted the washing practice over a 3-year period (Nishida 1987). Moreover, those who did so used a variety of different washing techniques, suggesting that the skill was acquired independently by each monkey (Galef 1988; Visalberghi and Fragaszy in press). Evaluating similar behavior among baboons and other species of macaques, Beck (1972, 1973) and Galef (1988) have argued that most novel skills and "cultural traditions" among monkeys are probably based on a combination of trial-and-error learning and social enhancement rather than purposeful imitation.

To date, the only strong suggestion of purposeful imitation in monkeys comes from an observation made by Marc Hauser (1988b) on the Amboseli vervets. Studying group A during a severe drought in 1984, Hauser observed the adult female Borgia dip a dry tortilis pod into a small pool of water that had formed in the hollow of a tree's trunk. Within 10 days, her offspring also adopted the practice, using a style of pod dipping that was almost identical to Borgia's.

Hauser's observation is striking at least in part because similar examples in other monkey species (or indeed vervets) are so rare. Visalberghi and Fragaszy (in press) suggest that monkeys may have difficulty in representing the task at hand and at recognizing the relation between actions and objects. Unlike chimpanzees, monkeys show little foresight and little ability to modify objects *in advance* of their use. Imitation may be uncommon at least in part because monkeys are unable to attribute purpose to others. Lacking a theory of mind, they may not recognize what others are trying to do.

Did they conclude monkeys lack theory of mind? How can one know for sure?

Chimpanzees and other apes seem more adept than monkeys at learning

Figure 8.4. A chimpanzee uses a stick to dip for ants at Gombe. Although different populations of chimpanzees use different methods to acquire food, it is not clear whether such local "cultural traditions" are transmitted through imitation. Photograph courtesy of Jim Moore/Anthro-Photo.

to use tools through observation, possibly because they are more adept at imputing purpose to others. In different areas of their range, chimpanzees use different techniques to acquire food or to cleanse their bodies, and these "cultural traditions" are passed on to other individuals within the local area (fig. 8.4; reviewed by Nishida 1987). At Gombe, for example, infant and juvenile chimpanzees often seem to mimic their mothers' efforts to dip for insects by breaking off small branches and poking them into holes. The infants' intense observation of their mothers, and their faithful though rather inept copying of their mothers' actions, give the impression of being purposeful (van Lawick-Goodall 1970, 1973).

Even in the case of chimpanzees, however, strong evidence for purposeful imitation and the rapid propagation of novel behavior is scant. Indeed, one experiment designed explicitly to test whether chimpanzees would imitate a knowledgeable model in learning how to use a tool concluded that imitation was probably *not* occurring (Tomasello et al. 1987). Juvenile chimpanzees who observed an adult using a tool to acquire food did learn to use the tool themselves, but they failed to copy or imitate the model precisely. On the other hand, more than just social enhancement seemed to be involved, since the chimpanzees did seem to recognize the tool as a tool, and they seemed to learn the tool's *function* from the model. The authors speculate that a mode of information transmission somewhere in between social enhancement and purposeful imitation was at work, in which the tool's function was understood, even if not all aspects of the problem were inferred or anticipated. We discuss tool use in more detail in chapter 9.

Social Relationships as Measures of Attribution

As we have seen, monkeys seem to be experts at reading each others' behavior; as yet, though, we have little evidence that they are equally expert at reading each others' minds. Part of this problem is methodological; under natural conditions, it is extremely difficult to distinguish between actions that result from knowledge of other individuals' states of mind and actions that result from knowledge of other individuals' behavior. Investigations of higher-order intentionality are easily confounded by possible audience effects, since it is difficult to present subjects with evidence of another animal's *knowledge* while simultaneously eliminating all visual or auditory evidence of the animal's physical *presence*. The problem is well illustrated by a series of experiments conducted by Keddy Hector, Seyfarth, and Raleigh (1989) that attempted to determine whether captive male vervets would change their behavior toward infants if the males perceived the infants' mothers to be watching them.

Among baboons, subordinate males sometimes attempt to establish a close bond with an adult female by interacting with that female's infant (chapter 2; see also Whitten 1987). This strategy potentially allows males to offset some of the competitive disadvantage brought about by their low dominance rank, since it can lead to a close bond between a male and a female that persists throughout the female's reproductive cycles. Close bonds between males and females are far less common among free-ranging vervet monkeys than among baboons (Andelman 1985), and affiliative interactions between vervet males and infants are almost nonexistent. In captivity, however, female vervets show a decided preference for particular males, and males also groom and carry infants at substantially higher rates than their wild counterparts (Keddy 1986; Raleigh and McGuire 1989). Keddy Hector's experiment tested whether or not the affiliative behavior shown by vervet males toward infants might constitute a reproductive strategy to influence female mate choice.

The vervet groups were housed in cages consisting of an outdoor and indoor enclosure connected by a chute similar to the one used in the "informing" experiment described earlier (see also chapter 5, p. 146). By manipulating doorways in this chute, it was possible to separate pairs of animals from the rest of the group for short periods of time. For each trial, a male was locked in the outdoor half of the enclosure with an infant while the rest of the group was locked inside. The mother was seated in the front of the chute at the edge of the outdoor enclosure. She was separated from the male and her infant either by a clear glass partition, a steel partition, or a one-way mirror that was transparent on one side. Neither the male nor the female could see each other in the steel door condition. In the one-way mirror condition, the female could see the male, but the male could not see the female.

All males, but particularly subordinate males, altered their behavior toward infants depending upon the perceived presence or absence of the mother. In the glass condition, when the mother appeared to be watching, males were more affiliative toward the infants—grooming, touching, and lipsmacking at them. By contrast, in the steel and one-way mirror conditions, when the mother appeared to be absent, males were more likely to threaten the infant. The mother's behavior under these different conditions also changed; mothers were less likely to threaten the male following the glass exposure, when the mother and the male could see each other, than following the one-way mirror exposure, when the mother could see the male but the male could not see her (Keddy Hector, Seyfarth, and Raleigh 1989).

It is impossible to conclude in these experiments that the males' behavior was altered by their attribution of knowledge to the mothers. Males might have been friendly to infants because they wanted to influence what the infants' mothers *thought.* Alternatively, the males' behavior might have simply been influenced by the mother's physical presence, much as alarm calls can be affected by the presence of an audience (chapters 5 and 7). If the males were influenced only by the presence of an audience, explanations based on states of mind are unnecessary. We cannot distinguish between these two alternatives, however, because in these experiments the mother's presence and the mother's knowledge were confounded.

In 1989, we designed an experiment that attempted to distinguish more precisely between an individual's apparent presence and her knowledge. Our aim was to investigate the extent to which monkeys attribute perspectives different from their own to each other. Do monkeys only take into account their own visual perspective when monitoring each other, or do they recognize that another individual's perspective, and the knowledge derived from it, can differ from their own? The experiment followed the same procedure as Keddy Hector's, but with one important modification: the orientation of the one-way mirror was *reversed,* so that the animals in the test arena could see the observer sitting in the chute even though the observer could not see them.

The subjects for this experiment were the same adult female and juvenile Japanese and rhesus macaques that we had used in the "informing" experiment described earlier. In both of these species, offspring acquire ranks similar to their mothers', and the ranks of younger animals often depend strongly on the support of their mothers and other female kin (chapter 2). Indeed, the juvenile offspring of dominant females may fall in rank to the members of normally subordinate matrilines if they are deprived of kin support (Chapais 1988a and 1988b).

Our experiment measured the influence of a dominant mother's apparent presence on the agonistic interactions of her offspring and a normally subordinate older female. To do this, we placed the offspring of a high-ranking female in a test arena with a subordinate adult female under each of three conditions: when the mother was visible behind a clear glass barrier; when the mother was invisible behind a dark opaque barrier; and when the mother was seated behind a one-way mirror. In this last case, the mother could be seen by her offspring and the subordinate female but she could not see them (fig. 8.5; Cheney and Seyfarth 1990). In other words, we attempted to control for the audience effect by separating the observer's apparent *presence* from her *knowledge*.

Under the glass condition, the subordinate female should have behaved much as she did under normal group conditions, showing little agonistic behavior toward the dominant offspring. In contrast, we predicted that the subordinate female would behave more agonistically when the mother was invisible behind the opaque barrier.

Figure 8.5. The design of experiments using clear glass, opaque glass, and one-way mirrors. Drawing by Cathy West.

There were two possible outcomes to the trials using one-way mirrors. If the subordinate female was influenced more by her audience's apparent presence than by its knowledge, her behavior under the one-way mirror condition should have been indistinguishable from her behavior under the glass condition. On the other hand, if the subordinate female was capable of distinguishing between her own visual perspective and the mother's visual perspective, she might have realized that the mother could not see what was occurring even though she was visible. If this were true, the subordinate female's behavior under the one-way mirror condition should have been indistinguishable from her behavior under the opaque condition.

Given the many anecdotal observations that suggest that monkeys do not distinguish easily between their own and other individuals' states of mind, we predicted that the subordinate female's behavior under the one-way mirror condition would be the same as her behavior under the glass condition. Only the apparent presence of an audience would affect her behavior; the audience' state of mind would be irrelevant.

Clearly, this experiment required that the monkeys become familiar with the properties of one-way mirrors. At least 4 weeks before the trials began, therefore, we placed a one-way mirror in each cage, at the entrance of the chute. We emphasize that this procedure only gave the monkeys the *opportunity* to learn that each side of the mirror provided a different visual perspective. If monkeys are incapable of comprehending that their own visual perspective, and hence the knowledge gained from what they see, can differ from somebody else's, they will never, by definition, understand how one-way mirrors work.

The results of the trials using the glass and opaque barriers supported many previous studies in suggesting that the dominance ranks of juveniles are to at least some extent dependent upon kin support. Juvenile subjects were supplanted and threatened significantly more by the normally subordinate females when their mothers were hidden behind the opaque barrier than when their mothers were visible behind the glass barrier (fig. 8.6). Conversely, the juveniles showed more agonistic behavior when their mothers were visible than when they were not. Subordinate females also avoided the juveniles most under the glass condition (Cheney and Seyfarth 1990).

Some aspects of the subjects' behavior were the same under both mirror and glass conditions, as would be predicted if we assume that monkeys do not distinguish between another animal's presence and her knowledge. For example, both subordinate and dominant subjects spent more time near the chute, looking at the mother, under the mirror and glass conditions than under the opaque condition.

According to other behavioral measures, however, subjects treated the mirror barriers as if they were opaque and therefore behaved as if they *could* distinguish between the observer's presence and her knowledge. Significantly more dominant juveniles showed agonistic behavior under the glass condition than under either the mirror or the opaque condition. In contrast, significantly more subordinate subjects showed agonistic behavior under the mirror and opaque conditions than under the glass condition (fig. 8.6). Subordinate subjects were also more likely to avoid their dominant companions under the glass condition than under either the mirror or the opaque condition (Cheney and Seyfarth 1990).

In summary, therefore, some of our results supported a "theory of mind" in monkeys, while others were inconclusive. Can we therefore conclude that the monkeys might have been capable of attributing ignorance to the observers, and of recognizing that the observers' visual perspective was different from their own? Not really, because we cannot rule out the more conservative and likely hypothesis that subjects were simply adept at monitoring the observers' apparent attentiveness. Rather than recognizing a mental state—the observer's ignorance—subjects may instead have been sensitive to her actions, orientation, and the direction of her gaze.

Anecdotes derived from field observations are subject to similarly ambiguous interpretations. When subordinate males copulate from behind bushes (fig. 8.7) or when females groom subordinate males only after they have first concealed themselves behind rocks (Kummer 1982), should we take this as evidence that monkeys recognize the difference between their own and other animals' visual perspectives? Human children as young as 3 years of age seem capable of recognizing this distinction, even though the distinction is not always made explicit (Donaldson 1978; Flavell, Ship-

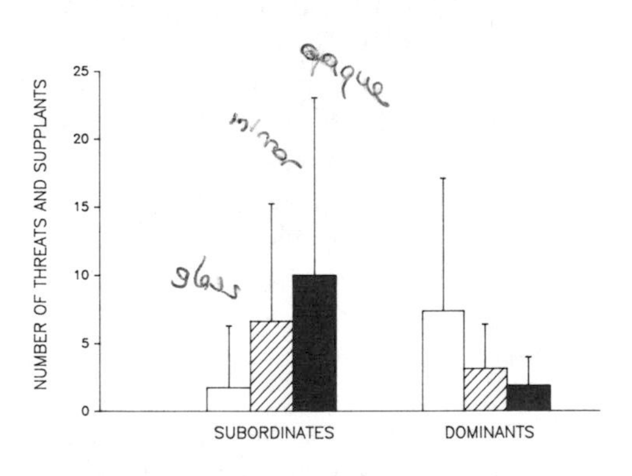

Figure 8.6. The number of threats and supplants given by subordinate and dominant subjects to their companions under the three test conditions. Eight pairs of subjects were tested under each condition. A significantly greater number of subordinate females showed more agonistic behavior under the opaque (solid) than under the glass (clear) condition (sign test, P = 0.031). Conversely, a significantly greater number of dominant juveniles showed more agonistic behavior under the glass and mirror (lined) conditions than under the opaque condition (glass versus opaque: P = 0.031; mirror versus opaque: P = 0.062).

Figure 8.7. A subordinate adult male, Bantam, checks the area around him to ensure that the dominant male is not looking before attempting to mate with female Aphro.

stead, and Croft 1978). Whether monkeys are also capable of such judgments, however, has not yet been demonstrated conclusively. Animals may simply learn that they can avoid attack if they conceal their actions from more dominant individuals.

It is, of course, by no means a trivial feat to adjust one's own behavior according to subtle variations in other individuals' orientation and direction of gaze. The ability certainly demands that monkeys recognize that attentiveness can strongly affect *actions*. It remains to be determined whether they also recognize that attentiveness can affect *knowledge*.

Compassion and Shared Emotions as Measures of Attribution

Pet owners often claim that their dogs or cats feel grief at the death or prolonged absence of a human companion, and these anecdotes are sometimes used as evidence for compassion and empathy in animals. Grief, compassion, and empathy, however, require quite different mental abilities and must be distinguished from each other.

Grief is an emotion, like fear; it does not imply any ability to impute mental states to others. It is certainly possible to feel grief or a sense of loss without ever recognizing that others experience the same emotions. Similarly, while an individual who behaves with compassion toward another must have some recognition of the other's needs or physical impairments, this recognition need not extend to the other animal's mental states. An animal who aids a wounded companion, for example, might recognize that his companion cannot walk properly without also knowing that his companion is experiencing pain. Strictly speaking, therefore, evidence for compassion does not constitute evidence for higher-order intentionality. It does, however, parallel Premack's and Woodruff's (1978) original experiments on the chimpanzee's theory of mind in implying that animals impute purposes and goals to others. Finally, empathy, at least as the term is applied to humans, requires that an individual be able to recognize emotions like grief or fear in his companions even if he is not experiencing these same emotions himself.

Despite the many accounts of close social bonds in monkeys, observers rarely describe their subjects as showing empathy or even compassion for each other. Monkeys almost never comfort each other after the loss of a close relative, look after sick group members, provide food for the old or disabled, or manifest any of the other forms of care that we like to think of as natural components of human interactions. Even when monkeys carry sick or dying infants, they do not treat the infants very differently from the way they treat healthy infants, and they show little evidence of recognizing their needs. Although monkeys do sometimes examine and groom the wounds of others, they appear to treat these disabilities as anomalies or objects of interest rather than as handicaps that require adjustments in their own behavior (fig. 8.8). Why should an animal who is capable of maintaining a close social bond for years at a time nevertheless be incapable of compassion? One possible explanation is that compassion demands some understanding of another's needs and purposes, an ability that monkeys may not possess if they cannot attribute states of mind to others.

Figure 8.8. Tycho and her sister Holborn examine Tycho's newborn infant. Although vervets will often carry dying or dead infants for several days, they do not seem to recognize physical disabilities in others and do not make adjustments in their own behavior to accommodate the needs of sick or handicapped individuals. Photograph by Marc Hauser.

Unlike monkeys, chimpanzees do sometimes appear to show compassion. For example, Jane Goodall (1986) recounts several examples of a female chimpanzee who climbed a tree to gather fruit for her mother when the mother was too old and ill to get the fruit herself. Chimpanzee mothers have also been observed to attend to the wounded or paralyzed limbs of their offspring, taking special care to arrange the disabled limbs gently (Goodall 1986). In the wild, chimpanzees care primarily for close kin. In captivity, however, even unrelated chimpanzees have been reported to help one another. The language-trained chimpanzee Washoe, for example, once scaled a steep restraining fence to rescue a juvenile that was drowning in a moat (cited by Goodall 1986).

But do chimpanzees ever show empathy? Goodall (1979, 1986) describes a number of instances in which chimpanzees apparently experienced mental depression, grief, and a sense of loss when a close relative died (see chapter 10). What is striking, however, is that in none of these descriptions is another animal ever reported to have consoled a grieving companion. Even though chimpanzees have mental states and feel grief, they seem not to attend to grief in *others* and as a result appear to show no empathy.

Clearly, different species' capacities for compassion and empathy will have to be investigated more thoroughly before we can conclude that such emotions demand a level of attribution that is lacking in all nonhuman primates or that is present in apes but not in monkeys. How many times do chimpanzees *fail* to show compassion for every time that they show it? In the absence of more systematic observations, even the intriguing examples from chimpanzees only provide hints of a potentially useful means of investigating other species' theories of mind. At the same time, these hints of compassion also present an evolutionary enigma. Is it possible that compassion is the by-product of an ability that evolved in the context of social competition?

We have argued that the pattern of redirected aggression among vervet monkeys suggests not only that animals recognize the close associates of others but also that they compare relationships and recognize the similarity between their own and other animals' close bonds (chapter 3). Do these data also suggest any ability to attribute emotions to others? If Newton fights with Tycho, and then later threatens Tycho's daughter, how do we rule out the possibility that Newton recognizes the emotions felt by Tycho toward her relatives and knows that one way to retaliate further against Tycho is to threaten her daughter? Similarly, if Newton's sister later threatens Tycho's sister, does she do so at least partially because she knows that this is an effective means of annoying Tycho and her kin? Even if recognition of the close bonds between Tycho and her relatives has originally been acquired through associative learning, why should either Newton or her sister threaten Tycho's kin *except* to retaliate against Tycho and assert the dominance of Newton's family over Tycho's? The crucial distinction here seems to be whether we conclude not just that Newton recognizes Tycho's close associates, but also that Newton recognizes whom Tycho *likes*. Functionally, the distinction is irrelevant. In terms of cognitive mechanisms, however, the distinction is crucial.

One interpretation of redirected aggression, then, suggests that Tycho experiences certain emotions toward her relatives and that other animals recognize these emotions. According to this view, in order to explain the pattern of redirected aggression in vervet monkeys we are forced to assume that the animals attribute motives and intentions to each other; yet there is no evidence that monkeys ever actually do this. Furthermore, the attribution of motives to others would seem to demand some awareness of motives in *oneself;* again, however, there is no evidence that monkeys have any awareness of their own thoughts (see below).

Several experiments with captive monkeys and chimpanzees have suggested that primates might have at least a rudimentary understanding of their companions' emotions and needs. In one of the pioneering studies of nonverbal communication in primates, Emil Menzel (1971, 1973) conducted a series of experiments in which one of six young chimpanzeess was taken into an outdoor enclosure in the absence of the others and shown some hidden item like food or a stuffed snake. The animal was then returned to the cage that held his group mates and, a few minutes later, the entire group was released. There was no indication that the first chimpanzee (termed the *leader*) communicated what he had seen to the others before being released; more often than not he just played and wrestled with his companions. Nevertheless, once released, the others behaved as if they knew not only *what* the leader had seen but also approximately *where* he had seen it. If the hidden item was some delectable bit of food, the entire group made a beeline to the approximate area, the rest of the group sometimes preceding the leader. If the hidden item was a stuffed snake or crocodile, the group would emerge with piloerected fur, approach the area cautiously, and even hit the hiding site with sticks. One of the larger males, Rocky, often monopolized the food cache once it was found, and one of the females, Belle, learned to avoid taking the group to the area when she was the leader. Nevertheless, Rocky was able to extrapolate from Belle's gaze where the food was located. When Belle was shown two hidden food sites, she would lead Rocky to the smaller one and then quickly run to the larger one while he ate.

Clearly, something was being communicated among the chimpanzees in these experiments, but precisely what remains unclear. Some aspect of the leader's behavior allowed the rest of the group to recognize when the hidden item was aversive, and their common distress upon release certainly suggests that the leader's emotions were communicated and shared. Unfortunately, Menzel's chimpanzees were highly dependent upon one another, so he was never able to release the rest of the group into the outdoor area without also releasing the animal who had already seen the hidden object. As a result, it was impossible to determine if the rest of the group were able to read the leader's emotions sufficiently well to know what was outside even if he did not accompany them. The chimpanzees were adept at reading humans' emotional expressions, however, and they would emerge from the holding cage in some agitation if a human had expressed fear of what was outside. It therefore seems likely that the rest of the group already knew quite a lot about what was hidden outside before their release.

Some experimental evidence also suggests that monkeys can learn to read other monkeys' emotional expressions. Miller (1967, 1971) tested the ability of rhesus macaques to "read" each other's facial expressions by requiring them to work together to avoid a shock. The monkeys were first trained individually to respond to two different stimuli on a lighted panel. When stimulus A appeared, the monkey had to pull a bar to avoid a shock. When stimulus B appeared, he had to pull a second bar to receive a food reward. After the monkeys had each learned to perform this task, Miller placed them in two separate cages. The monkey in the first cage could see the lighted panel but had no bars to push. The monkey in the second cage had access to the bars but not the lighted panel; instead he could see only the face of the first monkey, projected from a television screen. If the second monkey pressed the correct bar, both subjects received a food reward. If he pressed the wrong bar, both subjects received a mild shock. Miller found that when the subjects were familiar cage mates, the second monkey was correctly able to read the other's facial expressions in 80 to 90% of all trials. Monkeys who were unfamiliar with each other, however, fared poorly. The basis of the unfamiliar animals' poor performance is unclear. It is entirely possible, for example, that unfamiliar monkeys recognized each other's expressions but that they were for some reason unable to coordinate their activities.

In another series of experiments, laboratory rhesus macaques who observed wild-born monkeys' fear of snakes also developed a fear of snakes (Mineka and Cook 1988). These fear responses were to some extent independent of personal experience, because even monkeys who were unable to see the snake themselves became fearful if they saw other monkeys showing fear. This kind of vicarious fear did not persist for long periods of time, however, and monkeys who simply observed the fear responses of others soon ceased to show fear themselves.

But does the *recognition* of another individual's emotions constitute evidence for the ability to *attribute* emotions or even to feel compassion? In the absence of more definitive experiments, it is impossible to determine whether monkeys and apes become fearful after witnessing another individual's fear because they have simply learned to associate a particular facial expression with an aversive event or because they attribute states of mind to others. What seems more interesting is the fact that in all of tests described, the witnesses became fearful *themselves*. Even if they attribute emotions to others, in other words, monkeys and apes seem not to distinguish between their own and others' states of mind. Just as they seldom communicate about things that are not immediately present, they may not "mention" fear in the absence of that emotion.

Self-Awareness as a Measure of Attribution

The Role of Consciousness in Thinking

The ability to attribute mental states and perspectives to others would seem to require some degree of consciousness, or self-awareness. In the absence of some accessibility to his own mind, it is difficult to see how a monkey could distinguish between his own thoughts and beliefs and the thoughts and beliefs of others. Not surprisingly, though, evidence for consciousness in nonhuman primates is as inconsistent, patchy, and puzzling as the evidence for attribution.

It is not our intention to discuss in any detail the contentious issue of animal consciousness. To review the literature here would be merely to repeat, no doubt with far less success, Griffin's (1976, 1984) extensive review (see also comments by Mason 1976 and Campbell and Blake 1977). Nevertheless, even a cursory discussion of self-awareness demands some definitions.

Operational definitions of consciousness are slippery at best, primarily because self-recognition and self-awareness are multifaceted and can be manifested in different ways in different contexts. Although we use consciousness and self-awareness interchangeably here, we distinguish consciousness from self-recognition. Self-recognition is a more conservative term than consciousness and refers only to the ability to distinguish oneself from others without implying any awareness of so doing. There is ample evidence from studies of children, for example, that many aspects of self-recognition do not require active self-reflection (see below). Consciousness, however, is a kind of meta-self-awareness; it implies that the individual is aware of his own state of mind and can use this awareness to predict and explain the behavior of both himself and others.

Our distinction between self-recognition and consciousness is more explicit than some others' (e.g., Bunge 1980; Armstrong 1981; see discussion in Griffin 1984). Nevertheless, we believe that the distinction is both useful and important. It seems likely that few mental processes, of people no less than other animals, ever become accessible to consciousness. Hence it is easy to imagine that an animal could recognize his own place in a social group and even attribute some beliefs to others without having reflective access to his own mind. Higher-order intentionality, however, does require some degree of consciousness. To attribute false beliefs to others, for example, an individual must compare his own knowledge to that of others, an ability that demands at least some introspection. Throughout most of our discussion, we will restrict ourselves to the conservative term *self-recognition*, reserving the more inflationary *self-awareness* for examples that imply active access to one's own mind.

Resolving the role of consciousness in animal thinking would be much easier if there were some agreement about its role in human thinking. However, the supremacy and significance of consciousness for human thought processes were called into question as soon as Freud (and subsequently many others) drew attention to the variety of thoughts that remain unconscious and inaccessible to us. Today, many psychologists and philosophers echo Lashley (1956) in arguing that we can never be conscious of our mental processes, only of some of their products (see review by Weiskrantz 1985). For example, Dennett (1978b) and Jackendoff (1987) argue that the brain is a collection of modular subsystems, each adapted to perform different, largely unconscious, computational processes. Only some of the products of these computations become conscious, or accessible, to the individual, and not always in all modalities. Numerous studies of human decision-making have demonstrated that self-analysis can be wrong, misinformed, and even self-deceptive (e.g., Kahneman and Tversky 1982). Since we have unprivileged access to our own minds, consciousness cannot be regarded as always occupying a central role.

We do not yet understand precisely why or when some of our thought processes become conscious to us, and the degree to which conscious thoughts help to coordinate other mental processes remains unclear. Nevertheless, there does seem to be some concurrence that most mental processes remain unconscious and largely inaccessible, and that there is no reason why self-perception should be any less infallible than the perception of the outside world.

The function of consciousness also remains elusive. Humphrey (1986) has speculated that consciousness has evolved to allow individuals to see into others' minds and to predict the behavior of others on the basis of their own introspection. So, for example, a monkey might predict that she could retaliate effectively against an opponent by threatening her opponent's kin because she recognizes that similar retaliation against herself would make *her* angry. Consciousness, in this view, is an essential precondition to any ability to speculate about the mind of others (see also Markl 1985). Humphrey assigns consciousness a more central role in the mind than Jackendoff (1987). Even at the more intermediate, modality-specific, level advocated by Jackendoff, however, consciousness may serve the function of alerting the mind to what is being detected and what is being understood.

Whatever the precise definition or function of consciousness, there is no a priori reason to expect consciousness, any more than attribution, to be a discrete capacity that an organism either has or does not have. In children, the sense of self develops gradually, and manifestations of self-identity vary qualitatively with age (Guardo and Bohan 1971; Damon and Hart 1982; Anderson 1984b). Children under the age of 20 months, for example, usually fail to recognize their own mirror images, even though they may be capable of restricted self-descriptive statements and of recognizing their own physical limitations. Similarly, children will reliably identify their own place in a family or social group long before they are capable of much self-reflection about their own thoughts and moods (Damon and Hart 1982). Interestingly, children's knowledge of their own and other people's identities and personalities seems to develop in parallel stages and at similar rates (Damon and Hart 1982; Rotenberg 1982; reviewed in Flavell 1985:154).

Evidence for Self-Awareness in Monkeys and Apes

All of the ape language projects report some degree of self-recognition in their subjects. Patterson and Linden (1981), for example, describe numerous occasions when the gorilla Koko signed that she was sad, angry, happy, or afraid. Similarly, Kanzi, a bonobo trained by Savage-Rumbaugh and McDonald (1988), sometimes signed *bad* before doing something for which he would subsequently be punished. Indeed, most of the ape language projects have to some extent *presumed* self-recognition in their subjects by requiring them to use their own names when solving tests and answering questions. Given the many intriguing anecdotes that have emerged from these projects, it is rather surprising that systematic investigations of self-recognition in laboratory animals have been largely restricted to tests with mirrors.

It is not really clear what aspects of consciousness are reflected by tests with mirrors. Nevertheless, whatever it is that such tests reveal, they consistently suggest a qualitative difference between monkeys and apes. By a number of different measures, chimpanzees are clearly able to recognize their own images in a mirror (see reviews by Gallup 1982; Anderson 1984a). Chimpanzees whose foreheads have been daubed with paint, for example, will touch the relevant areas of their faces when they look into a mirror. Even after extensive experience with mirrors, however, monkeys typically attempt to interact with their mirror images (usually by threatening them); it is as if they treat these images as other monkeys.

Although studies with mirrors doubtless address some aspect of self-recognition, they reveal little about its function except under these rather restricted and artificial conditions. The negative results are even less illuminating, since they can never rule out the possibility that a particular species or individual might reveal some degree of self-recognition in another, more functionally relevant context. Gorillas, for example, "fail" the mirror test (Gallup 1982), even though they seem quite similar to chimpanzees and orangutans in other cognitive tests, and even though wild gorillas will look at and touch the surfaces of still water and camera lenses in ways that are substantially different from the ways they look at other gorillas (A. Harcourt, pers. comm.; see also Patterson and Linden 1981; Fossey 1983).

Tests with monkeys living in normal social groups also cast some doubt on the notion that the recognition of self in mirror images is an all-or-nothing phenomenon, present in some species but not in others. For example, when mirrors were presented to group-living Japanese macaques, the juveniles responded as if the images were other monkeys (although, unlike isolated laboratory animals, they seldom threatened their mirror images). Adults, however, typically stared at their images without attempting to interact with them in any way (Platt and Thompson 1985). Although the monkeys did not seem to treat their images as "themselves," they also failed to treat them as familiar or unfamiliar "others." For more discussion of the mixed responses of monkeys to mirrors see Gallup 1982; Eglash and Snowdon 1983; Anderson 1984a, 1986; and de Waal 1989.

Menzel, Savage-Rumbaugh, and Lawson (1985) investigated two chimpanzees' ability not only to recognize but also to *use* their mirror images. The chimpanzees had to reach through holes in a wall and to track mirror or video images of their hands to find a hidden object. Not only did they perform this task with apparent ease, but they were also able to discriminate between live video images and delayed tapes. Further, they could adjust their hand movements to accommodate novel orientations of the video camera. Similar experiments suggested that rhesus macaques did not even understand the problem, since they simply displayed at their images. The monkey findings in these experiments are somewhat puzzling, since other studies have shown that monkeys *can* learn to use mirror images to manipulate objects and to monitor the behavior of other animals (Itakura 1987a, 1987b; see also reviews in Gallup 1982; Anderson 1984a). Oddly, monkeys seem able to respond appropriately to mirrored information about their *environment,* but not about *themselves.*

Another realm of behavior suggestive of self-recognition and even of self-awareness is pretend play. Pretend play clearly requires some ability to distinguish between what is real and what is imagined, because it demands that individuals deliberately substitute something they have actively transformed for something they know to be real. Even 1-year-old children engage in pretend play. It seems to be one of the earliest manifestations of children's attempts to represent and manipulate their own and others' knowledge, and it seems to appear long before children are capable of recognizing false beliefs in others (Leslie 1987, 1988).

However, although an individual who engages in pretend play must entertain multiple representations of an object or an event, these representations are *shared* by all of the game's participants. To recognize someone else's false belief, on the other hand, an individual must consider multiple representations that *differ* among those involved. Perhaps as a result, a toddler who is as yet unable to lie successfully about the cookie crumbs on her face can nonetheless invite a teddy bear to a tea party. For the same reason, we might expect pretend play to be evident in nonhuman species whose theories of minds in other contexts seem rather incomplete.

There are a few anecdotal accounts of pretend play in apes. For example, Vicki, the chimpanzee raised by the Hayeses (1951), regularly played pretend games; she would pull an imaginary toy on an imaginary string around the house and even pretend to drop it into the toilet. Similarly, Kanzi, the young language-trained bonobo, not only regularly hid objects from his trainers but also pretended to hide and eat imaginary objects (Savage-Rumbaugh and McDonald 1988). Captive orangutans and gorillas have also been reported to pretend to hide or disguise themselves (Patterson and Linden 1981; Miles 1983). Pretend play, moreover, is not found only in animals raised by humans. A group of young bonobos studied by de Waal (1986c) at the San Diego Zoo regularly covered their eyes while chasing through a tree in an apparently deliberate effort to make the game more difficult.

But pretend play is provoking at least in part because we would probably be reluctant to ascribe higher-order intentionality or self-awareness to most forms of play. Many species of mammals, for example, adopt particular facial expressions when playing. These "play faces" signal that normally aggressive motor patterns are now not to be taken seriously. Similarly, older animals in species as diverse as dogs and baboons often apparently deliberately "scale down" the roughness of their play when playing with much younger animals (Fagen 1981). Does this indicate an ability to recognize the difference between one's own abilities and the abilities of younger animals? We may be inclined to aim Morgan's canon more charitably at baboons than at dogs, but in the absence of any more systematic evidence for attribution or self-awareness in either species it is difficult even to specify what our standards of parsimony should be.

If a child's understanding of other people's behavior reflects his understanding of himself, might we also use knowledge of other individuals' behavior as a means to measure self-recognition in nonhuman species? For chimpanzees and other apes, the answer may be yes. Premack (1986), for example, reports that two language-trained chimpanzees were actually slightly better at describing their own behavior than the behavior of others.

The mirror tests described earlier, however, suggest that the same may not be true of monkeys, since monkeys who regularly use mirror images to monitor the behavior of others nevertheless fail to recognize themselves (Gallup 1982; Anderson 1984a). Unfortunately for our methodological purposes, therefore, it may not be possible to correlate a monkey's knowledge of other individuals with self-recognition. In fact, in marked contrast to children and perhaps also chimpanzees, monkeys may be considerably better at describing and understanding others than they are at describing and understanding themselves.

Is that the end of the story, though? Once again, the tangle of inconclusive and contradictory evidence frustrates any definitive statements about the presence or absence of self-recognition in monkeys. Consider again the data on redirected aggression, reconciliation, and dominance rank presented in chapter 3. Monkeys, apparently, recognize both their own dominance ranks and the dominance ranks of others. They distinguish among the close associates of other individuals while simultaneously treating relationships within their own matrilines differently from relationships within other matrilines. Monkeys, in other words, seem to understand not only their own positions in the social group but also where others stand relative to each other. This ability would seem to demand that they be able to embed themselves and other group members within a network of social relationships while maintaining the distinction between their own social networks and the social networks of others (fig. 8.9). This is not to say that monkeys are aware of making these distinctions. As the evidence

Figure 8.9. Subadult female Shelley grooms her sister Carlyle. Although vervets may recognize their own place in the social network, they may not be aware of doing so.

from children suggests, an individual who is capable of some forms of self-recognition need not necessarily be able to reflect actively upon his emotions and his behavior (Damon and Hart 1982). There is no evidence that monkeys are *aware* of distinguishing their own relationships from the relationships of others. In this respect there may be more to the behavior of monkeys than meets their I.

What Do Animals Know About What They Know?

When a vervet monkey threatens the relative of a female who has previously threatened her *own* relative (see chapter 3), she appears to recognize the similarity between her own close relationships and the close relationships of others. This ability may be functionally equivalent to analogical reasoning. The chimpanzee Sarah, however, was able to solve analogical reasoning problems using a variety of stimuli (Premack 1983b). In contrast, the evidence from vervet monkeys has thus far been restricted to contexts involving familiar social companions. Indeed, as we will discuss in chapter 9, vervets seem quite poor at extending their social skills outside this familiar social domain, even when there appears to be strong selective pressure to do so.

To apply knowledge obtained in one domain to stimuli encountered in another, knowledge must be accessible (Rozin 1976); such accessibility is considerably facilitated if an individual is at least partially aware of what he knows. Do vervets and other monkeys know what they know? The question is clearly linked to issues of attribution and self-recognition, and it is not as arcane an issue as it might, at first, appear. Some human victims of amnesia, for example, can learn simple conditioned tasks while remaining unaware that they have learned anything. They cannot perform any task that requires reflection, matching, ordering, or reordering (Weiskrantz 1985).

An animal who knows what she knows necessarily has a theory of mind. However, just as the amnesiac's knowledge can be inaccessible to him, so might an animal perform quite complicated tasks or perceive complex distinctions between two stimuli without knowing or being aware of what she is doing. In chapters 5 and 6 we suggested that vervet monkeys might not be aware of classifying calls according to their meaning, or at least not aware in the same sense that Sarah was when she described what a blue triangle stood for. Here we discuss the question of knowledge about knowledge in more detail.

The dances of honey bees (*Apis mellifera*) are, at least functionally speaking, highly referential, designating very specific information about the direction of food, the abundance of food, and the distance from food to hive. Recruits are able to "read" the information contained in dances and hence find food with a high degree of precision. Do bees, however, *interpret* the information contained in dances, or simply react to it? Can they recognize false information?

To investigate this question, James and Carol Gould (1988) trained a number of marked bees to fly to a rowboat in the center of a lake, where they were provisioned with nectar. By keeping the quality of the nectar low, the Goulds were able to prevent the bees from dancing to other foragers at the hive during the training period. When training was complete, however, the quality of nectar was increased and marked bees now danced vigorously to potential recruits at the hive. Ordinarily, of course, nectar would never be found in the middle of a lake and any dance that conveyed information about the presence of food in this particular location would simply be wrong. The Goulds were interested in determining whether the recruits would interpret the dance as signaling implausible information and therefore ignore the dance or in some other way reveal that they knew the dance to be wrong (falling upon the liar and stinging her to death?). Interestingly, when the advertised food site was in the middle of the lake, no recruits responded. However, as the boat was moved closer to the shore the number of recruits increased substantially (Gould and Gould 1988).

The results of these and related experiments suggest that bees possess a well-developed mental map of their environs and that their interpretation of a dance is mediated by this map. If a dance denotes a food source whose location is at variance with their mental map, the dance is ignored. The experiments do not, however, address the question of whether bees are in any sense aware of the dancer's intentions.

Similar equivocal results have been obtained in tests using infant chimpanzees. Oden, Thompson, and Premack (1988) tested the ability of infant chimpanzeess to perceive a relational distinction using a habituation/dishabituation procedure. The infants were shown two objects that were either the same (AA) or different (AB). Next, the infants were shown two novel objects that were also either the same (CC) or different (CD). The experimenters reasoned that if the infants perceived the sameness or difference of *relations* between objects, they should respond for longer durations when they were presented with two heterogeneous sets of stimuli (e.g., AA then CD, or AB then CC) than when they were presented with two homogeneous sets of stimuli (e.g., AA then CC, or AB then CD). As predicted, the decrement in looking time was greater for the two sets of homogeneous objects than for the sets of heterogeneous objects.

The infants were then switched to a match-to-sample task and trained to match one of two alternative objects with a sample object. After reaching a high rate of accuracy with single objects, the infants were then asked to perform the match-to-sample task with relations instead of objects. Even though the stimuli used in this test were the same as those used in the habituation trials, none of the infants could solve the problem and all continued to perform at chance levels through 200 trials.

Although infant chimpanzees had no difficult in *perceiving* relational distinctions, therefore, they were unable to *use* these distinctions. Premack (1983b) argues that the ability to make use of relational distinctions may only emerge with language training or through the intervention of another species. A necessary first step in this research, however, would be to determine whether the same results obtain for adult chimpanzees.

The ability to solve problems by hypothesis rather than by simple associative processes allows knowledge to be more easily extended from one context to another. Many of the apparently insightful solutions to problems achieved by captive chimpanzees suggest that the animals have not only understood the nature of problems but also generalized their experiences to other problems. So, for example, the chimpanzees studied by Kohler (1925), Menzel (1972, 1973), and de Waal (1982) were able to use their experience with long sticks and poles to build and use probes and ladders spontaneously. We discuss these skills further in chapter 9.

Susan Essock-Vitale (1978) designed oddity tests for orangutans and two species of macaques that could be solved either through associative learning or through hypothesis formation. Tests were designed, moreover, so that the experimenter could distinguish between the two learning methods; individuals who had learned a specific discrimination through associative learning were predicted to have more difficulty transferring to a new discrimination problem than those who had learned through hypothesis formation. The orangutans used an abstract hypothesis from the outset, whereas the macaques began by solving the problem through association. The macaques' percentage of correct scores, however, was usually higher than the orangutans', and the monkeys also eventually learned to use an abstract hypothesis. These experiments emphasize that although individuals who use an hypothesis might know more about a given problem and might generalize more quickly to similar problems with novel stimuli, problem solving by association may nevertheless be as effective in achieving a particular goal (see also Yerkes 1916, who obtained similar results).

Problem solving by hypothesis or insight also allows an individual to predict the consequences of an act a number of steps into the future. A captive chimpanzee, Julia, was given a problem in which she had to open a series of transparent boxes, each containing a different key, in order to obtain the reward in the final box (Dohl 1968; Rensch and Dohl 1967). Only if she chose the box with the correct key at the beginning of the series could she embark upon the route that would eventually produce the reward. Julia was able to choose the correct initial box even in series that involved up to

10 boxes, as if she were able to anticipate and make judgments about her actions that were several steps removed from her immediate goal (cited by de Waal 1982 and Goodall 1986).

In a similar way, de Waal (1982) argues, the male chimpanzee Yeroen in the colony at the Arnhem Zoo was able to anticipate that his value as an alliance partner was more important to Nikki's dominance status than to the dominance status of a more powerful male, Luit. Yeroen therefore switched his allegiance to Nikki, even though his act initially resulted in increased aggression from Luit. Months later, however, Yeroen was able to reap subtle benefits from his alliance with Nikki through Nikki's greater tolerance of him. Yeroen had acted as if he could predict the consequences of his alliance with Nikki long before it resulted in any tangible rewards.

The mind's ability to extend information gained in one context to other, different contexts is a measure of its accessibility. Just as our thoughts and knowledge are not always accessible to consciousness, however, so are skills acquired in one context not always easily generalized to others (Rozin 1976). As we discuss in chapter 9, there is increasing evidence, both empirical and theoretical, that many of the mind's most impressive accomplishments are "domain specific" and used within a relatively narrow range of activities.

The Evolution of Attribution

We have argued that the complexity of primate social groups places strong selective pressure on the ability to recognize not only one's own social relationships but also the social relationships of others. Monkeys do seem adept at recognizing each others' matrilineal kin, dominance ranks, and friendships. They are also able to adjust their behavior according to who has reciprocated in the past, and they seem able to predict the consequences of their own behavior on the behavior of others. None of these abilities, however, requires the attribution of mental states. What is the adaptive value of higher-order intentionality?

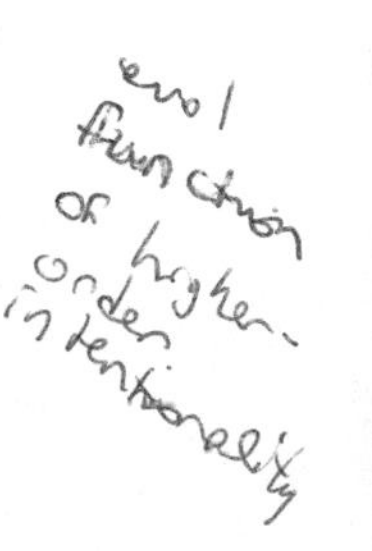

When we label behavior in monkeys, chimpanzees, or children as examples of second- or third-order intentionality, this is more than just an arid exercise in behavioral classification. As we have suggested, there are clear social (and perhaps also reproductive) advantages to be gained by an individual who can attribute mental states to others, recognize that such states can be different from his own, and recognize that what another individual thinks can have a causal effect on behavior.

For example, imagine a group of baboons in which individuals are extremely skilled at judging behavioral contingencies (if I do X, Y is likely to happen) but unable to identify the motives or knowledge of others or to recognize that these mental states can be different from their own. Imagine further that among these baboons, as among the baboons studied by Packer (1977), Smuts (1985), Bercovitch (1988) and others, adult males solicit each other for support in alliances. When male A, for instance, encounters male C in consortship with a sexually receptive female, he solicits support from male B, who may respond by joining him in chasing C away. Over time, some pairs of males exchange reciprocal alliances: A and B regularly cooperate to chase other males from females and each allows the other to consort with the female half of the time. In other pairs, however, support is not reciprocated: X helps Y gain access to a female but later, when X solicits Y, the latter refuses to help.

Under these conditions, males who are solicited as alliance partners will benefit if they can distinguish between an individual who seeks help and genuinely intends to reciprocate and one who seeks help but is unlikely to return the favor. This distinction, moreover, will not be an easy one to make, because selection will also favor males who solicit effectively, using the same patterns of behavior regardless of whether they intend to reciprocate or not. This is, in fact, what seems to occur in natural groups of baboons (e.g., Packer 1977; Noë 1986; Bercovitch 1988).

If male baboons are incapable of recognizing the motives of other animals—that is, if they are incapable of higher-order intentionality, direct experience is the only means by which they will be able to acquire information about honest and dishonest solicitors. And, in judging a solicitor's reliability, the more experiences they have had with him, the more accurate their assessment is likely to be; one interaction is unlikely to be sufficient. These baboons, in other words, will always be vulnerable to those who cheat on their first encounter, and they will remain vulnerable until they have gained enough experience, through sometimes costly interactions, to accurately assess their partners' reliability.

In addition, because these baboons base their judgments of another individual's reliability on his actions rather than his motives, the knowledge that one animal has about another is likely to be specific to a particular type of interaction. If male X has learned through experience that Y rarely reciprocates when forming alliances to obtain a female, he will know just that: Y rarely reciprocates when forming alliances to obtain a female. Such knowledge, however, will not necessarily make X skeptical about Y's reliability when their group is threatening another group or when it encounters a predator. Because their judgments of others rest on an assessment of behavioral contingencies, the knowledge possessed by these baboons is likely to be relatively context specific. As a result, they will be vulnerable to individuals who cooperate in one situation only to cheat in another.

Now imagine that into this group of nonintentional baboons comes a mutant male capable of attributing states of mind to others and of recognizing that these states of mind may be different from his own. The first point to make about this individual is that he recognizes a distinction between an animal's behavior and the motives that underlie it. As a result, he recognizes that however much a solicitor *seems* likely to reciprocate, this may not actually be his intention. Such knowledge will not necessarily make the mutant male any less vulnerable to cheaters on his first interaction with them, but it is certainly likely to make him more skeptical in subsequent interactions. More important, because the new male is attentive to motives and not just behavioral contingencies, he is more likely to generalize his knowledge of different individuals across a variety of different social contexts. When he encounters male Y cheating in the formation of alliances, for example, the mutant male's skepticism about this individual will extend not just to future alliances but also to Y's alarm calls, food calls, behavior during intergroup encounters, and so on. In short, the new male will have a competitive advantage over others in his group because, in being able to assess his companions' motives, he is better able to predict their behavior.

Consider another example. Suppose there exists a group of macaques in which one animal, like the famous Japanese macaque Imo, suddenly develops a new method for acquiring and preparing food. If the inventor deals only with behavioral contingencies, there is relatively little adaptive value to be gained from her discovery. She can, of course, feed herself better and raise healthier offspring. Other animals may learn that she, alone among their companions, is able to acquire this kind of food, and this may cause them to approach her when she is preparing the food and to handle the food as she does. Through such social enhancement, her offspring might eventually also acquire the skill. The female might also become a more attractive social partner to others, and her attractiveness might allow her to establish a relationship with a high-ranking female or male that might otherwise not have developed (Stammbach 1988b provides a good example).

However, if the inventor can attribute ignorance to others and understands that mental states can affect behavior, there is an immense amount to be gained. An inventor who possesses a theory of mind can selectively transmit her knowledge to kin, much as she can selectively distribute her grooming. She can also selectively *withhold* her knowledge from rivals, much as she selectively withholds other cooperative behavior like alliances. Finally, if the inventor can recognize the difference between her own knowledge and that of others, she need not depend on the relatively slow process of observational learning to transmit her skill but instead can engage in active pedagogy. Once again, an individual capable of attribution would seem to have a clear selective advantage over others.

We present these hypothetical examples in order to reemphasize that the existence (or lack) of a theory of mind, long recognized as an important watershed in children's cognitive development, also has considerable evolutionary significance. Once an individual recognizes that his companions not only behave but also think, desire, and believe about behavior, he becomes a much better social strategist, and he can use his knowledge much more skillfully to his own and his relatives' benefit.

Despite the apparent selective advantage of being able to attribute mental states to others, however, there is little evidence that monkeys do so. It seems unlikely that this failure is due to the lack of language; the capacities to inform, deceive, teach, and empathize may be enhanced by speech, but they certainly do not require it. Dennett (1987) suggests that higher-order intentionality has probably evolved as a result of selective pressures placed on individuals by their social environment, and he argues that the social environment in most primate species is probably too simple to require higher-order intentionality. Most species of nonhuman primates live in groups that forage and rest as cohesive social units. In only a few species, including humans, spider monkeys, and chimpanzees, do groups regularly split up into smaller parties or subunits whose composition is relatively fluid. As a result, Dennett argues, it is only in humans, and perhaps chimpanzees, that the web of social relationships becomes sufficiently entangled that higher-order intentionality really become essential. Vervets probably could not make use of most of the features of human language, including the intentional stance, because their world is so much simpler than ours. Since most monkey species do everything as a group, they live such a relentlessly public existence that there is not much novel information to impart, no secrets to reveal or withhold (Dennett 1987).

If the selective pressures favoring higher-order intentionality are absent in group-living monkeys, we might expect to find more evidence of a theory of mind in species that regularly split up into smaller subunits, including not just chimpanzees and spider monkeys (*Ateles* spp.) but also hamadryas baboons, dolphins, and elephants. Chimpanzees do indeed seem more skilled than monkeys at attributing states of mind to others, but to conclude that this is so because they can, if they want to, be alone, may beg a lot of interesting questions before they have been asked. It is even possible that we have misidentified our chicken and egg and that the ability to attribute states of minds to others is what *permits* social groups to become more fluid and less stable.

At least by most tests, gorillas too are intelligent, even though their cohesive groups rarely allow them to leave the public spotlight. Nevertheless, captive gorillas do seem to fail at least one measure of self-recognition (Gallup 1982). Can we link the failure to recognize a face in a mirror to the lack of self-recognition, the lack of self-recognition to the lack of a theory of mind, and the lack of a theory of mind to the lack of privacy? Unfortunately, it is often difficult to link many laboratory tests of self-recognition and higher-order intentionality in monkeys and apes to the animals' natural social behavior. It is easy to speculate about why a monkey who could attribute mental states to others might enjoy a selective advantage. This teleological approach, however, can only leave us wondering why such an apparently useful ability appears not to have evolved.

Summary

The study of attribution in monkeys and apes is still in its infancy and continues to rely heavily on intriguing but largely unsubstantiated anecdotes. While it is difficult to summarize an issue for which there is only a small amount of confusing and sometimes contradictory data, we will risk a few speculations.

Monkeys and apes do occasionally act as if they recognize that other individuals have beliefs, but even the most compelling examples can usually be explained in terms of learned behavioral contingencies, without recourse to higher-order intentionality. What little evidence there is suggests that apes, in particular, may have a theory of mind, but not one that allows them to differentiate clearly or easily among different theories or different minds. Indeed, we cannot even state conclusively that nonhuman primates attribute ignorance to each other. The problem may be at least partially related to the animals' inability to recognize and represent their own knowledge.

Many different examples, reflecting many different approaches to the same problem, all support the hypothesis that monkeys are unable to attribute mental states to others. First, while monkeys can easily learn the necessary steps to complete a task, they apparently find it more difficult to learn the roles of others, perhaps because they cannot impute motives to other individuals. Second, while they do attempt to deceive each other, monkeys' attempts at deception seem aimed more at altering their rivals' behavior than at affecting their rivals' thoughts. Third, although their vocalizations certainly function to alert others to the presence of food, danger, or each other, we have no evidence that monkeys ever communicate with the intent of changing a listener's mental state or of drawing the listener's attention to their own mental state. Monkeys do not adjust their behavior according to whether or not their audience is ignorant or informed, perhaps because they do not recognize that such mental states exist. Fourth, while monkeys are clearly able to acquire novel skills from others through observation, social enhancement, and trial-and-error learning, there is little evidence that they imitate each other, again perhaps because they are unable to impute motive. Fifth, monkeys do not teach each other. Again, we would argue that this lack of pedagogy reflects the animals' inability to distinguish between their own states of mind and the states of mind of others. Sixth, although monkeys experience emotions like fear and grief, they show no evidence of compassion or empathy and do not seem to recognize emotions in others. Finally, while monkeys are adept at recognizing their own position in a social network or dominance hierarchy, they show little self-awareness. This, too, is consistent with the view that monkeys do not know what they know and cannot reflect upon their knowledge, their emotions, or their beliefs.

Many of these generalizations may apply more to monkeys than to apes. Indeed, having gone so far as to suggest that monkeys, for the most part, lack a theory of mind, we will speculate further and predict that many of the most fundamental differences between the minds of monkeys and the minds of apes will ultimately be traced to the apes' superior skills in attributing states of mind to each other (see also discussion by Mason 1978a).

Although most of the data are anecdotal, there is strong suggestive evidence that chimpanzees, if not other apes, recognize that other individuals have beliefs and that their own behavior can affect those beliefs. Unlike monkeys, chimpanzees seem to understand each others' goals and motives. They deceive each other in more ways and in more contexts than monkeys, and they seem better than monkeys at recognizing both their own and other individuals' knowledge and limitations.

At the same time, however, chimpanzees may be like very young children in failing to attribute false beliefs to others. There is very little evidence that chimpanzees recognize a discrepancy between their own states of mind and the states of mind of others. They show little empathy for each other, and they do not explicitly teach each other. Although there is some suggestion that they recognize ignorance in others, we do not yet know if they ever deliberately take steps to rectify ignorance, or if they take into account their audience's knowledge when alerting others to the presence of another group or the discovery of a fruiting tree.

Shakespeare's *Romeo and Juliet* is a tragedy of ignorance and false beliefs of which only the audience is aware. The ultimate irony of the play comes when Romeo discovers the drugged and sleeping Juliet. Believing her to be dead, he laments that she looks beautiful enough to be alive. Romeo's suicide is tragic because it is based on a belief that the audience knows to be false. "If Romeo only knew what we know," we think, "this need not have happened." Tragedy arises, in the playwright Arthur Miller's (1949) words, when we are made to see the mismatch between a man's achievements and his aspirations, and to understand aspirations we must attribute mental states. But how would Romeo's death appear to an audience of monkeys, unable to distinguish between their own beliefs and Romeo's?

The probability that a monkey with a typewriter would produce the complete works of Shakespeare is one in many billions—even given thousands of years the right combinations of letters and spaces would simply never arise by chance (Dawkins 1986). The reason that monkeys would have to rely on chance has to do with their theories of mind. Even if a monkey could type and describe his characters' behavior, he could not reveal their minds. And without such attribution there could be no tragedy or comedy, no irony and no paradox.

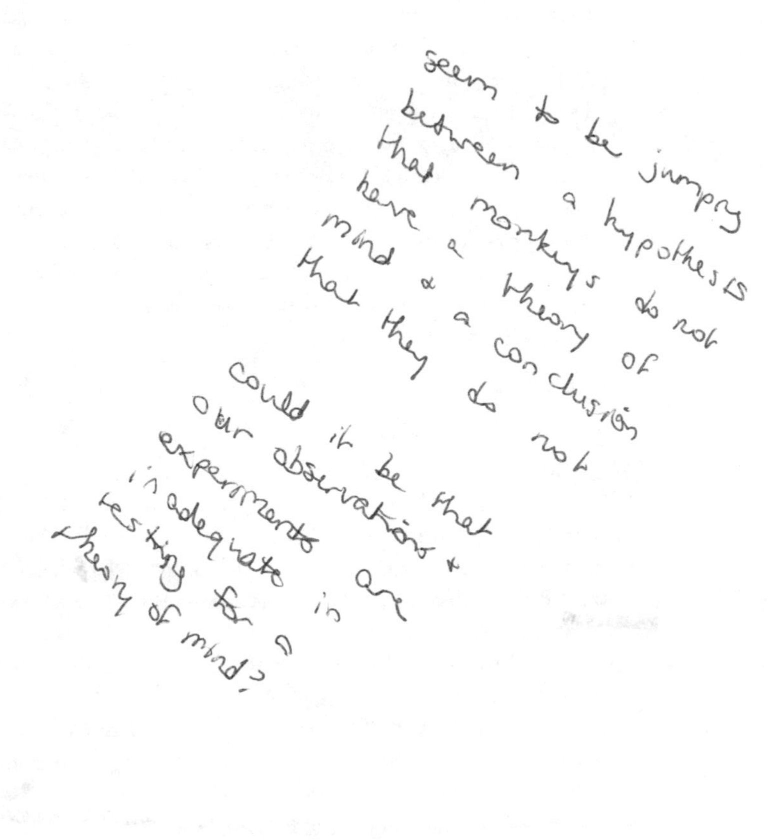

CHAPTER NINE | *SOCIAL AND NONSOCIAL INTELLIGENCE*

We arrived at our study area one day just after some lions had killed a buffalo. The lions had left the bloody, dismembered carcass in full view on the open plain and were resting, invisible from the road, in the shade of some dense acacia bushes about 20 m away. As we sat in the car deciding what to do next, a group of baboons began to approach. Although baboons normally give alarm calls when they see lions, and flee from them at close range, these baboons showed no apparent concern as they approached the carcass. As the group passed slowly by, some of the juveniles stopped to sniff the carcass. Then, suddenly, a few of the adults spotted the lions and began to alarm bark. All the baboons fled immediately toward the trees.

We watched this incident with great puzzlement. The baboons had often observed lions at a kill and must surely have had ample opportunity to learn that only a predator could be responsible for a freshly killed and dismembered buffalo carcass. And yet they had apparently needed to see the lions themselves; the fresh carcass alone had caused no obvious concern. Why did the baboons not seem to recognize the carcass as an indication of danger? Could they be such poor naturalists that they simply knew very little about the behavior of lions?

Comparing human intelligence with the intelligence of nonhuman primates is difficult for a number of reasons. First, primate intelligence has typically been measured by performance on learning tests. Comparatively little is known about the knowledge that monkeys and apes acquire naturally, in the absence of human intervention. More important, in the laboratory animal intelligence is generally tested with biologically arbitrary objects like tones or blinking lights. Many of the problems confronting nonhuman primates under natural conditions, however, derive from competitive and cooperative interactions with conspecifics. There is reason to believe that primates may reveal greater intelligence when dealing with each other than when they are dealing with what are, to them, irrelevant

good point

objects. In this chapter we examine what free-ranging vervet monkeys seem to have learned, without human training, about their physical environment. The hypothesis we wish to explore is that many of the cognitive skills exhibited by monkeys and apes have evolved as a result of selective pressures acting on individuals in the context of social interactions.

Primates tested with objects in the laboratory often face problems that are logically similar to the social problems confronting primates in the wild. Recall, for example, the experiments on transitivity described in chapter 3. Capuchin monkeys who have been taught to recognize an ordered, five-item series (like ABCDE) are easily able to order nonadjacent pairs and triplets (like AD or ACE) taken from that series. By contrast, pigeons apparently never learn to do this (D'Amato and Colombo 1988). Similarly, after considerable training with paired stimuli, squirrel monkeys (McGonigle and Chalmers 1977) and chimpanzees (Gillan 1981) seem able to make transitive inferences about the relative rankings of objects. These performances are paralleled by the behavior of monkeys and apes in the wild, where from a very young age individuals seem able to infer a dominance hierarchy among conspecifics from their observation of dyadic interactions (chapters 2 and 3).

To give another example, monkeys in captivity can be taught to classify objects according to similarity or oddity (chapter 3). However, they rarely do so spontaneously (that is, without some reward) or use any criteria other than perceptual similarity. In contrast, socially living monkeys regularly appear to classify individuals on the basis of kinship or close association. These classifications occur without human intervention, and in many cases they seem based on characteristics other than physical similarity, such as close behavioral association or family membership (chapters 3 and 6).

The skills exhibited by captive monkeys and apes, therefore, often seem to be duplicated in social groups. But while captive primates typically require extensive training, the performance of free-ranging primates in social interactions seems to emerge more spontaneously. One explanation for this difference is that primates in the laboratory are simply not motivated to perform in the absence of conspecifics under somewhat abnormal conditions. Research on captive primates has been plagued by motivational problems, and in many studies it has proved difficult to distinguish between a lack of ability and a lack of incentive to complete the task at hand (e.g., Terrace et al. 1979).

There is also, however, a qualitative difference between the stimuli encountered by monkeys and apes in the laboratory and the stimuli they confront in social groups. This difference suggests a provocative hypothesis: namely, that group life has exerted strong selective pressure on the ability of primates to form complex associations, to make transitive inferences,

hyp

and even to judge causal relations, but primarily when the stimuli are other primates. The same skills have not been favored outside the social domain. Monkeys and apes are more likely to solve at least some types of problems if they involve social stimuli, like conspecifics, than if they involve objects.

The hypothesis that certain features of primate intelligence have evolved to deal with the complexity of social interactions was first proposed by Alison Jolly in 1966 and elaborated more specifically by Nicholas Humphrey in 1976 (see also Chance and Mead 1953; Rozin 1976). According to these authors, natural selection has played a powerful role in shaping primates' ability to solve *social* problems. In Seligman's (1970) terms, primates are more "prepared" to deal with social stimuli than with those outside the social domain. Consequently, monkeys in the laboratory can be taught to classify objects according to some abstract criterion because this task requires the same skills that monkeys use whenever they classify animals according to membership in different kin groups or classify vocal signals into different functional categories. Likewise, when monkeys solve problems of transitive inference in the laboratory they are drawing upon skills they use daily in assessing other individuals' dominance ranks. Similarly, when captive chimpanzees solve technological problems that require foresight and an understanding of the consequences of past decisions, they are demonstrating abilities for which they have been preadapted as a result of the need to make equally strategic decisions about each other (de Waal 1982).

Before considering this argument in greater detail, we offer a few disclaimers. The hypothesis that some aspects of primate intelligence may be more highly developed in one domain than another is extremely speculative and largely untested. While we believe that the distinction between "social" and "nonsocial" activity is real and heuristically important, the boundary between these spheres of activity is poorly defined. Given such uncertainties, it seems important to specify what issues this hypothesis does *not* address.

The domain-specific view of primate intelligence posits that natural selection may have acted to favor some abilities in social interactions that are less easily extended or generalized to other contexts. The hypothesis does not, however, specify exactly how elaborate a monkey's social knowledge is or what processes underlie it, nor does it claim that social knowledge can never be extended to other spheres. Most important, it does not suggest that highly developed social skills necessarily imply impoverished abilities in other domains, for example, in remembering the location of food resources. Instead, the hypothesis argues that *certain types of problems* are solved more easily in the social domain than in other spheres of activity.

We focus on the distinction between social and nonsocial performance because we believe that scientists interested in primate intelligence have an obligation not just to describe the marvelous things their subjects do but also to specify where their subjects fail and where the intelligence of monkeys and apes differs most strikingly from our own. Already, in chapters 6 and 8, we have discussed some of the ways in which a nonhuman primate's understanding of his world may be at variance from our own. Contrasting a monkey's social behavior with his performance outside the social domain gives us another opportunity to specify both the richness and the limitations of the nonhuman primate mind.

Once again, our approach is not particularly novel. In recent years there has been a great deal of debate about the degree to which the minds of humans and other animals are modular, or domain specific, and adapted to perform certain tasks with greater skill than others. Below, we review some of these issues and also describe some data on the foraging behavior of birds and other animals that directly address the issue of domain-specific adaptations. We then return to the question of "social intelligence" in nonhuman primates and consider the hypothesis that there is a fundamental discrepancy between monkeys' understanding of their social and physical worlds.

The Modularity of Mind

Even though a honey bee is able to convey highly specific and accurate information about the location of food through her dance, she does not use the dance to convey information other than that concerned with food or the spatial location of objects. Bees do not, for example, use their dance to communicate about their nest mates. Similarly, although a salmon can "remember" the stream where he hatched so well that he is able to navigate back to the same site even several years later, he probably cannot remember the siblings who hatched with him. This may be because the honey bee's dance and the salmon's spatial memory are specialized adaptations that cannot be extended to other contexts (Sherry and Schacter 1987; Rozin and Schull 1988).

In the past, the behaviorist tradition that dominated laboratory psychology emphasized general learning processes and focused on associationist models that, with only minor modification, were thought to apply equally well regardless of the species or stimuli involved. Such general laws of learning were never very appealing to ethologists, whose evolutionary perspective by definition focused on species-specific adaptations to particular environments. Indeed, the earliest field studies of social development emphasized that many types of learning, such as filial imprinting, could *not* be explained in general associationist terms (e.g., Lorenz 1935, discussed in Lorenz 1975).

One of the best examples of a nonassociationist, species-specific learning process is provided by song learning in birds. In North America, song sparrows (*Melospizia melodia*) and swamp sparrows (*M. georgiana*) inhabit an environment where, as nestlings, they hear the songs of many species. Each sparrow, however, learns to sing only its own species' song. For swamp sparrows, the crucial parameter underlying this selective learning is the acoustic morphology of song notes. Swamp sparrows will learn to sing a song if it contains their own species' notes, even if these notes are presented in another species' temporal pattern (Marler and Peters 1977). For song sparrows, selective learning depends on both note morphology and temporal pattern (Marler and Peters 1981).

Although the view that learning is similar across most contexts and species persists in some comparative studies of animal intelligence (e.g., Macphail 1985), there is now widespread agreement that not all species are capable of learning the same thing and that at least some aspects of animal learning are shaped and constrained by species-specific predispositions (Seligman and Hager 1972, Hinde and Stevenson-Hinde 1973, Johnston 1981; Shettleworth 1984). Reinforcing the behavioral data is an increasing body of neurological evidence, from humans and other species, indicating that many patterns of behavior are controlled by specific areas in the brain.

Some of the most extreme examples of the mind's "modularity" come from human victims of strokes and tumors, who often suffer specific behavioral impairments as a result of damage to a particular part of the brain. At a more subtle level, even slight neuroanatomical differences between closely related populations can result in measurable changes in behavior that reflect the different environments in which the individuals have evolved. For example, eastern marsh wrens (*Cistothorus palustris*) typically have song repertoires composed of between 30 and 60 different song types, whereas the repertoires of western marsh wrens are approximately four times larger (Kroodsma and Verner 1987). The two subspecies also exhibit significant differences in the size of song control nuclei in the brain: western marsh wrens have larger song control areas (Kroodsma and Canady 1985). These neuroanatomical differences are apparently the result of different ecological and social selective pressures. Although both species use song to attract mates and drive away male competitors, western wrens are year-long residents in their territories and live at higher population densities, with the result that male-male competition is much more intense. Natural selection appears to have acted with greater intensity in western marsh wrens, favoring large song repertoires and the neural structures that allow them to be acquired (discussed by Kamil 1987).

The observations that animals are predisposed to learn some things more easily than others, and that specialized learning is often associated with specialized neural structures, have led to the view that many aspects of behavior are to some degree domain specific (e.g., Killeen 1985). One deliberately extreme proponent of this hypothesis is the philosopher Jerry Fodor (1983, 1985), who argues that the minds of humans and other animals are divided into modules that are "informationally encapsulated" and relatively discrete when it comes to communicating with other domains. In perception, for example, modules are sensitive to some types of information and insensitive to others. The perception of human speech is dealt with by a specific computational system adapted to perform a specific task; this system has little in common with those that deal with other tasks, such as the perception of music. According to Fodor (1985:4), the function of perception is "to propose to thought a representation of the world from which . . . irrelevant variability has been effectively filtered." Fodor's modules are computationally elaborate (grammar is a good example), domain specific, and involuntary in the sense that an individual cannot help but recognize a phoneme or understand a sentence (via the phoneme- or sentence-recognizing module), even when there is no reason to bother recognizing it (see also Rozin and Schull 1988). Fodor's modular mind is also hierarchical. He hypothesizes that the outputs of the mind's many modules feed into a "smarter" central processing system that is not modular, not encapsulated, and governs higher-order mental activities such as problem solving and decision making. This central processing system organizes and interprets information received from the more peripheral perceptual modules, and it is sometimes conscious and voluntary.

Fodor's model was originally formulated as a foil to the "general process" theory that learning is context independent. Some scientists, however, have criticized it for not being modular *enough* and for relying too heavily on higher, more central, cognitive processes. In particular, Ray Jackendoff (1987) uses evidence from research on human speech, vision, and the perception of music to argue not only that many human abilities are modular but also that modules communicate poorly, if at all, with other modules. He cites, for example, neurological and behavioral evidence that the perception and production of speech are poorly correlated with other "intelligent" activities. Many mentally retarded people who are otherwise incapable of memorizing complex structural rules nevertheless have little difficulty mastering grammar. Similarly, a musical composer whose iterated themes seem to demonstrate a detailed knowledge of mathematics may have difficulty with even simple sums (see also Gallistel and Cheng 1985; Gardner 1985).

In our discussion, we will refrain from using Fodor's modular vocabulary and instead use the less extreme term *domain specific* when referring to abilities that seem relatively restricted to particular contexts or stimuli. At least as commonly described, modules are truly discrete and relatively un-

communicative with other modules. In contrast, although domain-specific information may be coded in a particular way and applied to a finite number of contexts, in theory, rules learned in one domain can be abstracted and extended to other domains. By this view, learning is guided and facilitated by innate principles that organize different domains (Gelman 1987), but it need not be *restricted* to only one domain. Given the paucity of neurological data on the ways human and animal minds perceive and process information, it seems advisable to test the explanatory power of this less extreme model before accepting the more modular one.

Although at least some aspects of human behavior are doubtless relatively domain specific, human intelligence may nevertheless be more accessible than that of other animals. People can and do sometimes extend knowledge gained in one domain to another (Rozin 1976; Gardner 1985). We can, for example, use both inductive and deductive reasoning to derive simple geometric laws, and we can apply these laws either to calculate the length of an abstract shape or, more practically, to predict the angle at which a cue ball will carom off the side of a pool table.

Paul Rozin (1976) has proposed that increased "accessibility," or the ability to generalize knowledge from one context to another, is a crucial component of intelligence. According to Rozin, the evolution of human intelligence has involved an increase in the accessibility of what initially were domain-specific adaptations (Rozin and Schull 1988). If certain features of primate intelligence have indeed evolved to deal with social problems, then a crucial distinction between humans and other primates may be that humans are better able to generalize, or extend, skills used in social interactions to nonsocial domains. So, for example, if the ability to reason analogically first evolved as a result of the need to assess and compare different social relationships (see chapter 3), we might predict that vervet monkeys would have no problem dealing with analogies when the stimuli were other monkeys. They might flounder, however, if the stimuli were keys and locks or can openers and cans. Humans, in contrast, can learn to use analogical reasoning even when dealing with nonsocial or abstract stimuli.

The hypothesis that human intelligence is more accessible than other species' is difficult to test, because, as Rozin and Schull (1988) emphasize, the notion of accessibility is inherently vague. Like any negative result, failure to perform in a given domain could be due to many factors other than inaccessibility. As a result, the mechanisms that might underlie accessibility remain unclear. Although neurological and behavioral evidence suggest that many of our mental abilities are to some extent discrete and domain specific, there also seems to be at least some communication between domains. The extent to which this communication is a dialogue or a soliloquy remains a matter of debate.

Evidence for Domain-Specific Abilities in Animals

Although the social function of intelligence hypothesis has attracted considerable attention (see, for example, the chapters in Byrne and Whiten 1988a), there have been few attempts to test the hypothesis empirically. In fact, when one considers research on animals, almost the only support (however oblique) for domain-specific skills has come not from studies of social interactions but from studies of foraging behavior. The literature on foraging is of interest at least in part because it is one of the few areas of animal behavior in which field research has been complemented by experiments conducted in the laboratory. Here we will limit ourselves to a description of a few selected examples from the extensive literature on foraging behavior. For a more complete review, see Krebs and McCleery (1984); Kamil and Roitblat (1985); Stephens and Krebs (1986); Kamil, Krebs, and Pulliam (1987); and Gallistel (1989a, 1989b).

When a bird forages for food, it must make several decisions. First, it must evaluate and compare the nutritional value of the various types of food it encounters and consider the costs, in time and energy, of handling or capturing each item. Second, it must be able to decide when to abandon its present patch of food for an alternative one. To behave optimally, the bird must be able to compare its rate of return at its current patch with the average rate of return at other food patches, and it must also consider the cost of flying to an alternative patch. Together these decisions demand that the bird be able to represent features of its environment in quantitative terms, compare these values, and use them in quite sophisticated mathematical "calculations." Experiments and observations have demonstrated that the foraging behavior of many birds, fish, and insects conforms closely to what would be predicted by most optimization models. For example, when great tits are offered a choice of two perches, each of which gives access to mealworms after a different numbers of hops on the perch, the birds rapidly learn to spend most of their time at the perch offering the better ratio of reward to effort (Krebs, Kacelnik, and Taylor 1978).

One of the many decisions that an animal must make when feeding in a given food patch is when to abandon it and travel to a new, less depleted area. The animal can behave optimally only if it is able to compare the rate at which it is presently obtaining food with the average rate of reward in other patches. Several investigations have shown that birds are often remarkably good at comparing present intake rates with those experienced in the past. In one experiment, Cowie (1977) placed mealworms in plastic pots attached to artificial trees. The pots mimicked food patches, and travel time between patches was simulated by covering the pots with lids that were either easy or difficult to remove. When the lids were difficult to remove (that is, travel time was long), great tits spent longer at each pot (or patch), removing all the mealworms before moving on to another pot. Cowie compared the length of time spent by birds at each food patch with the values that would be predicted if the birds were comparing gain rates within each patch to the average gain rate for the whole environment. He found no difference between observed and expected probabilities (see also Krebs and McCleery 1984). Similarly, blue jays (*Cyanocitta cristata*) apparently use two simple rules to decide when to leave a patch: the number of prey they have already found and the length of time since they have found prey (or "run of bad luck"; Kamil, Yoerg, and Clements 1988). To use these rules, the birds must be able to compare depletion rates across patches, measure the time already spent in a patch or the amount of prey already captured, and track reward rates.

Animals also seem to use estimates of food depletion rates when "deciding" whether or not to defend their resources against intruders. In a study of golden-winged sunbirds (*Nectarinia reichenowi*) in Kenya, Gill and Wolf (1977) found that territorial defense depended on the rate of nectar renewal in flowers. When nectar levels were very high or very low, the birds abandoned their territories because the energetic costs of territory defense outweighed the benefits of controlling a fixed area. At intermediate levels of nectar replenishment, the benefits of territory defense were greater because defense allowed the flowers to replenish themselves without depletion by competitors. The costs of defense were offset by the decrease in foraging time required to find flowers with nectar. How the birds were able to measure depletion rates so accurately is not known (see also Davies and Houston 1981, 1984). Gallistel (1989a, 1989b) argues that the apparent ability of birds, rats, and even fish to calculate absolute and relative rates so accurately suggests that many animal species not only have mental representations of numbers and temporal intervals but also use these data to perform calculations analogous to division.

Experiments in operant conditioning (originally designed with entirely different issues in mind) provide further evidence that animals can learn to adjust their responses according to the rate and ratio of return. When presented with two food hoppers that offer different ratios of reward, pigeons soon peck almost exclusively at the hopper offering the higher ratio (Herrnstein and Vaughan 1980; see also Shettleworth 1984). Pigeons and rats can also count the pecks or bar presses required to obtain a food reward, even when they are interrupted (reviewed by Shettleworth 1984; Gallistel 1989a, 1989b). In one test of rats' counting ability, Davis and Bradford (1986) trained subjects to enter one of six different tunnels. The rats were able to use the tunnels' ordinal relationships to "count" forward or backward in order to enter the appropriate one. It is highly probable that these counting skills are adaptations that have evolved to allow these opportunistic foragers to monitor food intake and depletion rates more efficiently.

When captive rats or pigeons are presented with food hoppers (or "patches") that offer rewards at variable intervals, they soon learn to match the proportion of time spent at one hopper to the proportion of all rewards received at that hopper (Herrnstein and Vaughan 1980). In this case, matching the proportion of time spent at each hopper with the proportion of rewards received there is the best strategy to pursue because it provides a higher rate of return than simply concentrating on the better patch. Similarly, pigeons that are allowed to sample foraging "patches" (shuttle boxes) that provide a stable or a randomly fluctuating probability of reward succeed in tracking the changes in the fluctuating site to obtain a payoff that is close to optimum (Shettleworth et al. 1988).

Under natural conditions, the foraging behavior of birds also frequently approximates an optimal "matching" strategy. In a very simple but illuminating experiment, Harper (1982) tested the ability of mallard ducks (*Anas platyrhynchos*) at a village pond to estimate feeding rates. Two experimenters placed themselves approximately 20 feet apart at the pond's edge and threw bread to the ducks. One experimenter threw bread twice as fast as the other. Within minutes, the ducks had deployed themselves in such a way that there were twice as many ducks at the spot where the reward rate was twice as high. When the experimenters varied the size of the pieces of bread, so that larger pieces were being thrown at the site with the less productive throwing interval, the ducks rearranged themselves so that a higher proportion of animals concentrated at that site. Perhaps most interesting, this "ideal free distribution" of ducks (Fretwell and Lucas 1970) was achieved long before many of the subordinate ducks had obtained any bread. In other words, the ducks were able to estimate the rate of return at each throwing site through observation alone, without having to experience the feeding regimens themselves (see also Godin and Keenleyside 1984 for similar experiments on fish; and Gallistel 1989a, 1989b for reviews).

Some species are remarkably good at remembering and representing the spatial location of food. Birds that cache their food, like marsh tits (*Parus palustris*) and Clark's nutcrackers (*Nucifraga columbiana*), are able to remember the location of large numbers of storage sites for several days after they initially cached their food (Shettleworth and Krebs 1982; Kamil and Balda 1985; Balda, Kamil, and Grim 1987). The number of storage sites is by no means trivial; nutcrackers regularly hide seeds in as many as 7,500 separate caches. Somehow, however, they are able to "check off" sites that they have already visited to avoid revisiting caches that have been depleted (Balda and Turek 1984). The spatial memory of caching species may be considerably better developed than that of noncaching species like pigeons. For example, nutcrackers perform better than pigeons in mazes that require the birds to forage for previously hidden food items without revisiting sites that have already been exploited, and they remember the location of food for considerably longer periods than do pigeons (Balda and Kamil 1988; Sherry 1985; Spetch and Honig 1988; Shettleworth and Krebs 1986; see also Olton and Samuelson 1976 and Olton 1985 for rats). There are significant differences in the volume of the hippocampal complex between bird species that do and do not cache food, suggesting that ecological factors have exerted considerable selective pressure on the evolution of the specific neural structures associated with specialized memory capabilities (Krebs et al. 1989).

Ecological factors are not the only selective pressures favoring the evolution of spatial memory; in some species there are sex differences in spatial memory that can be linked to that species' social structure and mating system. Meadow voles (*Microtus pennsylvanicus*), for example, are polygynous rodents found in the eastern United States. In this species, males expand their ranges relative to those of females during the breeding season, a reproductive tactic that potentially allows males to encounter and monopolize more sexually receptive females. In contrast, prairie voles (*M. ochrogaster*), a related monogamous species, exhibit no sex differences in range size. In this species, each male mates with only one female and therefore cannot increase his reproductive success by increasing his exposure to females. Gaulin and Fitzgerald (1989) hypothesized that sexual selection favoring increased range size in male meadow voles might also have led to sex differences in spatial memory in this species, and they predicted that the polygynous meadow voles would exhibit more sex differences in spatial memory than the monogamous prairie voles. They tested this hypothesis with a series of experiments in which males and females of both species were required to shuttle back and forth between two food boxes at opposite ends of a maze in order to obtain food. As predicted, meadow voles showed signifiant sex differences in performance, whereas prairie voles did not. Male meadow voles made fewer errors than female meadow voles, while male and female prairie voles performed equally well.

Under both captive and free-ranging conditions, therefore, animals often appear to make complex calculations of handling time, rates of reward, and the spatial configuration of food items or territories (Gallistel 1989a, 1989b). Nevertheless, we do not typically think of most rodent, birds, and fish as particularly "intelligent," at least not in the human sense. Instead, the calculations made by animals to maximize their feeding efficiency seem to be more accurately regarded as species-distinct traits that have evolved under specific ecological pressures to serve a relatively narrow function and that are under relatively specific, perhaps even inflexible, neuronal control.

As yet, no experiments have investigated whether birds that cache food are also better than other species at remembering non-food-related experiences, such as the identities of siblings that dispersed from their natal area many years before. If we accept the hypothesis that the primary selective pressure favoring a highly developed memory in caching species is the need to remember the location of food storage sites, we would predict that food-caching birds would be unable to extend their abilities to nonforaging contexts and would be no better than other species at recognizing, for example, long-dispersed kin. Similarly, male meadow voles might be no more adept than female meadow voles at remembering whether they had already depleted a given food patch. How accessible is the spatial memory of a nutcracker or a meadow vole? It would be easy to investigate this question, and such studies would provide a much needed test of the modularity hypothesis.

Social Skills in Humans and Other Primates

Throughout this book, we have concentrated almost exclusively on social behavior, and we have focused in particular on the detailed knowledge that primates seem to have of their own and other animals' social relationships. Although many observers have been impressed by the social knowledge of primates, however, there have been almost no attempts to compare the types of abilities used by primates (or any other animals) in social interactions with those used in calculating rates of food rewards or the location of food in space and time. To investigate this question properly, we must first establish whether or not primates are in fact predisposed to attend to social stimuli. Second, we must establish whether a problem that involves social stimuli is more readily solved than a logically similar problem that involves nonsocial stimuli.

Not surprisingly, primates differentiate easily between their own species and the members of other species, even when these other species are quite closely related. For example, when individuals from five species of macaques were allowed to press levers to see slides of their own and other macaque species, they showed the most interest in looking at pictures of their own species (Fujita 1987). Primates also seem to distinguish *individuals* more readily within their own than within another species. Nicholas Humphrey (1974) examined the readiness of rhesus macaques to look at slides of other rhesus macaques by measuring the rate at which they pressed a lever that activated a slide projector. Although the monkeys rapidly habituated to slides of the same individual, they regained interest if they were shown slides of different individuals. By contrast, the monkeys quickly became bored by slides of pigs, even when the slide collection included portraits of different pigs. The monkeys treated other monkeys, but not other pigs, as distinct individuals.

This is not to suggest, however, that monkeys cannot learn to recognize the members of other species as individuals. After obtaining his preliminary results, Humphrey covered the monkeys' cages with pictures of pigs engaged in different piglike activities: black pigs jumping in mud, white pigs grazing peacefully, pink pigs sleeping, and so on. A few weeks later, he retested the monkeys with the pig slides. This time the monkeys behaved as they had when shown slides of other monkeys, responding to each slide of a different individual pig with renewed interest.

Some of the best evidence for the preeminence of social stimuli and the importance of social experience in primates comes from studies of social development. Monkeys (usually rhesus macaques) who are reared with inanimate surrogates are generally unable to interact normally with other monkeys as adults and show numerous sexual, maternal, and other social pathologies (see Harlow and Harlow 1965 and Suomi and Ripp 1983 for reviews). By contrast, many birds and nonprimate mammals, although group living and gregarious as adults, exhibit social behavior that is considerably less affected by periods of isolation during infancy.

The ability to engage in complex social interactions—for example, to recognize other animals' social relationships and relative dominance ranks—seems to depend even more strongly on experience with other conspecifics. In one study of captive rhesus macaques, Anderson and Mason (1974) compared the aggressive behavior of socially reared juveniles with that of juveniles who had been reared only with their mothers. When a socially reared juvenile redirected aggression to another animal following a fight with a more dominant opponent, he always threatened an animal who ranked lower than his opponent and never threatened one of his opponent's close associates. Juveniles who had been reared only with their mothers, however, seemed unable to assess other individuals' ranks and frequently threatened animals higher ranking than themselves (see also Mason 1978b).

Humans, too, seem predisposed to attend to members of their own species. Human infants show special sensitivity to human faces as opposed to inanimate visual stimuli (Sherrod 1981) and to speech sounds as opposed to other auditory stimuli (Eimas et al. 1971). Moreover, from a very early age human infants are sensitive to the distinction between animate and inanimate objects. Specifically, they note that animate objects, unlike inanimate ones, move on their own, perceive, and have feelings and intentions (reviewed in Gelman and Spelke 1981; Massey and Gelman 1988).

Since the distinction between animate and inanimate objects is drawn very early in life, children's understanding of complex phenomena often emerges first in the domain of social interaction. Take causality, for example. Though both animate and inanimate objects can act on one another and have causal relations, there is some evidence that children comprehend social causation before they understand causality in the physical domain. Fein (1972) showed children aged 4, 7, 11, and 15 years sequences of pictures, some depicting a causal sequence (like a child doing something good and then being rewarded) and some depicting a sequence of events that were causally unrelated. Some sequences depicted people; others depicted objects like blocks. When tested with pictures of people, children distinguished causal from noncausal sequences between the ages of 4 and 7 years. When tested with pictures of objects, the same distinction was not made until between 7 and 11 years (see also Leslie 1984). Of course, the ability to recognize causality in social contexts more easily than in nonsocial ones is also likely to depend on experience. Being social creatures, children naturally have had more experience with social interactions than with objects; experience in this cases poses a methodological problem that is difficult to overcome.

Studies of the earliest stages in language acquisition also suggest that social stimuli have a special salience for human infants. In a study of naming behavior in 2-year-olds, MacNamara showed subjects a toy block and a doll. He referred to one using a common noun ("This is a zav") and to the other using a proper noun ("This is Zav"). Common and proper nouns could be applied to either the block or the doll. After a brief play session, children were asked to perform some action using "the named object." When the named object was a doll they chose correctly, but if the named object had been a block, their selections were no different from chance. MacNamara concluded: "by the time the child comes to learn language, he has already learned that objects in certain categories are important as individuals, those in other categories are merely exemplars of the category. Person is the preeminent category of the first sort. . . ." (1982:30; see also Hood and Bloom 1979; Shatz, Wellman, and Silber 1983).

block vs doll

Although social stimuli may be particularly salient for young children, it would obviously be inaccurate to draw a strict dichotomy between cognitive development in social interactions and cognitive development outside the social domain. Clearly, normal social development is heavily dependent on many general cognitive skills and can be strongly affected by what a child learns outside the context of social interactions. In two sorts of pathological development, however, the dissociation between social and nonsocial performance is striking (Baron-Cohen, Leslie, and Frith 1986; Fein et al. 1986). On the one hand, children with Down's syndrome are surprisingly successful in social interactions, despite below-normal IQ scores and language skills. Their social smiles, attentiveness, and ability to express themselves through gesture are all well developed (e.g., Wing and Gould 1979; Sorce et al. 1982; Coggins et al. 1983). Children with Down's syndrome also respond positively to social reinforcers such as smiles or pats on the back. By contrast, children labeled as autistic, while in some cases possessing normal IQ scores, consistently show abnormalities in social behavior. They rarely make eye contact, show little interest in others, and are generally described as socially "aloof" (e.g., Rutter 1983; Lord 1984). The same social reinforcers that work well for children with Down's syndrome work poorly in the case of autism.

Baron-Cohen, Leslie, and Frith (1986) tested the ability of normal children, autistic children, and children with Down's syndrome to arrange pictures so that they depicted a predetermined narrative sequence. Across the three groups, autistic children had the highest scores on tests of nonverbal mental ability. In picture sequences that required no attribution of mental states to the objects or people displayed, the autistic children did as well or better than the others. In tests that required the attribution of mental states, however, autistic children performed significantly worse than normal children and those with Down's syndrome (see also Leslie 1988). In such experiments, as well as the clinical descriptions summarized previously, the separation between "social" and "nonsocial" performance is striking.

In their review of children's knowledge about animate and inanimate objects, Gelman and Spelke conclude, "Children's understanding of a given domain depends not only on the logical structure of the tasks used to assess competence and the level of development of some general set of cognitive structures. Their competence depends as well on the nature of the objects about which they must reason" (1981:44). The same may be true of adults. For example, when adults are given certain tests of logical reasoning, they seem to perform better when the problems are phrased in terms of social contracts and violations than when they are presented in a form that is divorced from social interaction.

The Wason selection task (Wason 1983) is a test of logical reasoning that asks subjects to determine whether a conditional rule has been violated. In a typical test, a subject is given four cards, each of which displays a number or a letter on one side and a letter or a number on the other. Given four cards labeled E, K, 4, and 7, the subject is asked to turn over the two cards that will allow him to determine whether there has been a violation of the rule "If a card has a vowel on one side, then it has an even number on the other." Only between 4 and 10% of naive subjects choose the correct combination: the true antecedent (E) and the false consequent (7) (Wason 1983). Subjects do considerably better, however, when the stimuli on the cards are not abstract but deal with a familiar situation, such as transportation by car or train to two different cities (Wason and Johnson-Laird 1972).

Previous experience with the stimuli in question, however, cannot entirely explain variation in the level of accuracy in the Wason test. In an explicit test of the hypothesis that humans have been selected to attend to the costs and benefits of social exchanges, Leda Cosmides (1989; Cosmides and Tooby 1989) designed a series of Wason tests to evaluate the ability of college students to detect a violation of a social contract. Cosmides made a special effort to design social situations that would be unfamiliar to her subjects. A typical test, for example, asked subjects to evaluate the problem "If a man eats cassava root, then he has a tattoo on his face." Over 70% of the subjects chose the correct combination of cards (correct antecedent, false consequent) when the problem was phrased in social contract terms. Cosmides concludes that people are adept at analyzing problems in terms of social costs and benefits, an ability to which they have been predisposed as a result of selection acting on the need to function in social groups that are simultaneously competitive and cooperative.

It is not at all surprising to find that humans are predisposed to attend to human faces or voices, since all animal species, regardless of whether or not they are social, are predisposed to compete, mate, and interact with conspecifics. What is of greater interest is the suggestion that humans may be predisposed to analyze at least some types of problems in terms of social exchange. Perhaps, however, this is also a tautology, since it is difficult to think of a myth, story, or novel in any culture that is not somehow concerned with these issues. In fact, as Humphrey (1976) suggests, our tendency to view problems in social terms is sometimes taken to absurd lengths. Our animistic ancestors placed gods in the heavens and on earth and remonstrated and negotiated with them about weather, crops, and personal fortune. Even now, in our more technologically oriented society, we continue to name hurricanes, to scream at recalcitrant cars, and to avoid walking under ladders. We may know that none of this makes sense, but somehow it is more comforting.

Comparing Social and Nonsocial Knowledge in Primates

Although many species of primates, including humans, may be predisposed to attend to social stimuli, it is more difficult to show that they manifest greater—or, more accurately, different—abilities in their social interactions than in their nonsocial behavior. In 1983, we began to test the "social intelligence" hypothesis by investigating vervet monkeys' knowledge of their physical environment. We tried to design a number of experiments with nonsocial stimuli that were formally comparable to ones we had previously used to examine the monkeys' understanding of social relationships. Now, however, our goal was to compare what monkeys knew about other vervets with what they knew about other *species*.

Our experiments are preliminary, highly speculative, and by no means as precisely controlled as laboratory tests. Nevertheless, they have at least two advantages. First, we avoid problems of motivation and human training by testing animals in their natural habitat, using stimuli that are biologically important to them. Every day, free-ranging primates encounter logically similar problems that involve both social and nonsocial objects and events, a serendipity that permits a direct test of performance in these two domains. Second, the monkeys regularly deal with objects in the external world that may be relevant or irrelevant to their survival. It is therefore possible to compare monkeys' social knowledge not just with their knowledge of biologically relevant objects but also with their knowledge of objects that are apparently unrelated to survival.

Two further caveats are important. First, as noted earlier, although we draw a distinction between social and nonsocial tasks, the border between these two domains is by no means clear. If it turns out that nonhuman primates do indeed perform some tasks better when they are dealing with social stimuli, we will need further research to establish the exact features that define this category. Second, in evaluating the observations and experiments presented here, we make no claims about the mechanisms underlying performance. Our tests define knowledge operationally; they measure only the responses that particular stimuli evoke and not the processes (mental or otherwise) that underlie such responses. Many of the results we describe could result from associative learning; they might also involve a more complex process that includes an understanding of the causal relations between different events. Our aim is not to argue for one of these alternatives. Instead, we use experiments to determine which of two stimuli seems more salient and to consider whether vervets attend more to some aspects of their environment than to others.

Relevant Aspects of Other Species' Behavior

For vervets, the behavior of other species falls along a continuum. At one extreme is behavior that is directly relevant to the monkeys' survival; the alarm calls of other species are an obvious example. Clearly, vervets ought to attend to these calls if they warn the monkeys of predators to which vervets are vulnerable. At another extreme is behavior by other species that seems irrelevant to the monkeys' well-being. This behavior includes, for example, the courtship calls of hornbills, the territorial displays of male hippopotami, nest building by vultures, and so on. How much do monkeys know about behavior that they observe every day but that is of no survival value to them?

Vervets obviously do attend to at least some aspects of other species' behavior. Recall, for example, their responses to the alarm calls of the superb starling (chapter 5). The superb starling, like many other birds, has two acoustically distinct alarm calls. One is a harsh, noisy call given in response to a wide variety of primarily terrestrial predators, including carnivores, snakes, and even vervets. The other, a clear, rising or falling tone, is given in response to raptors. When we played the starling's raptor alarm to vervets, they tended to look up. Although the monkeys' responses to the starling's terrestrial predator alarm were more varied, this alarm was the only starling call that regularly caused subjects to run to trees. Finally, the starling's song, presumably irrelevant to the vervets' survival, evoked no particular response or vigilance (chapter 5).

Over the years, we have noticed that vervets also respond to the alarm calls of many other species. The loud alarm snorts of ungulates like impalas (*Aepyceros melampus*), wildebeests (*Connochaetes taurinus*), zebras (*Equus burchelli*), Grant's gazelles (*Gazella granti*), and Thompson's gazelles (*Gazella thompsoni*) evoke vigilance and, frequently, flight into trees. Vervets also respond to the alarm calls of guinea fowls (*Numida mitrata*) and to the alarm barks of baboons. We do not know, however, how detailed their knowledge of these species' alarm calls is. Many different ungulates, for instance, give what seem to us to be similar-sounding alarm snorts. These species' alarm snorts also sound very much like the snorts given by males when they herd females during the mating season. We do not know if vervets distinguish among different species' snorts or if they can detect any differences between alarm snorts and rutting snorts. Figure 9.1 compares the alarm calls given by vervet monkeys, starlings, and impalas.

We also do not know whether vervets recognize that the alarm calls of different species are to some extent equivalent in the sense that they all denote the presence of danger. When different species' alarm calls occur separately, vervets certainly treat these calls as warning signals. But do the monkeys recognize that the alarm calls given by different species often have the same general referent and as a result are to some degree interchangeable? Do they recognize, for example, that a leopard could evoke both an impala's snort and a starling's terrestrial predator alarm call? Although this may seem a rather esoteric point, it is directly relevant to questions we raised earlier about the accessibility of knowledge and the monkeys' ability to know what they know.

In chapter 5, we described some experiments that examined whether vervets recognized the similarity between their own alarm calls and the alarm calls of starlings. We found that vervets treated starling raptor alarm calls and their own eagle alarm calls as having similar referents. Having habituated to one of the two alarm calls, vervets transferred their habituation to the other. By contrast, when tested with starling terrestrial predator alarms, vervets transferred habituation to both vervet leopard and vervet eagle alarms. We concluded that, to the monkeys, a starling's terrestrial predator alarm denoted a broader class of danger than the starling's raptor alarm.

We were also interested in determining whether vervet monkeys knew

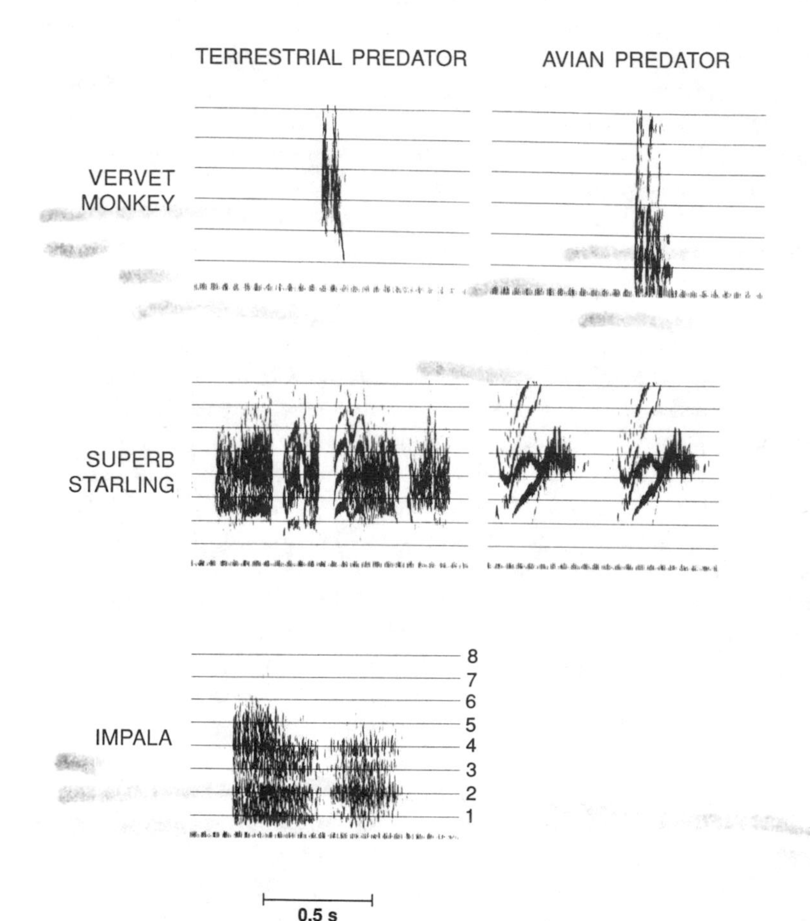

Figure 9.1. Spectrograms of alarm calls given by adult female vervet monkeys, starlings, and impalas to terrestrial and avian predators. Vervet alarm calls are taken from figure 4.4 and starling alarm calls are taken from figure 5.7. Impala alarm snorts were obtained from the British Library of Wildlife Sounds and were recorded from impalas in Kenya alarm calling at a lion.

enough about other species' alarm calls to recognize that two species *other than vervet monkeys* might have acoustically different alarm calls for the same class of predators. To investigate this issue, we replicated the habituation experiments described in chapter 5, this time using as stimuli the alarm snorts given by impalas and the terrestrial predator alarm calls of starlings. On day 1, we played a subject an impala snort (or a starling terrestrial predator alarm). Then, on day 2, the subject heard eight playbacks of the starling's (or the impala's) alarm, followed once again by the same call she had heard on day 1.

When vervets were played a vervet or a starling alarm after repeated exposure to the other species' alarm, they transferred habituation from one species' call to the other's (chapter 5). By contrast, vervets *failed* to transfer habituation between impala snorts and starling terrestrial predator alarms (fig. 9.2). Although all 10 vervet subjects rapidly habituated to repeated presentations of an impala alarm snort or a starling terrestrial predator alarm call, only five subjects transferred habituation from one species' call to the other's. In other words, even though the monkeys were being asked to compare two calls whose referents were rather similar, they behaved as if they did not recognize that starling calls and impala snorts referred to the same types of predators.

There are at least two possible explanations for this result. The first relates once again to the question of call specificity, since starling terrestrial alarms do seem to be less specific in the objects they denote than either vervet leopard alarms (chapter 5) or impala snorts. Although starlings give terrestrial predator alarm calls to a wide variety of predators, impala alarm snorts are given primarily in response to carnivores, including leopards, lions, hyenas, and jackals. Vervet monkeys may fail to transfer habituation between starling and impala alarms because, to the vervets, the two calls mean quite different things.

This is unlikely to be the sole explanation, however. After all, vervet leopard alarms are even more restricted in scope than impala alarm snorts,

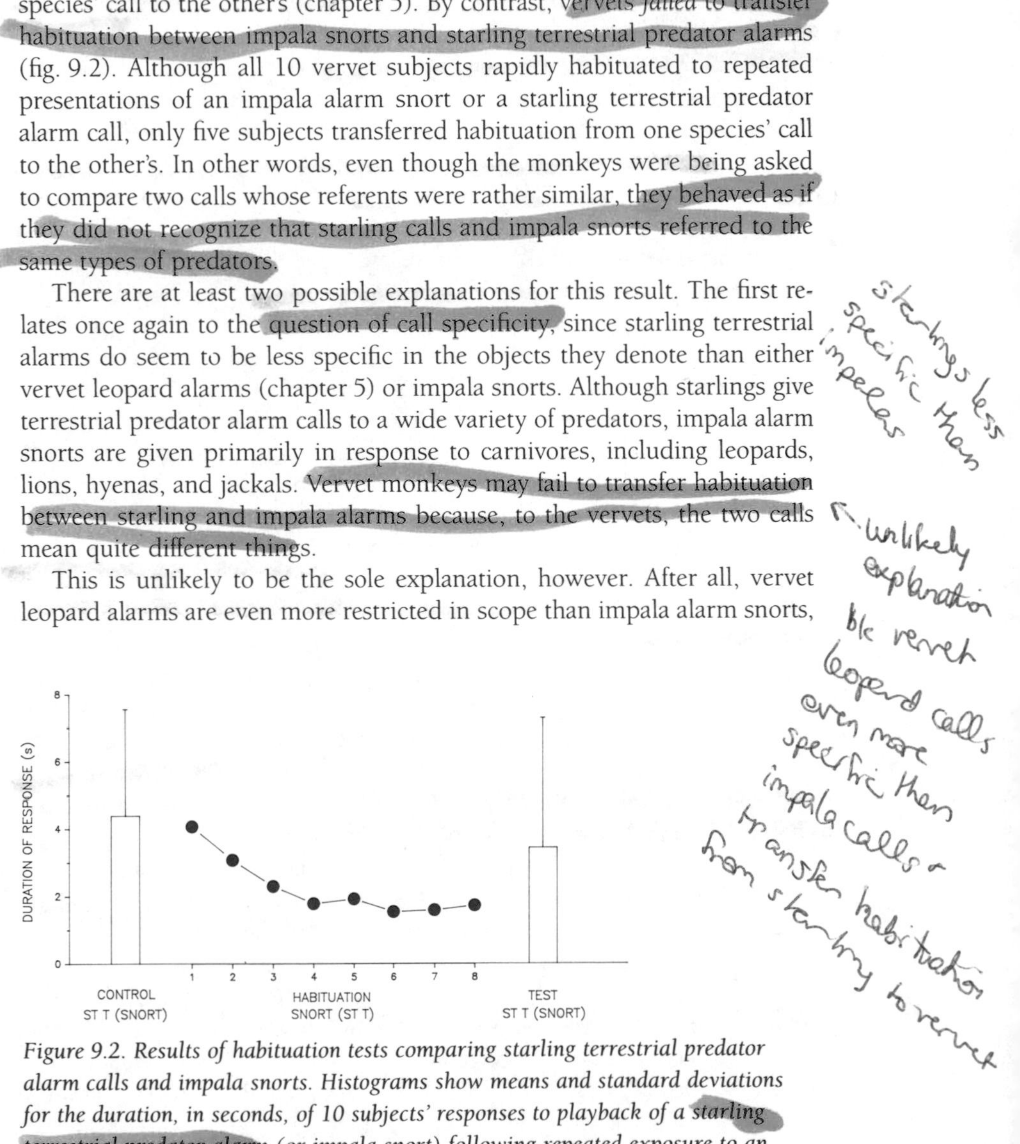

Figure 9.2. Results of habituation tests comparing starling terrestrial predator alarm calls and impala snorts. Histograms show means and standard deviations for the duration, in seconds, of 10 subjects' responses to playback of a starling terrestrial predator alarm (or impala snort) following repeated exposure to an impala snort (or starling terrestrial predator alarm) (test), compared with subjects' responses to the same alarm call in the absence of such exposure (control).

yet when the monkeys have habituated to either a starling's or a vervet's alarm call they transfer their habituation to the other species. Hence we offer a second, more speculative explanation, one based on the vervets' ability to assess aspects of the external world without reference to themselves. This explanation proposes that vervets learn the meaning of other species' alarm calls (that is, they learn what these calls denote) by comparing them with calls in their own repertoire. Vervets learn the meaning of starling raptor alarms, for example, by comparing the referents of these calls with the referents of their own eagle alarms. If this is true, then the monkeys may be unable to judge the relation between two *other* species' calls because this point of reference is absent. Vervets may simply be unable to regard the world through any perspective other than their own.

Even apes seem to have difficulty regarding problems from perspectives other than their own. Kohler's captive chimpanzees, for example, were easily able to guide *themselves* around barriers that were placed in their paths, but they had far more difficulty negotiating *items*, like food, around the same barriers. In one test, a banana was placed in a three-sided box outside the cage, which was oriented in such a way that the chimpanzee had first to push the banana away from himself before raking it toward him. Even though the chimpanzees all had experience with sticks, and even though all could solve detour problems that involved their *own* movements, most subjects were unable to complete this task (Kohler 1925). Failures such as these led Kohler to conclude that chimpanzees were unable to understand relations among objects in which they themselves did not serve as a reference point (see also Boakes 1984). If apes (and indeed young children) find it difficult to hold different, multiple representations of the same object or event (chapter 8), it may not be surprising that vervet monkeys can only interpret the calls of other species in relation to their own.

We were unable to test this hypothesis because the decreasing number of monkeys in Amboseli prevented us from conducting at least two crucial additional experiments. For example, because raptor alarm calls seem to be more specific and less ambiguous than terrestrial predator alarm calls, it would have been interesting to test whether vervets can recognize the similarity between a starling's raptor alarm and an acoustically different raptor alarm given by some small mammal like the slender mongoose. In addition, it would have been important to compare the vervets' assessment of impala snorts and vervet leopard alarm calls. If vervets judge snorts and their own leopard alarm calls as similar (by transferring habituation between the two calls), and if they also fail to recognize the similarity between the raptor alarms of a starling and of, say, a mongoose, then we would have more convincing support for the hypothesis that vervets assess the meaning of other species' calls largely by reference to the meaning of their own.

Apparently Irrelevant Aspects of Other Species' Behavior

As we noted earlier, the alarm calls of other species represent, for vervet monkeys, one end of a continuum of biologically relevant and irrelevant features of the environment. It is perhaps not surprising that vervets discriminate among such alarm calls, since they are so obviously important to the monkeys' survival. Can similar knowledge be demonstrated, however, for stimuli outside the social domain that are apparently unrelated to the monkeys' survival? This is an important question, because one striking feature of human intelligence is our inclination to accumulate information about the world that is not directly relevant to the getting and spending of daily life. Humans devote impressive amounts of time to the accumulation of bird life lists, baseball statistics, rock and roll trivia, and otherwise fairly useless information that they may never use or impart to anyone else. Can the same be said of vervet monkeys? Are vervets as good naturalists as they are primatologists?

To address this question we had first to identify two comparable features of the monkeys' environment—one social and presumably biologically relevant, the other nonsocial and apparently irrelevant to the monkeys' survival. We then designed experiments that could assess the monkeys' performance in these two domains. As a social, biologically relevant test, we used the cross-group recognition experiments described in chapter 3, which asked the monkeys how much they knew about the location of individuals in other vervet groups. As a nonsocial, apparently irrelevant test, we asked the monkeys how much they knew about the location of two species that neither compete nor interact with vervets in any obvious way.

In our earlier cross-group recognition experiments, we investigated vervets' knowledge of other groups' membership and ranges by playing the intergroup *wrr* call of a female, either from the true range of that female's group or from the range of another neighboring group. In these paired trials, subjects responded with significantly more vigilance to calls played from the inappropriate range than to calls played from the appropriate range (chapter 3). The monkeys appeared to recognize the group with which a particular female was associated, even though she was not a member of their own social group.

Our subsequent "nonsocial" tests followed the same experimental design but used as stimuli the calls of other species. Vervets were played the calls of two species that are habitually found in or near water, the hippopotamus (*Hippopotamus amphibius*) (fig. 9.3) and the black-winged stilt (*Himantopus himantopus*) (fig. 9.4). The hippopotamus' call is a territorial call, whereas the black-winged stilt's is a low-intensity alarm call given in response to a wide variety of potentially disturbing species. Neither the

Figure 9.3. Although hippopotami feed on land at night, they seldom emerge from the water during the day. The males' territorial calls can frequently be heard around waterholes.

hippopotamus nor the black-winged stilt attacks or competes with vervets, and both seem to be of little biological importance to the monkeys. At the same time, both species are so restricted to wet areas during the day that any indication of their presence in another habitat might be regarded, at least by humans familiar with the area, as anomalous. Black-winged stilts are never found away from water, and although hippos do emerge from water to feed on dry land, they do so only at night (Olivier and Laurie 1974).

We played hippo and stilt calls to vervets in paired trials, from either a swamp ("appropriate") or a dry woodland ("inappropriate") habitat. All subjects were members of groups whose ranges bordered both swamps and dry acacia woodlands, and all had regularly heard the calls of hippos and black-winged stilts when foraging near the swamps (Cheney and Seyfarth 1985a).

In marked contrast to the cross-group recognition experiments, the vervets did not respond differently to the two species' calls when they were played from swamps as opposed to dry woodland (fig. 9.5). Vervets typically showed little response to the playbacks of hippo vocalizations, regardless of the habitat from which the calls were played. Upon hearing stilt vocalizations, vervets often responded by looking in the direction of the loudspeaker, but the duration of response did not differ significantly between dry and wet habitats. There was some indication that the vervets treated the stilt's call as an alarm: 5 of 18 subjects looked up, and 3 subjects ran toward trees or stood bipedally when they heard the call. Again, how-

Figure 9.4. The black-winged stilt is a wading bird found only in and around waterholes.

ever, the monkeys did not respond more strongly in one habitat than another. Vervets responded to both hippo and stilt calls as if they did not recognize that calls played from a dry area were anomalous.

These negative results, of course, cannot distinguish between the failure to recognize an anomaly and the failure to respond to one. It is entirely possible, for example, that vervets know very well that hippos belong near water but simply do not respond to their calls. The negative results *are* of interest, however, when contrasted with similar experiments on cross-group recognition. Although vervets fail to respond to hippo or stilt calls coming from an inappropriate area, under comparable conditions they respond strongly to the calls of another vervet. The different performances are particularly striking given that the cross-group recognition trials asked

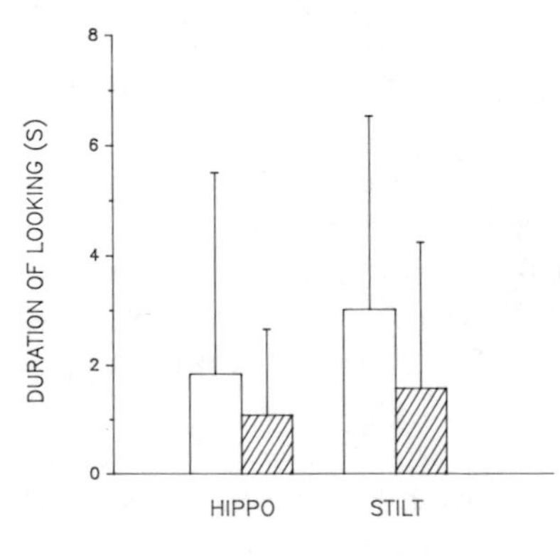

Figure 9.5. The duration, in seconds, that subjects looked toward the speaker after playback of hippopotamus and black-winged stilt vocalizations from wet (open histograms) and dry habitats (shaded histograms). Histograms show means and standard deviations for 10 subjects in hippo trials and 18 subjects in stilt trials.

subjects to assess the appropriate location of different *individuals*, whereas the hippo and stilt calls required only a gross understanding of the appropriate location of different *species*.

The cross-group recognition tests, however, investigated the monkeys' understanding of relevant social stimuli, whereas the hippo and stilt experiments did not. Knowledge of the identities of the individual members of other groups is crucial for territorial defense, while an understanding of the behavior of hippos and stilts has no apparent bearing on survival. The monkeys simply may not attend to those aspects of other species' behavior that are unimportant to their survival and reproduction, even when they encounter those species daily.

It seems likely that knowledgeable humans presented with the same experiments would, unlike vervets, respond strongly. However irrelevant the behavior of hippos and stilts may be to us, hearing these species call from a dry woodland presents a picture of the world that is at variance with our past experiences. In an analogous way, we might expect an ardent bird watcher to respond strongly to a wood thrush song coming from an open prairie or to a robin in Vermont singing in December. The fact that vervets do not respond as at least some humans might suggests that the monkeys do not perceive their world as we do ours and do not ask the same questions about it as we do.

Associations among Other Species

When interacting with conspecifics, vervets seem to have a clear understanding of the relationships that exist among other individuals. For example, vervets can associate the screams of particular juveniles with those juveniles' mothers (chapter 3; Cheney and Seyfarth 1980). Their recognition of social relationships is apparently based on observations of other individuals' behavior. To test whether vervets recognize similar associations outside the social domain, we tested their understanding of the relationships that exist among other species when one of these species is relevant to the monkeys' daily life.

Vervets regularly come into contact with Maasai tribesmen, who bring their livestock into Amboseli to graze during the dry season. The approach of Maasai causes the monkeys to give an acoustically distinct *strange human* alarm call and flee silently from the immediate area (chapter 4). Maasai in Amboseli are usually associated with cows, goats, and donkeys, which themselves pose no threat to vervets (fig. 9.6). Nevertheless, since these species are almost always accompanied by Maasai, their presence potentially signals the approach of danger.

If vervets have learned to associate Maasai and their livestock, we might expect them to respond to the approach of livestock almost as strongly as they respond to the Maasai themselves. There is, in fact, some anecdotal

Figure 9.6. A herd of cows emerges from a cloud of dust. During the dry season, the Maasai bring their livestock into Amboseli for water and grazing daily.

information to suggest that vervets do associate livestock with Maasai. One year during the rainy season when there had been no Maasai in the area for some months, Sandy Andelman, then a graduate student working on our project, heard the monkeys giving the distinctive *strange human* alarm call. Curious to see if Maasai were approaching, she looked around and saw only a herd of impalas. In the midst of the impalas, though, was a donkey that had strayed from a nearby Maasai settlement. The vervets had seen the donkey and had apparently associated the donkey with the approach of Maasai.

We decided to test the vervets' knowledge of the Maasai-livestock association through a series of paired experiments in which we played recordings of either cow or wildebeest vocalizations (see Cheney and Seyfarth 1985a for details). Cows and wildebeests are both grazers that do not compete with or threaten vervets (fig. 9.7). Both also give somewhat similar lowing calls. Cows, however, are associated with Maasai, whereas wildebeests are not, so we might expect the monkeys to respond more strongly to playbacks of cow vocalizations. This is in fact what we found. Playbacks of lowing cows caused subjects to look toward the speaker for significantly longer durations than did playbacks of lowing wildebeests (fig. 9.8). Their greater vigilance suggests that vervets associate cows with danger and that they respond to the apparent approach of cows as they would to the approach of Maasai themselves. Interestingly, this association appears not to be as explicit as it might be for humans confronted with the same problem, because in no case did playbacks of lowing cows cause vervets to utter a *strange human* alarm call.

Figure 9.7. Wildebeests are a common ungulate species in Amboseli.

Another auditory cue that signals the approach of Maasai cattle is the ringing of bells, since in any herd there are always a number of cows wearing bells. When we played recordings of cow bells to vervet monkeys, they responded as strongly to the bells as they did to the lowing calls. Playbacks of bells caused intense vigilance in the direction from which the bells were played, even when the sound came from over 100 m away.

Incidentally, these responses to cows and bells are by no means restricted to primates. Maasai in Amboseli occasionally spear rhinoceros and elephants, and both species respond strongly to playbacks of cows and bells. In fact, to the consternation of everyone involved, one of our playback experiments caused two rhinos to burst in panic-stricken flight from a nearby cluster of bushes.

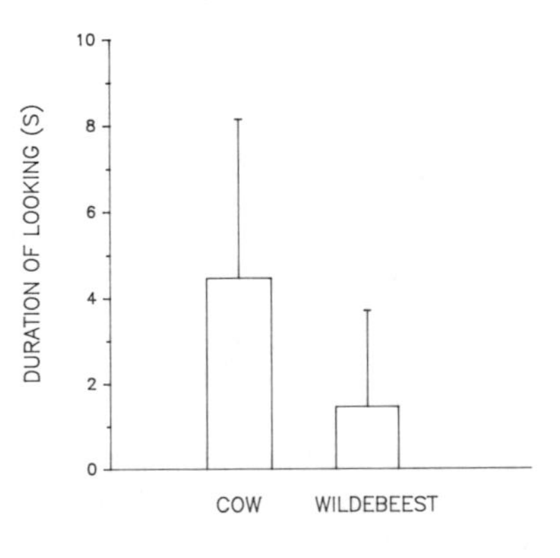

Figure 9.8. The duration, in seconds, that subjects looked toward the speaker after playback of cow and wildebeest vocalizations. Histograms show means and standard deviations for 19 subjects. The duration of response was significantly longer after playback of cow vocalizations (two-tailed Wilcoxon test, $P < 0.05$).

The vervets' strong response to ringing bells was not simply the result of a tendency to respond to any loud sound. Vervets who are habituated to human observers, for example, soon cease responding to the sound of vehicles. Instead, the monkeys seemed to associate bells with Maasai just as they did the lowing of cattle. We do not know, however, whether the monkeys recognize any difference between lowing and ringing. Do they understand that one sound is produced by an inanimate object and the other by an animal? We also know very little about the mental processes underlying the monkeys' responses to cows and bells. Clearly, the monkeys' responses could result simply from associative learning; it is also possible, though, that they reflect some more complex understanding of the relationship between two other species. One of the obvious weaknesses of playback experiments is that they define knowledge operationally, through the responses they evoke. Playbacks can reveal which of two stimuli is the more salient, but they can say nothing about the cognitive mechanisms underlying behavior.

Secondary Cues of Danger

The lowing of cows and the ringing of bells are auditory cues that signal the approach of danger. The vervets' response to these sounds suggests that they make use of these cues and that they have some knowledge about the association between two other species—cattle and Maasai. Oddly, however, much of the vervets' apparent knowledge of other species breaks down when we ask them what they know about the *visual* cues produced by other species.

Monkeys clearly rely on visual signs to locate food and avoid predators. Anyone who has ever traveled around tourist lodges in East Africa knows all too well that baboons and vervets are experts at associating cardboard boxes and tour buses with picnic food. Boxes, soda bottles, and even tour buses, however, are the packages in which food is delivered. They are not, therefore, secondary cues for food but the external manifestation of it. In contrast, an alarm call, whether produced by a vervet or another species, is a true secondary cue; it is not produced by the predator itself but signals the presence of a particular type of danger. Alarm calls are auditory secondary cues for predators. Can monkeys also make use of *visual* secondary cues that denote danger?

When leopards make a kill, they often drag their prey into trees, where they can feed without harassment from lions and hyenas. This behavior is peculiar to leopards, and humans soon learn to recognize that a fresh carcass in a tree reliably signals the proximity of a leopard. Each of the vervet groups in our study population had the opportunity to see leopards drag carcasses into trees. During all such cases when we were present they responded by giving repeated alarm calls, even when the leopard was already feeding on a carcass. We wanted to determine whether vervets knew enough about the behavior of leopards to recognize that, even in the absence of a leopard, a carcass in a tree signaled the same potential danger as did a leopard itself.

To carry out this experiment we first procured a limp, stuffed carcass of a Thompson's gazelle, a species that is frequently killed by leopards. (This aspect of the experiment presented one of the more bizarre methodological challenges of our entire study.) We then rose well before dawn and, while it was still dark, placed the carcass in a tree approximately 50 to 75 m from the monkeys' sleeping trees (fig. 9.9). The carcass was positioned in such a way as to mimic its placement by a leopard (in fact, our attempt fooled at least one tour bus driver into thinking that a leopard was in the area). At first light, we observed the behavior of the monkeys for 2 hours, noting at 5-minute intervals the direction of gaze of as many group members as could be seen. In total, we presented the carcass on five occasions, once to a group of baboons and four times to different groups of vervets. One of the vervet groups had seen a leopard with its carcass in a tree only 4 days earlier and had continued to alarm call even after the leopard ran away at our approach (see Cheney and Seyfarth 1985a for more details).

Despite all of the groups' experiences with leopards and carcasses in trees, neither the vervets nor the baboons gave alarm calls at the sight of the carcass alone. In fact, there was not even any increased vigilance in the direction of the carcass over what might have been expected by chance. In all cases, the monkeys behaved as if they did not recognize that a carcass in a tree signaled the proximity of a leopard.

It is, of course, possible that the carcass was somehow not as realistic as we (and the tour bus driver) imagined it to be and that the monkeys simply

Figure 9.9. The stuffed carcass of a Thompson's gazelle placed in a tree near the vervets' sleeping site.

did not recognize the carcass as the prey of a leopard. Nevertheless, the animals' apparent failure even to regard the carcass as an unusual object suggests that they do not pay very much attention to changes in their habitat, even when such a change seems to be more than obvious. Recall, for instance, that the baboons in the example we cited at the beginning of this chapter walked right around the bloody buffalo carcass. In this respect, it would be of interest to determine how vervets or baboons would respond to a massive but largely irrelevant change in their environment. How would they respond, for example, if we were somehow able to move Kilimanjaro so that it appeared to the north rather than the south, or if we could cause the sun to rise in the west? We expect that the monkeys would remain utterly unfazed. However much we might alarm and discombobulate the local human population, the vervets would probably react with the kind of aplomb that stems from profound ignorance.

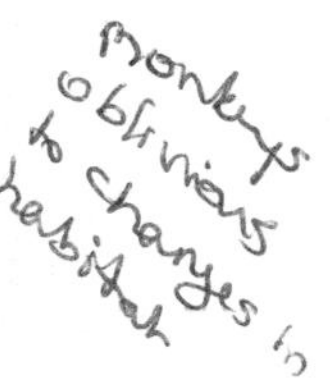

As a further test of monkeys' knowledge of visual secondary cues, we tested the vervets' recognition of python tracks. Pythons in Amboseli frequently prey on vervet monkeys (Cheney et al. 1988), and when vervets encounter a python they give alarm calls and closely monitor its movements through the area (chapter 4; Seyfarth, Cheney, and Marler 1980b). Pythons lay distinct, wide, straight tracks that cannot be mistaken for those of any other species and that are easily recognized by us and local humans (fig. 9.10). It is possible to determine the freshness of a python track by noting the clarity of its outline and whether any other animals have walked across it. In fact, on many occasions we have been able to find a python by following its track into a nearby bush. Vervets in Amboseli have had ample opportunity to watch and alarm call at pythons as they laid down tracks

Figure 9.10. Pythons lay wide, straight tracks that cannot be mistaken for the tracks of any other species.

and then disappeared into nearby bushes. Have the monkeys learned that a fresh python track represents potential danger?

To investigate this issue we relied on both observation and experiments (see Cheney and Seyfarth 1985a). We noted all those occasions when a python laid down a track in the dust when there were no monkeys in the area. We then observed the monkeys as they approached the track on their foraging route. In no case did individuals show vigilance or change their behavior when they approached and crossed the track. Indeed, on several occasions we watched in utter mystification as a vervet calmly followed a python track right into a bush, only to leap away in shocked horror when he encountered the snake there. We also attempted five replications of real python tracks by laying down an artificial track in an area that the monkeys were approaching. Again, the animals showed no increased vigilance toward the track, behaving as if they did not recognize that it signaled danger.

Finally, consider the inconsistent responses of vervets to the secondary cues denoting the approach of Maasai herdsmen. In Amboseli, a herd of cows or goats produces a thick, dense cloud of dust that humans readily learn to distinguish from the lighter, more dispersed dust clouds produced by wild herds of zebras, wildebeests, or elephants (fig. 9.11). As we have discussed, vervets respond strongly not only to the sight of cows, donkeys, and goats but also to the lowing, braying, bleating, and bell ringing produced by these species. Strangely, though, they never seem to recognize the distinctive clouds of dust that signal these animals' approach. Day after day, vervets will impassively watch an approaching dust cloud produced by livestock, only to give surprised alarm calls when, invariably, the cows and goats emerge.

Why should we find these curious failures to detect secondary cues? It seems unlikely that the monkeys rely on olfactory cues in place of visual or auditory information. Unlike many other mammals, the olfaction of primates is relatively poorly developed, and, at least at distances of over a meter, the vervets' sense of smell is probably quite similar to our own. For each of the three examples we have given, one could argue that a particular cue is not sufficiently salient, or encountered in close enough temporal association, or reliable enough to be used as a consistent predictor of danger. Taken together, however, these objections have the hollow ring of unsatisfactory, post hoc explanations. We are left with the distinct impression that monkeys are ignoring cues that they might logically be expected to pick up.

The Relative Importance of Auditory and Visual Cues

The secondary cues associated with vervets' predators can be divided roughly into four categories, depending on whether they are animate or in-

Figure 9.11. Herds of livestock produce large, dense clouds of dust that often appear before the livestock themselves are seen and can easily be distinguished from the more dispersed dust clouds produced by wild ungulates.

animate, auditory or visual. Animate auditory cues include the lowing of cows and the alarm calls of other species. Animate visual cues include the sight of cattle or other livestock. Inanimate auditory cues include the ringing of cow bells, whereas inanimate visual cues include clouds of dust, python tracks, and carcasses. Of these four types of secondary cues, vervets seem to attend to all but the last. Why might this be the case?

Assuming for the moment that our categories are at least heuristically valid, there are at least three explanations for the vervets' apparent lack of attentiveness to inanimate visual cues. First, auditory cues may be more salient than visual ones. Auditory signals have a more rapid onset time, and it has been shown that rats are more likely to associate sudden events with other sudden events, and gradual events with other gradual events (Testa 1974). It may therefore be easier for vervets to associate and respond to secondary cues of imminent danger when these cues are auditory. This explanation is limited, however, because it fails to explain why, in the first instance, natural selection should have favored different sensitivities in visual and auditory domains.

Another salient feature of auditory cues is their close temporal association with the stimuli that evoke them. In contrast, visual cues are not always closely associated with the stimuli that originally gave rise to them. A python track, for example, could have been laid down hours before the monkeys encountered it. Humans find it easier to place a causal interpretation on an event when the stimuli involved are spatially and temporally contiguous (Michotte 1963). Perhaps the lack of consistent spatial and temporal contiguity between, say, pythons and their tracks or leopards and car-

casses in trees makes it difficult for vervets to recognize that these visual stimuli provide information about a nearby predator. Explanations based on the salience of stimuli and their temporal association are not, however, very convincing. Clouds of dust, for example, appear in close temporal association with the approach of Maasai, and carcasses in trees and python tracks frequently *do* signal immediate danger. Leopards typically remain for several days in the area where they have recently made a kill, and vervets often encounter pythons in bushes when they have failed to attend to tracks.

We estimate that predation accounts for at least 70% of all deaths among vervet monkeys in Amboseli, and both leopards and pythons frequently prey on vervets (Cheney et al. 1988; Isbell 1990). As a result, one might expect strong selective pressure favoring monkeys who can associate these sometimes temporally discontinuous cues, especially because they do have the opportunity to observe the predators creating them. Research on learned taste aversion has shown that animals can learn to associate the taste of certain foods with illness, even when these two stimuli are widely separated in time. Rats and coyotes, for example, learn after only one trial to avoid food that has previously made them sick, even if they became sick hours after eating (Garcia and Koelling 1966). Taste aversion research has also shown, however, that some temporally discontinuous stimuli are more easily paired than others, and this may also be true for the visual cues left by predators.

An alternative argument, which grows out of the domain-specific hypothesis described earlier, begins by suggesting that the vervets' use of visual and auditory secondary cues first evolved to deal with social problems. Inconsistencies in the monkeys' use of visual and auditory signals when dealing with other species result from inconsistencies in their use of visual and auditory signals when communicating with each other.

Consider, for example, the different ways in which monkeys use auditory and visual signals in their social interactions. Vervets use vocal signals both in the presence and in the absence of face-to-face contact. A monkey can give a grunt in full view of her intended recipient, or she can call to another when the two animals are completely out of sight of each other. If monkeys are foraging in dense bush, a vocalization can signal that another group is approaching or that a snake is nearby without any supporting visual information. In this respect, vocal signals are not only expressive but also denotative.

In contrast, the only denotative visual signals used by vervet monkeys are other vervet monkeys. The approach of a high-ranking juvenile can signal to a subordinate female that the juvenile's mother will probably soon also appear, but the use of visual secondary cues apparently does not extend to other species or to inanimate objects. Vervets do not create visual signs, nor do they use visual signals (except other vervets) referentially, to provide information about something that is not present. The monkeys do not, for example, follow each others' tracks when foraging, look for the tracks made by other groups when patrolling a territorial boundary, or visually mark aspects of their physical environment to denote, for instance, rank or group membership. Because it has evolved to meet social needs, the monkeys' system of visual communication is ill-suited to solving certain problems outside the social domain. In their social interactions, the monkeys may never have *needed* to recognize that a visual cue can denote some absent referent. As a result, when they confront a python's track or a carcass in a tree, the monkeys do not make the connection between these stimuli and the predators that put them there.

Some of these generalizations concerning the lack of denotative visual cues may be less applicable to apes than to monkeys. At Gombe in Tanzania, for example, chimpanzees make sleeping nests each night, and when the members of one group make incursions into the range of another they sometimes make aggressive displays upon encountering their neighbors' empty nests (Goodall, et al. 1979). Chimpanzees therefore react to nests as if they denote other chimpanzees, even when the latter are absent. Similarly, solitary male mountain gorillas sometimes seem deliberately to follow the trails made by other gorillas when tracking a group (K. Stewart, pers. comm.). Given these observations, it would be interesting to see whether apes performed better than vervets on the tests we have described. As we discuss below, there are good reasons to believe that the intelligence of apes in general may be less domain specific, and more "accessible," than that of monkeys.

A final alternative explanation for the vervets' inability to recognize visual cues as signs of danger is based on the assumption that these cues require some understanding of causality. At present, we do not know whether monkeys understand the relation between cause and effect, because the only systematic investigations of causal inference in nonhuman primates have been carried out on chimpanzees.

There is strong suggestive evidence that chimpanzees do have some understanding of causality. Premack (1976) presented a number of language-trained chimpanzees with a choice of alternative agents to complete a sequence of actions and found that they chose the correct agent with a high degree of accuracy. For example, when the chimpanzees were shown a whole and a cut apple and allowed to choose among a knife, a glass of water, and a pencil, the animals selected the knife. Premack argues that the chimpanzees could not have been guided by previous associations between particular objects and agents because some of the objects were more closely associated with one of the incorrect agents. When shown a whole and a severed sponge, for example, the chimpanzees chose a knife rather than a glass of water, even though sponges and glasses of water were more closely associated with each other than were sponges and knives. The numerous examples of tool use by free-ranging chimpanzees also suggest some understanding of cause and effect (see Nishida 1987 for a review; see also pp. 296–298). Could a chimpanzee find and prepare a stick to fish for termites without understanding the causal relation between the right kind of stick and the acquisition of termites?

In their social interactions, monkeys often behave as if they have some understanding of causality. When a juvenile baboon avoids an altercation with the daughter of a high-ranking female, she may do so because she recognizes that a fight will cause that individual's mother to intervene. Similarly, Kummer's female hamadryas baboon, described in chapter 7, may groom a subordinate male out of sight of the dominant male because she knows that the sight of her interacting with another male will cause him to chase her (Kummer 1982). In most cases, however, the monkeys' behavior could also be based entirely on associative learning and involve no understanding at all of causal inference. We simply have no evidence at the moment that would allow us to distinguish clearly between these two alternatives.

Even if, for purposes of argument, we accept the view that monkeys understand causality in their social interactions, this understanding is apparently not extended to the nonsocial environment. For example, just as vervet monkeys seem to ignore the secondary cues left by predators, so do baboons at Gombe ignore the methods of termite fishing developed by chimpanzees. Despite ample opportunity to observe chimpanzees, baboons do not imitate them and instead feed on termites only when the insects emerge after rainfall (discussed by Premack 1976). These observations suggest either that baboons and other monkeys lack the ability to make *any* causal inferences or that causal inference in monkeys is more restricted to social interactions than it is in apes. Of course, the chimpanzees' notion of causality where termites are concerned is still less refined than that of local humans, who, recognizing that termites emerge after rain, simulate rain by pouring water or urinating on the termite mound. This act of deception requires that humans manipulate the termites' "beliefs," something chimpanzees may not be able to do.

Cooperation and Reciprocity

Cooperative alliances among humans are often characterized by the exchange of goods or services. Significantly, such exchange can involve either reciprocal social interactions or the exchange of objects. Humans, for example, may exchange cooperative behavioral support (as in a political alliance), individuals (the exchange of spouses between two villages), or material goods (the donation of money or food to cement an agreement).

When interacting with both kin and nonkin, monkeys and apes also sometimes exchange one form of cooperative behavior, such as grooming, for another, such as an alliance or tolerance at a feeding site (chapters 2 and 3). Primates seem able to remember past interactions and to adjust their cooperative acts, depending on who has behaved altruistically toward them in the past. Curiously, though, while monkeys and apes often reciprocate cooperative *acts,* under natural conditions reciprocity rarely involves the use or exchange of *objects.* The two frequently mentioned exceptions to this rule, cooperative hunting and the sharing of meat by chimpanzees and perhaps also baboons, are worth closer examination.

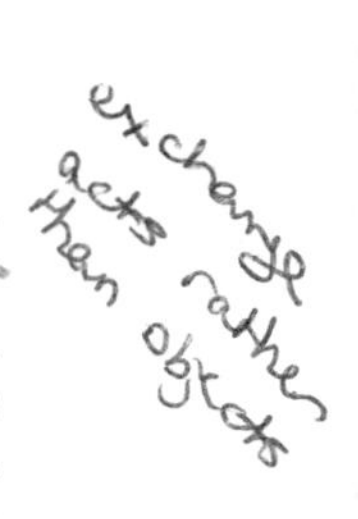

It is true that male baboons and chimpanzees sometimes engage in joint hunts that result in some distribution of the prey among hunters and uninvolved bystanders. Among baboons, however, hunts seem to consist of a number of independent, though simultaneous, efforts rather than genuinely cooperative ventures (fig. 9.12; Kummer 1968; Altmann and Altmann 1970; Strum 1981). Chimpanzees' hunts more often involve concerted action and even, occasionally, some division of labor. At Gombe, chimpanzees have been observed to recruit others before hunting elusive or dangerous prey like baboons or wild pigs, almost as though they recognize that other hunters are crucial to their success (Goodall 1986). Occasionally, one individual will chase the prey while two or more other males position themselves to block potential escape routes (Wrangham 1975; Busse 1978;

Figure 9.12. A male baboon eats a Thompson's gazelle calf that he has captured as juvenile baboons look on. Baboon hunts are rarely cooperative, and hunters rarely share their prey. Photograph by Cynthia Moss.

Teleki 1981; Goodall 1986). Not all chimpanzee hunts, however, are cooperative, and this division of labor has been observed only a handful of times. Most important, chimpanzees seldom willingly share their prey. Males do sometimes give meat to their close relatives, a sexually receptive female, or someone with whom they have recently groomed. More often, however, those who obtain a portion of the kill do so by grabbing, fighting, begging, and scavenging rather than through sharing or active exchange. Recent observations of chimpanzee hunts in the Tai forest of the Ivory Coast suggest that cooperation and sharing may in fact be more common than the Gombe data imply (Boesch and Boesch 1989). Nevertheless, chimpanzee hunting behavior still appears to be no more cooperative than the hunting behavior of wolves, wild dogs, or even some spiders (Packer and Ruttan 1988).

Reciprocity in nonhuman primates therefore occurs more commonly in the form of social interactions, such as grooming and alliances, than in the exchange of material goods. Before we conclude that animals in general differ from humans in restricting their cooperative acts primarily to social interactions, however, a number of caveats deserve attention.

First, the relative rarity of food sharing among free-ranging monkeys and apes may result at least partially from the fact that, except for meat, their food is simply not worth sharing. Nonhuman primates feed primarily on leaves and fruit, and these items are typically distributed in a way that prevents their being easily monopolized by one individual. There may therefore be little benefit in acquiring food directly from another. Individuals might derive greater benefit through tolerance at a particular feeding site or fruiting tree, and in fact grooming, copulation, and other affinitive interactions do sometimes permit subordinate individuals to feed near dominants (chapter 3; Wrangham 1975; Weisbard and Goy 1976; Smuts 1985; Goodall 1986). Furthermore, although nonhuman primates seldom exchange material goods for future beneficial acts, such patterns of exchange *do* occur in other animal species. In the courtship displays of many birds and insects, for example, the male offers food to his mate, and females apparently select mates at least partly on the basis of the "gifts" they offer. Finally, although nonhuman primates seldom share food and never provision others who are absent from the immediate area, many species of carnivores (including vampire bats, foxes, jackals, coyotes, wolves, hyenas, and wild dogs) do bring food to a central den or gathering point where young and other individuals are fed.

In some animal species, therefore, cooperative behavior includes the exchange of material goods. We do not yet know, however, whether such patterns of exchange are at all modifiable. Although humans can readily substitute a behavioral altruistic act for a material one, similar flexibility in the "currency" of reciprocal acts has never been convincingly documented in other animals. More research is clearly needed before cooperation and reciprocity in monkeys and apes are fully understood. For the moment, though, it seems safe to suggest that reciprocity among monkeys and apes is manifested more commonly through social interactions than through the exchange of material goods.

The Importance of Ecological Factors

The Challenge of Finding and Exploiting Food

Some of the food exploited by nonhuman primates, particularly ripe fruit, is both spatially and temporally dispersed. Many species of monkeys and apes range over large geographical areas and are skilled at remembering the location and phenological patterns of food and water (Clutton-Brock 1977; Rodman 1977; Wrangham 1977; Sigg 1980; Sigg and Stolba 1981). As a result, it is sometimes hypothesized that ecological pressures have played the primary role in the evolution of primate intelligence (e.g., Clutton-Brock and Harvey 1980; Milton 1981, 1988).

Using as evidence the correlation between brain size and diet, Milton (1988) argues that the primary selective pressure favoring large brain size in many primates, including early hominids, has been the cost of procuring high-quality food. Species that depend on fruit, which is difficult to locate in both space and time, must be able to remember the location of widely dispersed, ephemeral food resources. Within primates, as within other taxa, there is a positive correlation between brain size, range size, and particularly diet; frugivorous genera, which feed primarily on widely dispersed fruit, typically have larger brains and larger ranges than folivorous genera, which feed on the more abundant leaves. In contrast, Milton argues, across species there is no correlation between the complexity of social systems and brain size.

The relation between foraging and brain size should not, however, be considered in a primatological vacuum. After all, many species of animals other than primates display quite complex foraging skills. Birds, fish, and rodents are able to estimate patch size, to compare food quality and handling time, to remember where they have stored food, to compare rates of return at different feeding sites, and even to count (see Gallistel 1989a, 1989b for reviews). Across taxa, these abilities are not closely correlated with gross brain size. Indeed, foraging skills comparable to those of great tits or nutcrackers have not yet been documented in monkeys or even apes. In general, it has proved difficult to test optimal foraging models on nonhuman primates, at least partially because ecological measurements are often complicated by intervening social variables. For example, because feeding bouts can be interrupted by the approach of a more dominant individual, the time that a monkey spends at a given food patch does not necessarily provide any accurate indication of the patch's quality (see e.g., Post, Hausfater, and McCuskey 1980; Stacey 1986; van Noordwijk and van Schaik 1987).

Just as correlations between relative brain size and food quality may be overly simplistic, it may be equally misleading to draw conclusions about the selective pressures favoring large brain size from correlations with social structure. What seems important here is not so much the general features of a species' social system but the types of interactions that occur once a given system has evolved. Zebras and gorillas both live in one-male groups characterized by female dispersal, and both zebras and gorillas feed on evenly dispersed, relatively low-quality food. Nevertheless, gorillas have considerably larger brain/body weight ratios than zebras; these differences cannot easily be explained by food type or the gross outlines of the two species' social structures.

We must also be wary of any attempt to rank social systems according to what must ultimately remain a subjective assessment of social complexity. Do species that live in fission/fusion groups, such as chimpanzees and spider monkeys, really have to remember more individuals and social alliances than species that live in more cohesive groups? At the moment, we simply do not know.

Although ecological factors have undoubtedly contributed to the evolution of primate intelligence, nonhuman primates do not appear to manipulate objects in their physical environment to solve ecological problems with as much sophistication as they manipulate each other to solve social problems. The challenge of exploiting widely dispersed and ephemeral food sources may therefore have led to increased intelligence not simply because food collection itself became more difficult, but also because ecological complexity placed increased selection pressure on *social* skills, including the ability to cooperate to defend resources, to detect nonreciprocating cheaters, and to communicate about resources that are displaced in time and space. Milton (1988) makes the same point when she argues that the high quality diet of early hominids probably favored the evolution of increased communicative and social skills.

We do not mean to imply that primates never face challenging ecological problems. Whereas vervet territories seldom exceed 1 km^2, chimpanzee communities patrol areas of over 30 km^2, and some baboon groups regularly cover areas of over 100 km^2 (Cheney 1987; J. Altmann, pers. comm.). Often the animals seem to have a detailed knowledge of the location of food and water over enormous areas. For example, in Sigg's and Stolba's (1981) study of hamadryas baboons in Ethiopia, the monkeys unerringly remembered and found even rarely used waterholes within their 28 km^2 range. Their ability to locate waterholes did not seem to depend on topographical cues, since the animals used different, flexible routes to water that minimized the distance traveled. Well-used waterholes were approached from a variety of directions, suggesting that the baboons had a "mental map" of familiar areas. In these familiar areas, the group would often split up into small foraging units and then reunite at a waterhole, as if the animals all recognized a common goal. In contrast, the baboons stuck to more rigid routes when in less familiar areas (see also Boesch and Boesch 1984; Sigg 1986; Jolly 1988).

These results demonstrate that the distinction between social and nonsocial knowledge is not a simple one. When one group of hamadryas baboons rejoins another at a distant waterhole, the animals are undoubtedly drawing on both their knowledge of the habitat and their knowledge of familiar social companions. With this point in mind, we should reiterate that the argument about domain-specific intelligence does *not* aim to oppose one comprehensive ecological argument against an equally comprehensive social one. The evolution of specialized skills to solve social problems by no means precludes the evolution of other, equally specialized skills to solve problems in other contexts. Rather, the hypothesis posits that *specific* abilities have evolved to cope with specific social or ecological demands. As a result, abilities that are manifested in one context may not always be generalized to another. So, for example, while a male baboon might have no difficulty in assessing the relative ranks of other males, he might be unable to rank the relative amounts of water in a series of containers. Similarly, if the need to exploit widely dispersed and ephemeral waterholes has favored the evolution of complex spatial memory in hamadryas baboons, we should not necessarily expect hamadryas baboons to be better than other monkeys at remembering the genealogies of band members. Instead, we argue that the monkeys' intelligence is largely inaccessible to them; hence knowledge gained in one domain is not necessarily extended to another. Thus far, of course, this is only an hypothesis. It remains to be determined how flexible monkeys are in extending social (or foraging) skills to other contexts.

Tool Use

The few experiments we have described are the only ones to have attempted to compare a primate's social knowledge with her performance outside the social domain. This is unfortunate, because it suggests that primatology continues to focus on the marvelous things that monkeys and apes do, with few parallel attempts to determine the limits of their intelligence. The lack of any comparison between social and nonsocial performance is particularly unfortunate in the case of the great apes, because it seems likely that the intelligence of apes (like that of humans) is less domain specific and more accessible than that of monkeys. Anecdotal accounts of chimpanzees, for example, suggest that they readily distinguish between animate and inanimate objects. The captive chimpanzee Vicki, for instance, was able to sort pictures of animate and inanimate objects into distinct categories without previous training (Hayes and Nissen 1971). Whether a monkey would be capable of performing a similar classification is not known.

Chimpanzees also construct and use tools more often than any other nonhuman primate species. Tool use is relevant to our current discussion because, across the animal kingdom, the frequent use or manufacture of tools is undeniably associated with larger brain/body weight ratios. Tool use is substantially greater among nonhuman primates than among other orders. Perhaps more important, using or manufacturing tools involves behavior that is not easily classified as social or nonsocial. On the one hand, tools themselves are manufactured from objects like branches or stones that seem to fall outside the social domain. On the other hand, using tools and learning about them can be an intensely social affair, as Jane Goodall makes clear in her descriptions of the ways in which infant and juvenile chimpanzees appear to mimic their mothers when fashioning sticks to fish for termites and ants (van Lawick-Goodall 1970, 1973). Since tool use is at least potentially linked to social learning and imitation, it is tempting to speculate that it represents a breakdown in the barriers between social and nonsocial domains.

Wild primates use tools in three main contexts (Beck 1974, 1980; Passingham 1982), the most common of which appears to be threatening or attacking intruders. All apes and many forest-dwelling monkeys drop twigs and branches onto human observers, whereas gorillas often throw vegetation as part of their chest-beating display. In one bizarre but resourceful instance of polyspecific symbiosis, a capuchin monkey even threw a squirrel monkey at a human observer (Boinski 1988). In most cases, it is difficult to determine whether such "tools" are deliberately aimed. For example, when chacma baboons dislodged stones from a cliff onto humans, they did so not only when the observers were directly under the cliff but also when the observers were too far away to be struck (Hamilton, Buskirk, and Buskirk 1975). At Gombe, where chimpanzees and baboons competed for food, chimpanzees threw objects at baboons but rarely hit them (Goodall 1968). Baboons, however, never threw objects at chimpanzees.

Tool use also occurs in the context of bodily care. Chimpanzees wipe blood and feces from their hair with leaves (McGrew and Tutin 1973); orangutans construct crude shelters to protect themselves from rain (Mac-

Kinnon 1974). The use of "tools" to cleanse the body seems to be rare in monkeys, but a baboon has been reported to wipe blood from its face with a maize cob (Goodall, van Lawick, and Packer 1973).

A third context of tool use involves the acquisition and preparation of food. Chimpanzees use at least five types of tools to obtain termites, ants, and honey, and to break open intractable nuts (see Nishida 1987 for a review). In captivity, chimpanzees have used poles to construct ladders (Menzel 1973; de Waal 1982). These observations are of interest at least in part because, as we discussed in chapter 8, they seem to indicate a shared motive, and the representation of a goal.

The use of crude hammers and anvils by chimpanzees in the Ivory Coast to break open nuts illustrates the animals' foresight in selecting the appropriate materials and their detailed spatial memory of their home range (Boesch and Boesch 1983, 1984). At the start of a typical nut-smashing session, a chimpanzee will collect as many nuts of a particular species as he can carry and take them to a broad, flat rock or surface root that serves as an anvil. If the nuts are a particularly hard species, the chimpanzee will travel (often more than 40 m) to find a stone anvil. The animal's route suggests that he has a mental map of the area that allows him to compare distances between trees and stones in order to minimize the distance traveled (Boesch and Boesch 1984). Often a wooden club or stone (the hammer) will already be lying next to the anvil, but on other occasions the chimp will first locate a hammer and then carry it to the anvil along with the nuts.

In three respects, data on primate tool use support laboratory investigations of differences in intelligence among different primate species (reviewed by Passingham 1982; Essock-Vitale and Seyfarth 1987). First, just as prosimians generally perform worse than other primates on many tests, they also exhibit much less tool use. Second, just as apes almost always do at least as well, and often better, than monkeys on intelligence tests, they also use tools more often and in more different ways than monkeys. This is particularly true of tool making, as opposed to tool use. Chimpanzees who fish for termites and use hammers to crack open nuts are the only primates that select particular objects as tools, modify them appropriately, and do so in a way that shows foresight. Finally, even within primate families, tool use is associated with large brain size. Both chimpanzees and capuchins have larger brain/body weight ratios than other members of their families, and they also use tools far more often. In captivity, for example, capuchins spontaneously use and even manufacture tools to obtain food. In contrast, squirrel monkeys never learn to use tools, even when they are housed in the same cages as capuchins (Westergaard and Fragaszy 1987). Although we should be wary of suggesting that large brain size is the result of any specific ecological or social factor, it is certainly possible that ecological

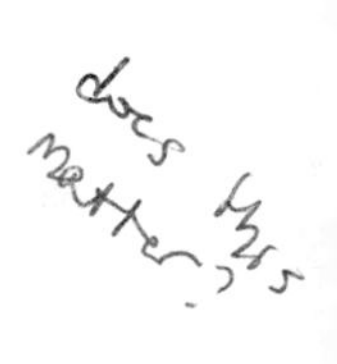

pressures have favored the evolution of larger brains in these species (see also Parker and Gibson 1977 and Boinski 1988 for discussions). On the other hand, as is true of any hypothesis that relies on correlational data, we could also argue the opposite. Perhaps social pressures favoring abilities like planning and foresight have, in turn, allowed some primate species to exploit tools more easily than others.

Comparisons of tool use in different primate species are by no means completely straightforward, however. For example, reports of tool using and making are far more numerous for chimpanzees than for orangutans, even though orangutans perform as well as chimpanzees on many learning tests, and even though no one who has ever worked with captive orangutans would deny their extraordinary skill at manipulating objects or escaping from their cages (e.g., Wright 1972).

What remains striking and puzzling is how relatively impoverished the tool use of even chimpanzees and capuchins is. In captivity, capuchins readily learn to use tools. Despite many thousands of hours of observation in the wild, however, there has been only one account of capuchins using sticks for purposes other than threat displays (Boinski 1988). As Hans Kummer (1982) has noted, the paucity of such observations contrasts markedly with the many instances in which primates apparently use each other as "social tools" to achieve a particular result (see also Kummer 1968; Chance and Jolly 1970; chapters 3, 7, and 8).

Social Skills in Nonprimate Species

Although we have concentrated throughout this book on the social behavior of primates, we certainly do not mean to suggest that social skills are unique to monkeys and apes. If the social environment has favored the evolution of some aspects of learning and memory, like the ability to identify the kin of other group members, then we might expect to find similar abilities in many other group-living birds and mammals (Kamil 1987). For example, long-term research by Cynthia Moss and Joyce Poole has shown that elephants live in a multilevel social community that apparently demands quite sophisticated abilities to observe and remember the social relationships of other individuals (Moss and Poole 1983; Moss 1988). Females spend most of their lives with their close female kin. Different families, however, also regularly associate to form distinct "bond groups" (Moss 1988). Specific bond groups, in turn, regularly associate with each other to form clans. Finally, clans sometimes come together to form massive herds of over 500 individuals, all of whom appear to recognize one another both as individuals and as members of particular families. There is increasing evidence, moreover, that elephants not only recognize the specific association patterns of different families, but also know where each family group typically ranges (Moss 1988; Poole, in prep.). Among primates, this sort of multilevel social organization is known to occur only among hamadryas baboons and perhaps also drills (*Papio leucophaeus*).

Like other observers, we have used alliances as a measure of social skills in nonhuman primates (e.g., de Waal 1982; Smuts 1985; Harcourt 1988). Monkeys and apes seem to be masters at choosing those alliance partners who are most likely to do them some good, and they do seem to differ from other animal species in choosing alliance partners on the basis of their quality or potential worth, such as their dominance rank, rather than simply on the basis of reciprocity (Harcourt 1988).

One of the most important features of human alliances, however, is that they are not restricted to individuals but also extend to groups. Two clans unite against a third; two nations sign a treaty to defend themselves against a third. Among monkeys and apes, intergroup alliances also occur, although in a far more limited way. Members of two vervet matrilines in the same group will behave as rivals until another group makes an incursion into their territory, when both matrilines will unite to drive out the invading group. Similarly, two units of hamadryas baboons that normally interact at low rates will nevertheless act in concert to chase away a third (Abegglen 1984).

Human groups, however, are able to sustain alliances even when they rarely come into contact with each other. It was not necessary, for example, for Germany and the Ottoman Empire to associate at high rates, much less to like each other, in order to form an alliance against France and Great Britain in 1914. In contrast, intergroup alliances in nonhuman primates seem to be limited to units that already associate, at least spatially, at high rates. Two matrilines from the same vervet group may ally themselves against another group, but two distinct groups that normally inhabit different ranges will never form an alliance against a third. The same is probably also true of hamadryas baboons, since interunit and interclan alliances seem to be restricted to groups that sleep and travel together frequently.

Intergroup alliances also occur in nonprimate species. Male dolphins form consistent and predictable bonds with other males, and allied males cooperate to herd females away from rivals (Connor, in press). Occasionally, two allied groups will act in concert to drive away a third band of males. Similarly, among elephants, there is some suggestion that matrilines occasionally unite to threaten a third family group (Moss 1988; Poole, in prep.). Although we do not know whether intergroup alliances in these species also occur primarily among groups that associate at high rates, it remains possible that intergroup alliances in at least some nonprimate species are more complex than those found in monkeys and apes. In any case, the very existence of intergroup alliances in other mammals reminds us that social skills are by no means unique to primates, and that, in fact, some species may exhibit social behavior that is rare or nonexistent in monkeys and apes.

As a final example, consider the white-fronted bee-eater (*Merops bullockoides*), a colonial, communally breeding bird living throughout much of East Africa (Emlen 1981; Hegner 1982; Hegner, Emlen, and Demong 1982; Emlen 1984; Hegner and Emlen 1987; Emlen and Wrege 1988). Bee-eaters nest in holes dug into the sides of riverbanks, forming large colonies made up of kin and nonkin. Although they are monogamous breeders, close relatives nest in adjacent holes and enter each others' holes at will. Nonkin, in contrast, are excluded from breeding holes. The birds appear to distinguish among kin according to their degree of genetic relatedness and reproductive value. If a breeding pair loses its clutch, it will often help its kin to rear offspring, and individuals are more likely to aid close kin than more distantly related individuals. Groups of relatives defend a feeding territory that can be located as many as 3 km away from the nesting site. Each clan group recognizes the boundaries of its territory, and although pairs tend to feed in separate locations within the territory, all the members of a clan will cooperate to drive nonrelatives from the territory. The birds also seem to discriminate between unrelated territorial neighbors and nonneighbors.

Bee-eaters, therefore, display both highly developed spatial memory in their foraging behavior and primatelike social skills. Not only do they remember the location and boundaries of distant feeding territories, but they also seem to distinguish kin from nonkin, close relatives from nonrelatives, and unrelated neighbors from nonneighbors. As yet, it is impossible to say whether bee-eaters approach primates in recognizing both their own social relationships and the social relationships of others, because the relevant observations and experiments have not been conducted. We also do not know whether any of the specific skills that bee-eaters use in foraging are also manifested in social interactions. Do bee-eaters, for example, discriminate the location of nonrelatives' breeding holes with the same acuity that they distinguish the boundaries of their neighbors' territories? If bee-eaters (or any other species) failed to extend an ability like spatial memory from a foraging to a social context, this would further support the domain-specific view of intelligence.

Summary

In this chapter we have continued our attempt to specify the limits of nonhuman primate intelligence. Having suggested, in chapter 8, that the ability of monkeys to attribute mental states to each other is limited, we have argued here that many of the social skills exhibited by monkeys may be inaccessible to them. Natural selection has acted on at least some features of primate intelligence with particular force in the domain of social interactions. As a result, the skills we observe when primates interact with each other and make judgments about social relationships are not always apparent in nonsocial contexts.

Social groups of primates are composed of individuals of varying ages, ranks, and degrees of genetic relatedness. To compete successfully, an individual must know not only her own social position but also the social positions of others. Group life, in other words, exerts strong selective pressures on the ability to classify individuals into subgroups, to make transitive inferences, and perhaps even to make same/different judgments about relations among fellow group members. We suggest that the ability of monkeys to classify objects in captivity, as well as the ability of captive apes to understand the relation "X is different from Y, but both are members of the class Z," first emerged in the animals' need to classify each other in this way.

Despite frequent opportunity and often strong selective pressure, however, monkeys perform less well when dealing with other animal species or with inanimate objects. Vervet monkeys do not seem to attend to many aspects of their physical environment, even when to do so would be adaptive. Although the monkeys do recognize and respond to the different alarm calls given by birds and nonprimate mammals, they appear to ignore the visual and behavioral cues associated with some predators. Apparently, they do not recognize the relationship between a python and its track, nor do they understand that a carcass in a tree indicates a leopard's proximity, even though they have had ample opportunity to learn such associations.

Similarly, although vervets and other primates exhibit many forms of cooperation and reciprocity in their social interactions, comparable behavior using nonsocial currency (for example, food sharing) is relatively rare. Monkeys behave altruistically and form alliances to achieve social goals, but they seldom cooperate to learn new ways of exploiting food.

Finally, vervet monkeys are poor naturalists. They seem disinclined to collect information about their environment when that information is not directly relevant to their own survival. Vervets do not seem to know that hippos stay in water during the daytime or that particular shore birds do not occur in dry woodlands. These data are perhaps not surprising, but they do point out a potential difference between monkeys and human beings, who are naturally curious about much of their environment and who engage in many activities that have little practical value to survival.

These observations suggest that there is, in nonhuman and perhaps also human primates, an evolutionary predisposition that makes it easier for individuals to understand relations among conspecifics than to understand similar relations among things. Among humans, the predisposition is more subtle, but nevertheless apparent in the earliest years of childhood, when infants seem remarkably sensitive to the emotions, behavior, and social relations of other people while remaining ignorant of much of the world around them.

Monkeys have a kind of laser-beam intelligence. While they solve social problems with little difficulty or training, they often flounder when confronted with the same problems outside the social domain. They do not always generalize their social abilities to other species or to inanimate objects, and in this sense their skills seem relatively restricted. Apparently, the animals do not know what they know and cannot apply their knowledge to problems in other domains.

It has been suggested that one of the most important features of human intelligence is its *accessibility:* a skill acquired in one context can be extended, or generalized, to another. We use analogical reasoning not only to classify different types of kinship relations but also to solve rather irrelevant problems on college entrance exams; we use our skills in classification not only when talking about people but also when arguing vociferously about taxonomic relations among other species. If some of our intellectual abilities did indeed appear first in the context of social interaction, then one of the primary distinctions between our own and other primates' intelligence may lie not so much in any specific skill as in our ability to extend knowledge gained in one context to new and different ones. "Only connect" admonished E. M. Forster in *Howard's End.* It may be peculiarly human advice.

CHAPTER TEN | HOW MONKEYS SEE THE WORLD

On October 1, 1972, the London *Sunday Times* printed the obituary of Flo, an adult female chimpanzee who had lived in the Gombe National Park, Tanzania, and who had been studied by Jane Goodall for over 11 years. An excerpt from this obituary reads as follows: "Flo has contributed much to science. She and her large family have provided a wealth of information about chimpanzee behavior—infant development, family relationships, aggression, dominance, sex. . . . But this should not be the final word. It is true that her life was worthwhile because it enriched human understanding. But even if no one had studied the chimpanzees at Gombe, Flo's life, rich and full of vigour and love, would still have had a meaning and a significance in the patterns of things."

With this final sentence we return to the problem with which this book began. If no human observer had interpreted Flo's life, could we still say that she had knowledge, motives, beliefs, and desires, and that her life was full of vigor and love? Do such mental states really exist in the mind of any animal? Or are they artifacts, invented by ethologists as the best means of describing what they have seen? When we watch nonhuman primates and analyze their social behavior, do we have their minds or ours under the microscope?

We began this book by describing the social behavior of vervet monkeys and suggesting that vervet behavior is best understood if we attribute knowledge, motives, and strategies to the individuals involved. Watching monkeys, one is tempted to treat them like tiny humans, not only because they look rather like us, but also because features of their social organization—close bonds among kin and status striving, for example—look like simplified versions of our own. More important, anthropomorphizing *works*: attributing motives and strategies to animals is often the best way for an observer to predict what an individual is likely to do next.

Descriptions of social behavior in anthropomorphic terms do not, however, constitute an explanation. After all, one way to describe and predict the behavior of an automatic teller machine is to assume that it wants to help you do your banking, but this motive is presumably not a central part of the machine's operation. A second goal of our book, therefore, has been to dissect the knowledge and motives that make monkeys do what they do. Are the mechanisms that govern behavior and communication in vervet monkeys—superficially so human—really similar to our own? If so, can we prove it? If not, what are the differences? What can we do that monkeys cannot, and how does this make their lives—and their view of the world—different from our own?

Perceiving Social Relationships

Many of the mechanisms that underlie human social behavior occur so automatically that we take them for granted. Without giving it much thought, for example, we recognize individuals, observe their behavior and, on the basis of many different types of interactions, abstract and describe the relationships that two, three, or more individuals have with one another. We then compare relationships in at least two distinct ways. In the simplest comparison we contrast one relationship directly with another: Tom and Mary are much closer than Steve and Shirley, who aren't getting along very well these days. In more complex comparisons we classify relationships into types, like lovers, friends, or enemies, and then analyze the extent to which a particular bond conforms to expectations given the class of relationships into which it falls: Claire has good relations with her parents but is unusually close to her grandmother. Finally, in an attempt to understand *why* relationships are the way they are—why they differ from one another and why they may or may not conform to type—we search for the motives (usually mental states like desires, fears, hopes, or dreams) that cause individuals to behave the way they do: in science, relations between senior professors (silverbacks) and their various students often *look* the same, as both parties strive to keep up the appearance of relaxed colleagiality. But some students dislike their mentors, whereas others are genuinely close, both academically and personally. In such cases, our understanding of relationships is based ultimately on the thoughts, motives, and mental states we attribute to others.

Compared with our relatively full, coherent picture of human social relationships, the vervet monkey's view of relations among other vervets seems like Picasso's cubist guitar: many of the parts are there and some of them even fit together, but the elements are in disarray and the net effect is that of a puzzle that has not been assembled correctly.

Certainly vervet monkeys, like many other animals, recognize individuals and take note of the interactions that occur among them. Moreover, from their observation of different types of interactions, by a process we do not yet fully understand, the monkeys may also create in their minds a number of representations that describe different sorts of social relationship: mother-offspring relationships, relationships among kin, friendships between males and females. We suspect that such representations exist because experiments with captive monkeys have shown that primates compare relationships according to the *types* of bond they instantiate and not just according to the particular individuals involved. They judge mother-offspring pairs as the same even if one involves a mother and her adult daughter and the other involves a mother and her infant son (Dasser 1988a). Furthermore, we hypothesize that monkeys make use of such representations in their daily lives because a vervet monkey is more likely to threaten another individual if one of her own close relatives and one of her opponent's close relatives have recently fought (Cheney and Seyfarth 1986, 1989). Such behavior is difficult to explain without assuming that vervets recognize some similarity between their own close bonds and the close bonds of others.

Although monkeys may have mental representations of social relationships, however, these representations are probably not as abstract and flexible as our own. Language, for example, allows us to label different types of relationships, to specify the criteria by which we include a bond in one class or another, and to discuss types of relationships in general, abstract terms that are independent of the individuals we had in mind when we first organized relationships into types. By contrast, there is as yet no evidence that monkeys' knowledge of social relationships is accessible to them or that their classification of social bonds is abstract enough to include unfamiliar individuals or social structures. We suspect that a monkey raised among families A, B, C, D, and E, could tell us that the bond between mother A and infant A1 is the same sort of bond as that between mother C and infant C1. We do not know, however, if the same monkey could deduce that bonds just like this occur in many other vervet groups and even in many other species.

Perceiving States of Mind

Humans not only classify social relationships into types but also examine the motives and strategies of others in an attempt to explain *why* some relationships are alike and others different. In our effort to understand behavior, it seems likely, as Nicholas Humphrey (1983) suggests, that humans often use introspection as a guide. Knowing that our own actions are often caused by particular mental states, we look for the same processes in others.

In contrast, we have as yet no evidence that monkeys are aware of their own knowledge or attribute mental states to others. Monkeys are doubtless excellent at monitoring and predicting each other's actions, and they probably have little difficulty recognizing that monkey X's actions can have a particular effect on Y's behavior. It seems unlikely, however, that monkeys take

analogies

into account each others' thoughts, motives, or beliefs when they assess what other individuals are likely to do next.

Many of these generalizations seem to apply less to apes than to monkeys. More than monkeys, chimpanzees seem to recognize thoughts as agents of actions, and much of their behavior seems designed to alter or control other individuals' states of minds. At the same time, however, even apes seem to have difficulty recognizing ignorance and false beliefs in others; although they seem to understand that other individuals have beliefs, they do not distinguish easily between different beliefs or even no beliefs.

Why is the attribution of mental states so important? Consider what one *cannot* do if one is unable to attribute mental states to others or recognize that not all individuals share the same beliefs, motives, and knowledge.

To begin with, there can be no pedagogy in the sense that we know it among humans. To teach, one must recognize a difference between one's own knowledge and someone else's knowledge and then take explicit steps to redress this imbalance. Without attribution, instruction cannot even begin, because those with knowledge do not realize that the information possessed by others can be quite different from their own. The transmission of information about predators by vervet monkeys provides a good example. Recall that although vervet infants seem to be born with a general inclination to give eagle alarm calls to things flying in the air, they often make mistakes and warn others unnecessarily about the approach of, say, pigeons (chapter 4). When infants correctly give an alarm call to a genuine predator, adults often respond by giving alarm calls themselves. Since "second alarms" by adults occur much more often when the infant is right than when she is wrong, the data suggest that adults may be "teaching" infants about which birds are really dangerous. In fact, however, adults do not go out of their way to help infants learn: they never correct infants when the infants make mistakes and, as for selective reinforcement, adults are just as likely to give second alarms after another adult has called correctly as they are after a correct alarm by an infant (Seyfarth and Cheney 1986).

Infant vervets, then, are left to learn by observation alone, without explicit tutelage from adults. This reliance on observational learning is widespread among animals (Nishida 1987; Galef 1988) and severely constrains the rate of cultural transmission (Boyd and Richerson 1985). In our view, it can ultimately be traced to the failure to recognize that the knowledge possessed by others can be different from one's own.

Deception, in the human sense, also requires the attribution of mental states to oneself and to others. When we lie to someone we recognize a distinction between our own and another person's thoughts, and we depend on the fact that a person's beliefs can affect his subsequent behavior. If we are correct in arguing that monkeys live in a world without attribution, it would also seem to follow that they live in a world without deception. The issue, however, is more complicated than this.

The behavior and vocal signals of many different species often *function* to deceive or mislead others (chapter 7). A review of the evidence, however, raises doubts about the flexibility of animal deception and provides little evidence for the attribution of mental states to others. Great tits, for example, give apparently deceptive alarm calls at feeding perches, and they are skillful enough to vary their false alarm calls depending upon who is nearby. If the birds at the feeding perch are lower ranking than the signaler, false alarm calls are rarely given, presumably because the caller can simply supplant his rivals by approaching. When higher-ranking birds are present and a supplant is not possible, however, lower-ranking birds do give false alarm calls (Moller 1988). There is, then, some flexibility in the use of deceptive alarms by great tits; however, the limits of great tit deception are equally striking. We have no evidence, for example, that the birds use any other signals to deceive each other or that they use deceptive signals in any other social context (chapter 7). Even in the case of nonhuman primates, we have little evidence that individuals ever act to manipulate each others' *beliefs*, as opposed to each others' *behavior*.

Results to date suggest, therefore, that deception in animals is less flexible than human deception. We suspect that many of the differences between human and nonhuman deception, like the differences between human and nonhuman pedagogy, ultimately derive from the failure of most animal species to attribute mental states to others. Human deception rests on the assumption that other individuals have mental states, that these mental states affect behavior, and that other individuals' mental states can be manipulated to one's own benefit. Armed with this general theory, humans can modify deceptive behavior widely, within a single context or from one context to the next. By contrast, animal deception seems to rest primarily on the recognition of certain behavioral contingencies: if I do this, he will do that. It must be learned and relearned from one circumstance to the next. As a result, each species can provide us with one or two examples of "deceptive" behavior but no one species, except possibly chimpanzees, exhibits the flexibility, modifiability, and variety of lies, deceit, and half-truths that are so easy to find among humans.

association & lacking flexibility

Finally, an inability to attribute mental states limits the extent to which nonhuman primates can express empathy or compassion toward one another. When his mother Flo died, the young male chimpanzee Flint exhibited many of the behavioral patterns we associate with grief in humans. He avoided others, stopped eating, and spent many hours a day sitting in a hunched posture, rocking back and forth (Goodall 1979). Eventually he died. It is clear from this and other descriptions of death in chimpanzees

(Goodall 1971, 1979) that these animals experience grief and a sense of loss when someone close to them dies. Equally striking, however, is the absence of sympathy among other chimpanzees. Although chimpanzees have mental states and grieve at the loss of close friends, they do not seem to recognize the same mental states in others. As a result, they are unable to share their sorrow or show empathy toward them.

In sum, many fundamental differences in social behavior between human and nonhuman primates depend on the presence, or lack, of a *theory of mind:* whether individuals can recognize their own knowledge and attribute mental states to others. Apparently, monkeys see the world as composed of things that act, not things that think and feel. Although they are acutely sensitive to other animals' behavior, they know little about the knowledge or motives that cause animals to do what they do. In a monkey's world, the knowledge possessed by an individual exists in a kind of vacuum: the individual does not know what he knows and cannot recognize knowledge (or the lack of it) in others.

The Perception of Word Meaning

Few things are more difficult than to specify precisely and exhaustively what a word or sentence means. When Caliban says, in Act V of Shakespeare's *Tempest,* "What a thrice-double ass Was I, to take this drunkard for a god, And worship this dull fool!" the speech conveys subtly different shades of meaning to each character on stage and to each person listening in the audience, many of whom will find still more interpretations when they read the play's reviews in the next morning's papers. Part of the problem, of course, is tied up in the question of attribution: to interpret Caliban's remarks, we not only need information about his social interactions during the past few weeks, but we also have to be mind readers and interpret Caliban's interactions from his *own* perspective. Finally, our analysis will be completely stymied if we fail to recognize that Caliban's remarks cannot be taken literally.

Even casual remarks that require no recourse to mental states or metaphor can often be interpreted in as many different ways as there are listeners, because meaning depends not only on external referents but also on the speaker's behavior, his personal idiosyncrasies, and the context in which the remarks are given. As he starts walking toward the door, a man says, "Oh! It's just beginning to rain." This simple statement conveys information about a particular external referent, the man's disposition toward that referent, and the likelihood that he will engage in certain behavior, like picking up an umbrella or going back to the closet for a coat. The man's audience might also draw a number of other conclusions if they knew, for example, that the country was in the midst of a drought, that the man liked to go out in the rain, that he did not own an umbrella, or that he had been planning to play tennis within the hour.

One lesson to be learned from the study of language, therefore, is that we should not expect to specify exhaustively what the vocalizations of nonhuman primates mean, because any call correlated with a specific external referent cannot help but be correlated also with behavior and context. We can, however, chip away at the problem by eliminating some of the more extreme hypotheses that have been offered in the past. Although previous studies ruled out the possibility that monkeys, apes, or indeed any other animal ever signaled *about* things, there is now clear evidence that the alarm calls, grunts, *wrrs,* and chutters of vervet monkeys (as well as many calls in other species) provide listeners with information about objects and events in the environment. Two sorts of evidence support this view. First, playback of a call in the absence of its supposed referent generally elicits the same response as would be elicited by the referent itself. Without any other supporting information, for example, leopard alarm calls elicit the same response as would a leopard itself and grunts to another group, *wrrs,* and chutters elicit the same response as would a neighboring group (chapter 4). In this sense, vervet calls function to denote objects or events. Second, and perhaps more important, when monkeys in habituation experiments are asked to compare two vocalizations, they do so not just according to the calls' acoustic properties but also according to their referents. Vervets, in other words, treat vocalizations not just as physical entities but as sounds that represent things (chapter 5).

To repeat, however, our emphasis on the representational function of primate vocalizations should not be taken as an exhaustive characterization of what each call means. Without doubt, leopard alarms, grunts, *wrrs,* and chutters, in addition to denoting external objects and events, provide listeners with information about the signaler's level of excitement, the immediacy of the danger, what the signaler is likely to do next, and so on (e.g., Smith 1977, 1990). We emphasize the referential function of vocalizations not because it is the only function but because it was previously thought not to exist and because it may shed some light on the mental processes that underlie vervet communication and behavior.

Since vervet vocalizations convey many sorts of information, our assessment of call meaning will invariably be imprecise. It is presently unclear, for example, whether a vervet's alarm call should be glossed as a word (Leopard!) or as a proposition that includes information about both a referent and the speaker's attitude or disposition toward it (Leopard! Head for the trees!). Specifying precisely what a call means is difficult in at least one other respect. To describe the referents of leopard alarm calls fully, we need to know both the objects that elicit leopard alarms (positive instances of the class) and objects that elicit a different vocalization or no vocalization at all (negative instances). The meaning of any one call, therefore, is relative and depends on the meaning of others. With an enormous repertoire of words, like our own, we can specify meaning quite precisely; with a small repertoire like the vervets', meaning is inevitably imprecise.

There are important parallels between the processes that underlie vervet monkeys' social behavior and the processes that underlie their vocal communication. In social behavior the monkeys seem to begin by observing the social interactions of others. Then, using this raw material, they recognize relationships, classify relationships into types, and adopt a behavioral strategy based on this classification. Similarly, in their vocal communication the monkeys seem to begin by distinguishing among different calls. They then learn the meaning of each call and classify and compare vocalizations according to their referents.

Vervet monkey vocalizations deserve the label *semantic* because they function to denote objects and events in the environment. Vervet calls are also semantic in the stronger sense that their production and interpretation depend on the mental states of the signaler and the recipient. Vocalizations generate representations in both signaler and recipient and are compared and responded to on the basis of these representations. To a vervet, the world is composed of two fundamentally different sorts of things: objects, like leopards, snakes, or other groups, and vocalizations, which serve as representations of these objects. Monkeys respond to objects according to their physical features; they respond to vocalizations according to the things for which they stand. While vocalizations are semantic in this stronger sense, however, vervet calls seem not to be semantic in the strongest sense of being given with an intent to modify the mental states of others. Though representations play a crucial role in the monkeys' use and interpretation of calls, there is no evidence that monkeys (or even apes) ever communicate with the intent of changing a receiver's mental state or of drawing the receiver's attention to the sender's own mental state (Grice 1957; Dennett 1978a). Monkeys cannot communicate with an intent to modify the mental states of others because, lacking attribution, they do not recognize that such mental states exist.

Domain-Specific Intelligence

Many animals are specialists, performing skills with much greater sophistication in some contexts than in others. Honey bees have extraordinarily complex ways of communicating with each other about food, beavers build elaborate dams, and arctic terns are superb celestial navigators. Yet none of these species displays comparable abilities outside its one area of expertise. Natural selection, it appears, has acted not on *general* skills but on behavior in more narrowly defined ecological domains.

Throughout this book we have focused on the skills displayed by vervets and other monkeys in the domain of social interactions. A complete understanding of how monkeys see the world, however, requires that we place their social behavior in a broader context, contrasting it with, for example, their feeding behavior and their knowledge of other species.

Although far more research remains to be done, there is some support for the suggestion that monkeys, too, are specialists and that at least some of their skills in social interactions are not extended to other contexts. Within their social groups, monkeys seem to recognize not only their own but also other individuals' dominance ranks and social bonds. Monkeys also seem to compare social relationships, performing crude analogies to recognize, for example, that different individuals have similar social relationships. In contrast, vervet monkeys seem to know remarkably little about the environment in which they live. Even in circumstances when it would obviously be adaptive for them to do so, vervets have not learned to associate predators with their secondary visual cues. They seem inattentive, for example, to the tracks of snakes or the hunting behavior of leopards. In tests that directly compare the monkeys' abilities using two different sorts of stimuli, vervets perform far better if the question concerns other vervets than if the question concerns birds, hippos, or predators.

Conversely, primates undoubtedly use some skills when locating and exploiting food resources that they do not extend to social interactions. Hamadryas baboons, chimpanzees, and probably many other species seem to have mental maps, or representations, of their home range, which they use to take the most efficient route to water, fruit, or tools. It remains to be seen whether similar mental maps are used or even needed in social interactions. To argue that the mental representations used in feeding are more or less abstract or complex than those used in social interactions is specious. The intelligence shown by monkeys in the social domain is not superior to that shown in the nonsocial domain; it is simply different.

The question of domain-specific knowledge is linked closely to issues of consciousness and the attribution of mental states. An individual who cannot reflect upon his own knowledge to form hypotheses about what he knows will almost by definition be unable to extend knowledge from one context to another. If, as seems likely, monkeys do not know what they know, then much of their knowledge may be highly compartmentalized and inaccessible to them.

Despite this post hoc rationalization for the monkeys' differing performance in different domains, the vervets' apparent failure to attend to biologically important features of their environment remains surprising and unsatisfying. Can it really be more important, in terms of survival, to develop a conceptual understanding of social relationships than to recognize the track of a python? For reasons that are not entirely clear, vervets seem to have some adaptive specializations (Rozin and Schull 1988) in the domain of social behavior that are not, but should be, extended to their interactions with other species. We will make little progress in understanding the minds of nonhuman primates until we investigate these issues more thoroughly.

Summary

When we study the social behavior of monkeys we are tempted to anthropomorphize and treat them as if they were human. This is not entirely inappropriate. Like the primatologists who study them, vervet monkeys observe social interactions and draw generalizations about the types of relationships that exist among individuals. The monkeys also use sounds to represent things and compare different vocalizations according to their meaning.

There are, however, many ways in which a vervet's view of her world is very different from our own. Though a monkey may make use of abstract concepts and have motives, beliefs, and desires, her mental states are not accessible to her: she does not know what she knows. Further, monkeys seem unable to attribute mental states to others or to recognize that others' behavior is also caused by motives, beliefs, and desires.

The inability to examine one's own mental states or to attribute mentality to others severely constrains the ability of monkeys to transmit information, deceive, or feel empathy with one another. It also limits the extent to which monkey vocalizations can be called semantic. True, calls function to denote objects and events in the environment and, like words, are caused by the mental states of those who use them. Unlike our language, however, the vocalizations of monkeys are not given with an intent to modify the mental states of others. Though monkeys are skilled observers of each others' *behavior,* they seem to be far less astute observers of each others' *minds,* and they seldom seem to proceed beyond other animals' actions to analyze the motives underlying their behavior. *We* attribute motives, plans, and strategies to the animals, but they, for the most part, do not.

This chapter ties all together well.

APPENDIX

The primary aim of this demographic record is to present a complete genealogy of the adult females and juveniles in our three main study groups. It also, however, has a more frivolous purpose.

Anyone who has ever studied animals in their natural habitat recognizes that even the most stimulating project includes moments of unrelieved tedium. An objective sampling regime cannot always concentrate on fighting, grooming, or copulating animals but must also include the sleeping pregnant females and the vacant wanderings of the peripheral male. Such sampling sessions invariably find the observer gazing off into the middle distance, musing quietly about past and future foibles.

Our system for naming vervets was originally devised to enliven the many hours we and our colleagues spent watching sleeping or solitary monkeys. And, since infant vervets are notoriously difficult to sex, we attempted to avoid potential bias by assigning names that were not specific to either sex.

At the beginning of our study we assigned all females, males, and juveniles in each group to a particular theme, such as scandals, prisons, enigmas, "countries that aren't countries," and so on. (The adult females in group A when we began our study were an exception.) Giving each age-sex class a theme was not original; in her earlier study of baboons, for example, Thelma Rowell (1968) named the adult females in her group after characters in Jane Austen's novel *Emma*. Each successive year, the infants in a given group and birth season were assigned a new theme, devised by whomever was studying the monkeys at the time. Each theme, therefore, has its own creator, and many people in addition to ourselves are responsible for the names listed below. We leave it to the interested reader to decipher each year's theme in each group.

The far left column lists the adult females in each group as of March, 1977; females are arranged in order of descending dominance rank. The offspring of each female are listed beneath her name and are arranged according to the birth season (October to February) in which they were born. Males are indicated by a ♂ sign. Dates beneath animals refer to their year of death. For older adult females, minimum estimated birth dates are provided. * = transfers into group C as a result of a group fusion. ? = for infants, sex unknown; for males, year in which a male was last seen after transferring to another group.

GROUP A

Year Born	1975	1976	1977	1978	1979	1980	1981	1982	1983	1984	1985	1986	1987
Disney (1970–81)	♂ Bobby V (1984)		♂ Savarin (1978)		♂ Dixie (1980)								
Lady Bountiful (1970–80)	LaBelle (1977)		Escoffier (1978)		Little (1980)								
Escoffier										Snoopy		Unnamed (1986)	
Borgia (1969–87)	Leslie (1986)		♂ Claiborne (1986?)		♂ Bantam		♂ Goofy (1986)	♂ Brass Monkey (1983)	♂ Earl Grey (1989)	♂ Wood-stock (1985)		♂ Strawberry (1987)	
Leslie						Widmer-pool (1984)		Lead Balloon (1983)	♂ Morning Thunder (1984)	♂ Pig Pen (1987)		? Telluride (1986)	
Madame Pompadour (1971–78)		♂ Waidus (1977)	♂ Egg McMuffin (1978)										
Anastasia (1970–80)	♂ Del Shannon (1982)		♂ Julian (1986)		♂ Alaking (1980)								

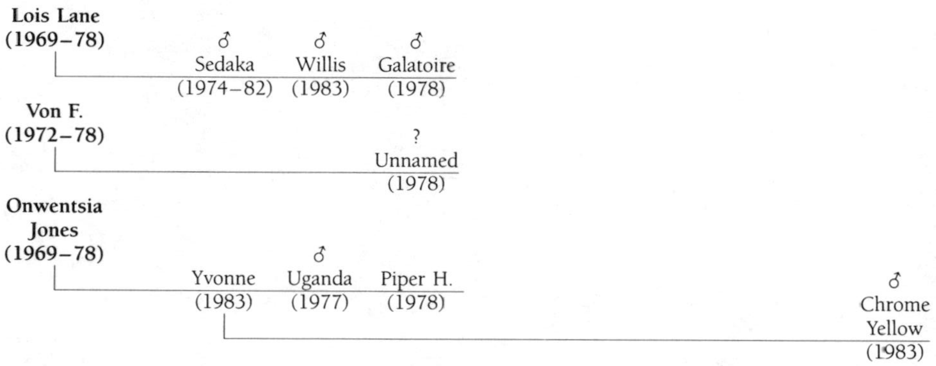

Lois Lane
(1969–78)

♂ Sedaka (1974–82)
♂ Willis (1983)
♂ Galatoire (1978)

Von F.
(1972–78)

? Unnamed (1978)

Onwentsia Jones
(1969–78)

Yvonne (1983)
- ♂ Chrome Yellow (1983)

♂ Uganda (1977)
Piper H. (1978)

GROUP B

Year Born	1975	1976	1977	1978	1979	1980	1981	1982	1983	1984	1985	1986	1987	1988
Bokassa (1974–80)														
Somoza (1971–80			♂ Galileo (1980)		Satin Sheets (1980)	Picadilly (1983)	Sing Sing (1983)							
Amin (1972–87)		♂ Copernicus (1983)	♂ Halley (1980)		Aphro (1988)	♂ Knightsbridge (1981)	Attica (1982)	♂ Big Apple (1983)	♂ Peter Piper (1984)	♂ Betamint (1986)	♂ Acushnet (1989)			
(Aphro)									Unique New York (1987)	♂ Condiment (1985)	Bismarck (1987)		♂ Unnamed (1987)	
Franco (1973–81)			♂ Einstein (1980)		Feather (1980)	Victoria (1981)								
Duvalier (1972–81)			Tycho (1987)		♂ Diddle (1980)	Holborn (1984)								
(Tycho)								? Traffic Jam (1982)	Woodchuck (1986)	♂ Predicament (1985)				

Marcos (1973–89)

Newton (1987) — ♂ Mazolla (1980) — Charing Cross — Wormwood Scrubs (1987) — Cereal Monogamy (1983) — Toy Boat (1986) — Emolument (1986) — Andrea Doria — ♂ Mistral

Alcatraz (1982) — Nut Case — Fig Plucker (1985) — ♂ Tournament (1987) — Lusitania (1986) — ♂ Baby Krebs (1987)

Sentiment (1986) — ♂ Naldo (1987) — Chinook (1988)

♂ Tory Canyon (1987)

Alizee (1989)

Pinochet (1972–85)

♂ Kepler (1979) — ♂ Pernod (1984?) — ♂ Playboy (1980) — Paddington (1980) — Pepper Tick (1983) — ♂ Sea Shells (1987) — ♂ Parliament (1986)

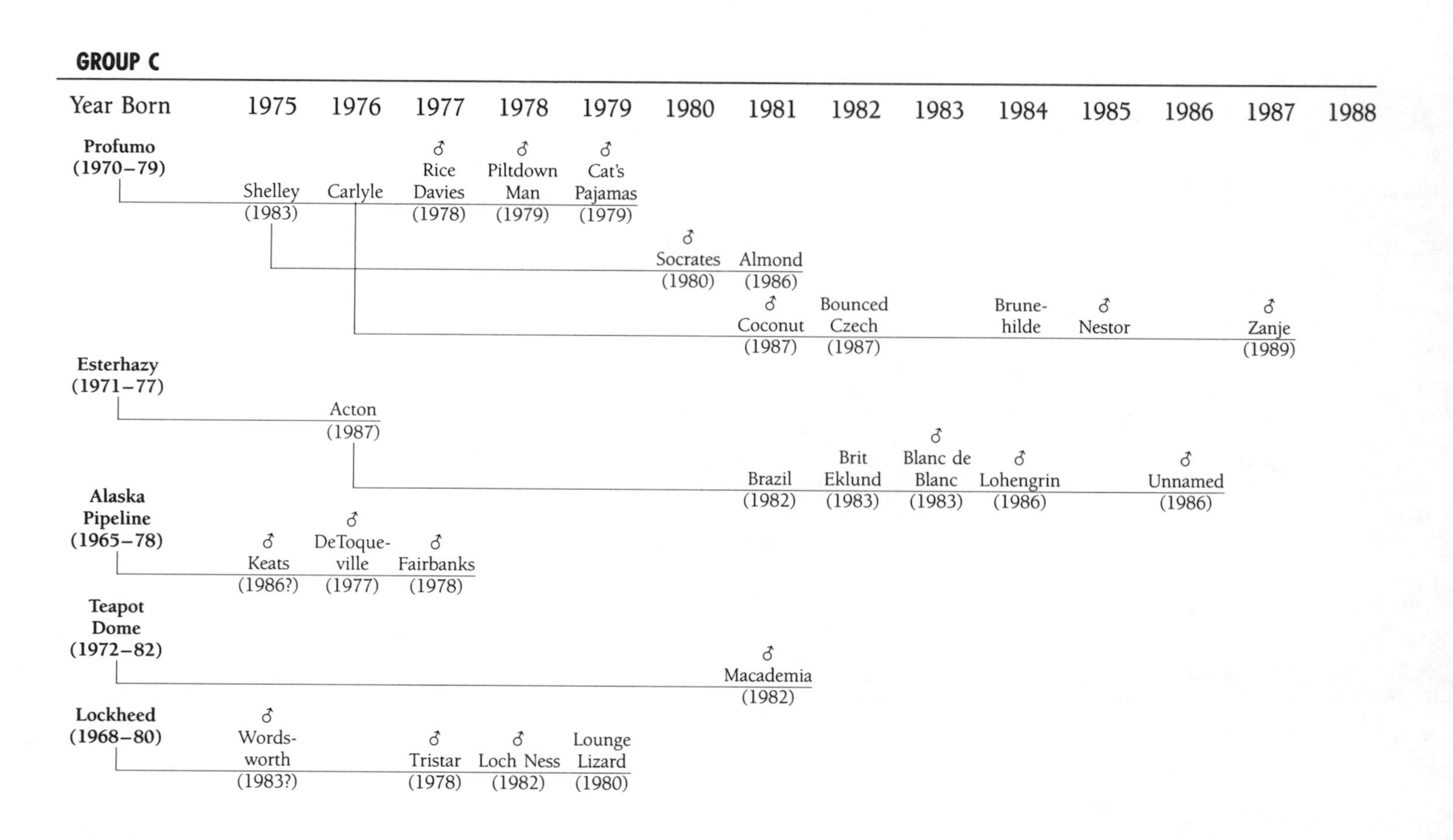
GROUP C
Year Born
1975
1976
1977
1978
1979
1980
1981
1982
1983
1984
1985
1986
1987
1988
Profumo (1970–79)
Shelley (1983)
Carlyle
♂ Rice Davies (1978)
♂ Piltdown Man (1979)
♂ Cat's Pajamas (1979)
♂ Socrates (1980)
Almond (1986)
♂ Coconut (1987)
Bounced Czech (1987)
Brune-hilde
♂ Nestor
♂ Zanje (1989)
Esterhazy (1971–77)
Acton (1987)
Brazil (1982)
Brit Eklund (1983)
♂ Blanc de Blanc (1983)
♂ Lohengrin (1986)
♂ Unnamed (1986)
Alaska Pipeline (1965–78)
♂ Keats (1986?)
♂ DeToque-ville (1977)
♂ Fairbanks (1978)
Teapot Dome (1972–82)
♂ Macademia (1982)
Lockheed (1968–80)
♂ Words-worth (1983?)
♂ Tristar (1978)
♂ Loch Ness (1982)
Lounge Lizard (1980)

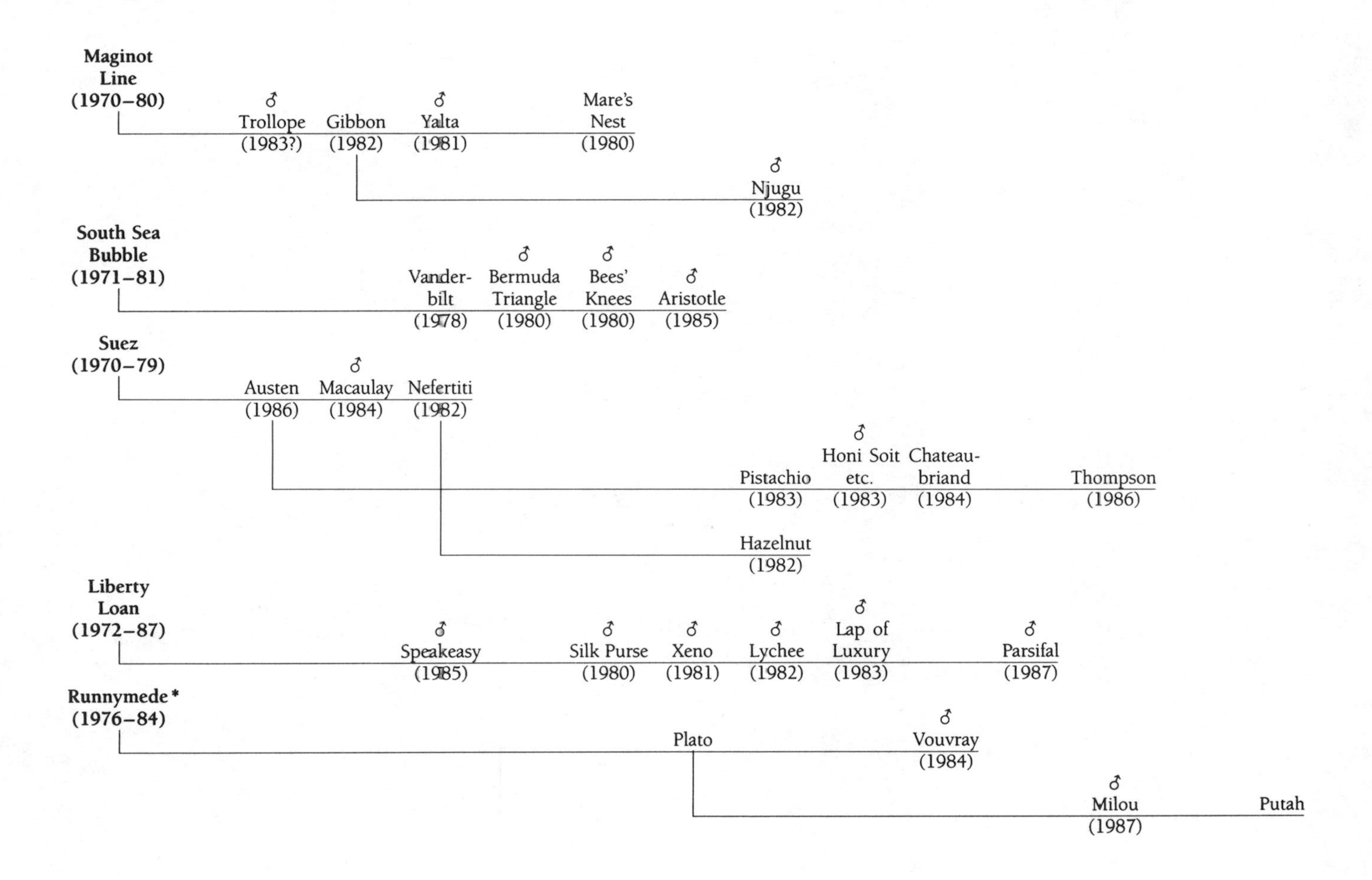

Maginot Line (1970–80)
♂ Trollope (1983?)
Gibbon (1982)
♂ Yalta (1981)
Mare's Nest (1980)
♂ Njugu (1982)
South Sea Bubble (1971–81)
Vander-bilt (1978)
♂ Bermuda Triangle (1980)
♂ Bees' Knees (1980)
♂ Aristotle (1985)
Suez (1970–79)
Austen (1986)
♂ Macaulay (1984)
Nefertiti (1982)
Pistachio (1983)
♂ Honi Soit etc. (1983)
Chateau-briand (1984)
Thompson (1986)
Hazelnut (1982)
Liberty Loan (1972–87)
♂ Speakeasy (1985)
♂ Silk Purse (1980)
♂ Xeno (1981)
♂ Lychee (1982)
♂ Lap of Luxury (1983)
♂ Parsifal (1987)
Runnymede* (1976–84)
Plato
♂ Vouvray (1984)
♂ Milou (1987)
Putah

References

Abegglen, J. J. 1984. *On socialization in hamadryas baboons.* Cranbury, N.J.: Associated University Presses.

Abercrombie, D. 1967. *Elements of general phonetics.* Edinburgh: Edinburgh University Press.

Alatalo, R. V., A. Carlson, A. Lundberg, and S. Ulstrand. 1981. The conflict between male polygamy and female monogamy: The case of the pied flycatcher *Ficedula hypoleuca. Am. Nat.* 117:738–53.

Alatalo, R. V., A. Lundberg, and K. Stahlbrandt. 1984. Female mate choice in the pied flycatcher *Ficedula hypoleuca. Behav. Ecol. Sociobiol.* 14:253–61.

Alexander, R. D. 1987. *The biology of moral systems.* Hawthorne, N.Y.: Aldine.

Allen, C. 1989. Philosophical issues in cognitive ethology. Ph. D. diss., University of California, Los Angeles.

Altmann, J. 1974. Observational study of behaviour: Sampling methods. *Behaviour* 49:227–65.

———. 1979. Age cohorts as paternal sibships. *Behav. Ecol. Sociobiol.* 6:161–9.

———. 1980. *Baboon mothers and infants.* Cambridge, Mass.: Harvard University Press.

Altmann, J., S. A. Altmann, and G. Hausfater. 1978. Primate infant's effects on mother's future reproduction. *Science* 201:1028–30.

———. 1988. Determinants of reproductive success in savannah baboons (*Papio cynocephalus*). In *Reproductive success,* ed. T. H. Clutton-Brock. Chicago: University of Chicago Press.

Altmann, S. A. 1967. The structure of primate social communication. In *Social communication among primates,* ed. S. A. Altmann. Chicago: University of Chicago Press.

Altmann, S. A., and J. Altmann. 1970. *Baboon ecology.* Chicago: University of Chicago Press.

Andelman, S. J. 1985. Ecology and reproductive strategies of vervet monkeys (*Cercopithecus aethiops*) in Amboseli National Park, Kenya. Ph. D. diss., University of Washington.

———. 1987. Ecological and social determinants of cercopithecine mating patterns. In *Ecological aspects of vertebrate social evolution,* ed. D. I. Rubinstein and R. W. Wrangham. Princeton, N.J.: Princeton University Press.

Anderson, J. R. 1984a. Monkeys with mirrors: Some questions for primate psychology. *Int. J. Primatol.* 5:81–98.

———. 1984b. The development of self-recognition: A review. *Develop. Psychobiol.* 17:35–49.

———. 1986. Infant stumptailed macaques reared with mirrors or peers: Social responsiveness, attachment, and adjustment. *Primates* 27:63–82.

Anderson, C. O., and W. A. Mason. 1974. Early experience and complexity of social organization in groups of young rhesus monkeys. *J. Comp. Physiol. Psychol.* 87:681–90.

Andersson, M. 1980. Why are there so many threat displays? *J. Theor. Biol.* 86: 773–81.

Anglin, J. M. 1977. *Word, object and conceptual development.* New York: W. W. Norton.

Armstrong, D. M. 1981. *The nature of mind and other essays.* Ithaca, N.Y.: Cornell University Press.

Armstrong, S. L., L. R Gleitman, and H. Gleitman. 1983. What some concepts might not be. *Cognition* 13:263–308.

Astington, J. W.; P. L. Harris, and D. R. Olson, ed. 1988. *Developing theories of mind.* Cambridge: Cambridge University Press.

Axelrod, R., and W. D. Hamilton. 1981. The evolution of cooperation. *Science* 211:1390–96.

Bachmann, C., and H. Kummer. 1980. Male assessment of female choice in hamadryas baboons. *Behav. Ecol. Sociobiol.* 6:315–21.

Balda, R. P., and A. C. Kamil. 1988. The spatial memory of Clark's nutcrackers (*Nucifraga columbiana*) in an analogue of the radial arm maze. *Anim. Learn. Behav.* 16:116–22.

Balda, R. P., A. C. Kamil, and K. Grim. 1987. Revisits to emptied cache sites by nutcrackers. *Anim. Behav.* 34:1289–98.

Balda, R. P., and R. J. Turek. 1984. The cache-recovery system as an example of memory capabilities in Clark's nutcracker. In *Animal cognition,* ed. H. L. Roitblat, T. G. Bever, and H. S. Terrace. Hillsdale, N.J.: Lawrence Erlbaum Associates.

Baron-Cohen, S., A. M. Leslie, and V. Frith. 1986. Mechanistic, behavioural and intentional understanding of picture stories in autistic children. *Br. J. Develop. Psych.* 4:113–25.

Bateson, P. P. G. 1980. Optimal outbreeding and the development of sexual preferences in Japanese quail. *Z. Tierpsychol.* 53:231–44.

Beck, B. B. 1972. Tool use in captive hamadryas baboons. *Primates:* 13:276–96.

———. 1973. Cooperative tool use by captive hamadryas baboons. *Science* 182: 594–7.

———. 1974. Baboons, chimpanzees, and tools. *J. Hum. Evol.* 3:509–16.

———. 1980 *Animal tool behavior.* New York: Garland Press.

Beer, C. 1990. From folk psychology to cognitive ethology. In *Cognitive ethology: The minds of other animals (essays in honor of Donald R. Griffin),* ed. C. A. Ristau. Hillsdale, N.J.: Lawrence Erlbaum Associates.

Bennett, J. 1976. *Linguistic behaviour*. Cambridge: Cambridge University Press.

Bercovitch, F. 1988. Coalitions, cooperation, and reproductive success among adult male baboons. *Anim. Behav.* 36:1198–1209.

Berman, C. M. 1980. Early agonistic experience and rank acquisition among free-ranging infant rhesus monkeys. *Int. J. Primatol.* 1:153–70.

———. 1982. The ontogeny of social relationships with group companions among free-ranging infant rhesus monkeys I. Social networks and differentiation. *Anim. Behav.* 30:149–62.

———. 1983. Effects of being orphaned: a detailed case study of an infant rhesus. In *Primate social relationships: An integrated approach,* ed. R. A. Hinde. Oxford: Blackwell.

Bitterman, M. E. 1965. Phyletic differences in learning. *Am. Psychol.* 20:396–410.

Boakes, R. 1984. *From Darwin to behaviorism: Psychology and the minds of animals.* Cambridge: Cambridge University Press.

Boesch, C., and H. Boesch. 1983. Optimisation of nut-cracking with natural hammers by wild chimpanzees. *Behaviour* 83:265–86.

———. 1984. Mental map in wild chimpanzees: An analysis of hammer transports for nut cracking. *Primates* 25:160–70.

———. 1989. Hunting behavior of wild chimpanzees in the Tai National Park. *Am. J. Phys. Anthrop.* 78:547–73.

Boinski, S. 1988. Use of a club by a wild white-faced capuchin (*Cebus capucinus*) to attack a venomous snake (*Bothrops asper*). *Am. J. Primatol.* 14:177–80.

Boyd, R. 1988. Is the repeated prisoner's dilemma a good model of reciprocal altruism? *Ethol. Sociobiol.* 9:211–22.

Boyd, R., and P. Richerson. 1985. *Culture and the evolutionary process.* Chicago: University of Chicago Press.

Boysen, S. T., and G. G. Berntson. 1989. Numerical competence in a chimpanzee (*Pan troglodytes*). *J. Comp. Psychol.* 103:23–31.

Bradshaw, J. L., and N. C. Nettleton. 1981. The nature of hemispheric specialization in man. *Behav. Brain Sci.*: 4:51–91.

Breslow, L. 1981. Reevaluation of the literature on the development of transitive inferences. *Psych. Bull.* 89:325–51.

Bretherton, I., and M. Beeghley. 1982. Talking about internal states: The acquisition of an explicit theory of mind. *Develop. Psych.* 18:906–21.

Bretherton, I., S. McNew, and M. Beeghley-Smith. 1981. Early person knowledge as expressed in gestural and verbal communication: When do infants acquire a 'theory of mind'? In *Infant social cognition,* ed. M. E. Lamb and L. R. Sherrod. Hillsdale, N.J.: Lawrence Erlbaum Associates.

Brogden, W. J. 1939. Sensory pre-conditioning. *J. Exp. Psychol.* 25:323–32.

Brooks, R. J., and J. B. Falls, 1975. Individual recognition by song in white-throated sparrows. III. Song features used in individual recognition. *Can. J. Zool.* 53:1749–61.

Brown, L. 1966. Observations on some Kenya eagles. *Ibis* 102:285–97.

Brown, L. H., and D. Amadon. 1968. *Eagles, hawks, and falcons of the world.* New York: McGraw-Hill Book Co.

Brown, R. 1973. *A first language: The early stages.* Cambridge, Mass.: Harvard University Press.

Bryant, P. E., and T. Trabasso. 1971. Transitive inferences and memory in young children. *Nature* 240:456–8.

Bunge, M. 1980. *The mind-body problem, a psychological approach.* New York: Pergamon.

Busse, C. D. 1978. Do chimpanzees hunt cooperatively? *Am. Nat.* 112:767–70.

Busse, C. D., and W. J. Hamilton, III. 1981. Infant carrying by male chacma baboons. *Science* 212:1281–83.

Byrne, R., and A. Whiten, ed. 1988a. *Machiavellian intelligence: Social expertise and the evolution of intellect in monkeys, apes, and humans.* Oxford: Oxford University Press.

———. 1988b. Tactical deception of familiar individuals in baboons. In *Machiavellian intelligence: Social expertise and the evolution of intellect in monkeys, apes, and humans,* ed. R. W. Byrne and A. Whiten. Oxford: Oxford University Press.

———. 1988c. Towards the next generation in data quality: A new survey of primate tactical deception. *Behav. Brain Sci.* 11:267–73.

———. 1990. Computation and mindreading in primate tactical deception. In *Natural theories of mind,* ed. A. Whiten. Oxford: Blackwell Scientific.

Campbell, D. T., and R. Blake. 1977. Animal awareness? *Am. Sci.* 65:146–7.

Capaldi, E. J., and D. J. Miller. 1988. Counting in rats: Its functional significance and the independent cognitive processes which comprise it. *J. Exp. Psychol. Anim. Behav. Proc.* 14:3–17.

Cargile, J. 1970. A note on 'iterated knowing'. *Analysis* 30:151–5.

Carpenter, C. R. 1942. Sexual behaviour of free-ranging rhesus monkeys, *Macaca mulatta.* I. Specimens, procedures, and behavioural characteristics of estrus. *J. Comp. Psychol.* 33:113–42.

Caryl, P. G. 1979. Communication by agonistic displays: What can game theory contribute to ethology? *Behaviour* 68:136–69.

Chance, M. R. A. 1961. The nature and special features of the instinctive social bond of primates. In *Social life of early man,* ed. S. L. Washburn. New York: Viking Fund Publications.

Chance, M. R. A., G. Emory, and R. Payne. 1977. Status referents in long-tailed macaques (*Macaca fascicularis*): Precursors and effects of a female rebellion. *Primates* 18:611–32.

Chance, M. R. A., and C. Jolly. 1970. *Social groups of monkeys, apes, and men.* London: Jonathan Cape.

Chance, M. R. A., and A. P. Mead. 1953. Social behavior and primate evolution. *Symp. Soc. Exp. Bio., Evol.* 7:395–439.

Chapais, B. 1981. The adaptiveness of social relationships among adult rhesus monkeys. Ph. D. diss., University of Cambridge.

———. 1983. Dominance, relatedness and the structure of female relationships in rhesus monkeys. In *Primate social relationships: An integrated approach,* ed. R. A. Hinde. Oxford: Blackwell Scientific.

———. 1988a. Experimental matrilineal inheritance of rank in female Japanese macaques. *Anim. Behav.* 36:1025–37.

———. 1988b. Rank maintenance in female Japanese macaques: Experimental evidence for social dependency. *Behaviour* 104:41–59.

Cheney, D. L. 1977. The acquisition of rank and the development of reciprocal alliances among free-ranging immature baboons. *Behav. Ecol. Sociobiol.* 2: 303–18.

———. 1978. Interactions of immature male and female baboons with adult females. *Anim. Behav.* 26:389–408.

———. 1981. Inter-group encounters among free-ranging vervet monkeys. *Folia primatol.* 35:124–46.

———. 1983a. Extra-familial alliances among vervet monkeys. In *Primate social relationships: An integrated approach,* ed. R. A. Hinde. Oxford: Blackwell Scientific.

———. 1983b. Proximate and ultimate factors related to the distribution of male migration. In *Primate social relationships: An integrated approach,* ed. R. A. Hinde. Oxford: Blackwell Scientific.

———. 1984. Category formation in vervet monkeys. In *The meaning of primate signals,* ed. R. Harre and V. Reynolds. Cambridge: Cambridge University Press.

———. 1987. Interactions and relationships between groups. In *Primate societies,* ed. B. B. Smuts, D. L. Cheney, R. M. Seyfarth, R. W. Wrangham, and T. T. Struhsaker. Chicago: University of Chicago Press.

Cheney, D. L., P. C. Lee, and R. M. Seyfarth. 1981. Behavioral correlates of nonrandom mortality among free-ranging adult female vervet monkeys. *Behav. Ecol. Sociobiol.* 9:153–61.

Cheney, D. L., and R. M. Seyfarth. 1977. Behaviour of adult and immature male baboons during inter-group encounters. *Nature* 269:404–6.

———. 1980. Vocal recognition in free-ranging vervet monkeys. *Anim. Behav.* 28:362–7.

———. 1981. Selective forces affecting the predator alarm calls of vervet monkeys. *Behaviour* 76:25–61.

———. 1982a. How vervet monkeys perceive their grunts: Field playback experiments. *Anim. Behav.* 30:739–51.

———. 1982b. Recognition of individuals within and between groups of free-ranging vervet monkeys. *Am. Zool.* 22:519–29.

———. 1982c. Social knowledge in nonhuman primates. Paper presented at the IXth meeting of the International Primatological Society, Atlanta, Georgia, August 1982.

———. 1983. Non-random dispersal in free-ranging vervet monkeys: Social and genetic consequences. *Am. Nat.* 122:392–412.

———. 1985a. Social and non-social knowledge in vervet monkeys. In *Animal intelligence,* ed. L. Weiskrantz. Oxford: Clarendon Press.

———. 1985b. Vervet monkey alarm calls: Manipulation through shared information? *Behaviour* 93:150–66.

———. 1986. The recognition of social alliances among vervet monkeys. *Anim. Behav.* 34:1722–31.

———. 1987. The influence of intergroup competition on the survival and reproduction of female vervet monkeys. *Behav. Ecol. Sociobiol.* 21:375–86.

———. 1988. Assessment of meaning and the detection of unreliable signals by vervet monkeys. *Anim. Behav.* 36:477–86.

———. 1989. Reconciliation and redirected aggression in vervet monkeys, *Cercopithecus aethiops. Behaviour* 110:258–75.

———. 1990. Truth and deception in animal communication. In *Cognitive ethology: The minds of other animals (essays in honor of Donald R. Griffin),* ed. C. A. Ristau. Hillsdale, N.J.: Lawrence Erlbaum Associates.

——. 1990. Attending to behaviour versus attending to knowledge: Examining monkeys' attribution of mental states. *Anim. Behav.* 40:742–53.

Cheney, D. L., R. M. Seyfarth, S. J. Andelman, and P. C. Lee. 1988. Reproductive success in vervet monkeys. In *Reproductive success,* ed. T. H. Clutton-Brock. Chicago: University of Chicago Press.

Cheney, D. L., R. M. Seyfarth, and B. B. Smuts. 1986. Social relationships and social cognition in nonhuman primates. *Science* 234:1361–66.

Cheney, D. L., R. M. Seyfarth, B. B. Smuts, and R. W. Wrangham. 1987. The study of primate societies. In *Primate societies,* ed. B. B. Smuts, D. L. Cheney, R. M. Seyfarth, R. W. Wrangham, and T. T. Struhsaker. Chicago: University of Chicago Press.

Chevalier-Skolnikoff, S. 1989. Spontaneous tool use and sensorimotor intelligence in *Cebus* compared with other monkeys and apes. *Behav. Brain Sci.* 12:561–88.

Chivers, D. J., and J. MacKinnon. 1977. On the behavior of siamang after playback of their calls. *Primates* 18:943–8.

Chomsky, N. 1972. *Language and mind.* New York: Harcourt, Brace, Jovanovich.

Church, R. M., and W. H. Meck. 1984. The numerical attributes of stimuli. In *Animal cognition,* ed. H. L. Roitblat, T. G. Bever, and H. S. Terrace. Hillsdale, N.J.: Lawrence Erlbaum Associates.

Churchland, P. M. 1984. *Matter and consciousness.* Cambridge, Mass.: MIT/Bradford Books.

Clark, E. 1973. What's in a word? On the child's acquisition of semantics in his first language. In *Cognitive development and the acquisition of language,* ed. T. Moore. New York: Academic Press.

Clutton-Brock, T. H. 1977. Some aspects of intraspecific variation in feeding and ranging behaviour in primates. In *Primate ecology: Studies of feeding and ranging behaviour in lemurs, monkeys and apes,* ed. T. H. Clutton-Brock. London: Academic Press.

———. 1988. Reproductive success. In *Reproductive success,* ed. T. H. Clutton-Brock. Chicago: University of Chicago Press.

Clutton-Brock, T. H., and P. H. Harvey. 1980. Primates, brains and behaviour. *J. Zool., London* 190:309–23.

Coggins, T. E., R. L. Carpenter, and N. O. Owings. 1983. Examining early intentional communication in Down's syndrome and nonretarded children. *Br. J. Disord. Comm.* 18:98–106.

Collins, D. A. 1981. Social behaviour and patterns of mating among adult yellow baboons (*Papio c. cynocephalus* L. 1766). Ph. D. diss., Cambridge University.

Colvin, J. 1983a. Description of sibling and peer relationships among immature male rhesus monkeys. In *Primate social relationships: An integrated approach,* ed. R. A. Hinde. Oxford: Blackwell Scientific.

———. 1983b. Influences of the social situation on male emigration. In *Primate social relationships: An integrated approach,* ed. R. A. Hinde. Oxford: Blackwell Scientific.

Cords, M. 1988. Resolution of aggressive conflicts by immature long-tail macaques, *M. fascicularis. Anim. Behav.* 36:1124–36.

Cosmides, L. 1989. The logic of social exchange: Has natural selection shaped how humans reason? *Cognition* 31:187–276.

Cosmides, L., and J. Tooby. 1989. Evolutionary psychology and the generation of culture. II. Case study: A computational theory of social exchange. *Ethol. Sociobiol.* 10:51–97.

Cowie, R. J. 1977. Optimal foraging in great tits (*Parus major*). *Nature* 268:137–9.

Curtiss, S. 1977. *Genie: A linguistic study of a modern-day 'wild child'.* New York: Academic Press.

Daanje, A. 1941. Uber das Verhalten der Haussperlinge. *Ardea* 30:1–42.

D'Amato, M., and M. Colombo. 1988. Representation of serial order in monkeys (*Cebus apella*). *J. Exp. Psychol. Anim. Behav. Proc.* 14, 131–9.

———. 1989. Serial learning with wild card items by monkeys (*Cebus apella*): Implications for knowledge of ordinal position. *J. Comp. Psychol.* 103:252–61.

———. 1990. The symbolic distance effect in monkeys (Cebus apella). *Anim. Learn. Behav.* 18:133–40.

D'Amato, M., and D. P. Salmon. 1984. Cognitive processes in *Cebus* monkeys. In *Animal cognition,* ed. H. Roitblat, T. G. Bever, and H. S. Terrace. Hillsdale, N.J.: Lawrence Erlbaum Associates.

D'Amato, M., D. P. Salmon, and M. Colombo. 1985. Extent and limits of the matching concept in monkeys (*Cebus apella*). *J. Exp. Psychol. Anim. Behav. Proc.* 11:35–51.

D'Amato, M., D. P. Salmon, E. Loukas, and A. Tomie. 1985. Symmetry and transitivity of conditional relations in monkeys (*Cebus apella*) and pigeons (*Columba livia*). *J. Exp. Ana. Behav.* 44:35–47.

D'Amato, M., and P. van Sant. 1988. The person concept in monkeys (*Cebus apella*). *J. Exp. Psychol. Anim. Behav. Proc.* 14:43–55.

Damon, W., and D. Hart. 1982. The development of self-understanding from infancy through adolescence. *Child Develop.* 53:841–64.

Darwin, C. 1859. *The origin of species.* London: Murray.

Dasser, V. 1985. Cognitive complexity in primate social relationships. In *Social relationships and cognitive development,* ed. R. A. Hinde, A. Perret-Clermont, and J. Stevenson Hinde. Oxford: Oxford University Press.

———. 1988a. A social concept in Java monkeys. *Anim. Behav.* 36:225–30.

———. 1988b. Mapping social concepts in monkeys. In *Machiavellian intelligence: Social expertise and the evolution of intellect in monkeys, apes, and humans,* ed. R. W. Byrne and A. Whiten. Oxford: Oxford University Press.

Datta, S. B. 1983a. Patterns of agonistic interference. In *Primate social relationships: An integrated approach,* ed. R. A. Hinde. Oxford: Blackwell Scientific.

———. 1983b. Relative power and the acquisition of rank. In *Primate social relationships: An integrated approach,* ed. R. A. Hinde. Oxford: Blackwell Scientific.

———. 1983c. Relative power and the maintenance of dominance. In *Primate social relationships: An integrated approach,* ed. R. A. Hinde. Oxford: Blackwell Scientific.

Davies, N. B., and M. Brooke. 1988. Cuckoos versus reed warblers: Adaptations and counteradaptations. *Anim. Behav.* 36:262–84.

Davies, N. B., and T. R. Halliday. 1978. Deep croaks and fighting assessment in toads, *Bufo bufo. Nature* 274:683–5.

———. 1979. Competitive mate searching in common toads, *Bufo bufo. Anim. Behav.* 27:1253–67.

Davies, N. B., and A. I. Houston. 1981. Owners and satellites: The economics of territory defence in the pied wagtail, *Motacilla alba. J. Anim. Ecol.* 53:895–912.

———. 1984. Territory economics. In *Behavioural ecology: An evolutionary approach,* ed. J. R. Krebs and N. B. Davies. Oxford: Blackwell Scientific.

Davis, H., and S. A. Bradford. 1986. Counting behaviour by rats in a simulated natural environment. *Ethology* 73:265–80.

Davis, R. T., R. W. Leary, D. A. Stevens, and R. F. Thompson. 1967. Learning and perception of oddity problems by lemurs and seven species of monkey. *Primates* 8:311–22.

Dawkins, R. 1986. *The blind watchmaker.* New York: W. W. Norton.

Dawkins, R., and J. R. Krebs. 1978. Animal signals: Information or manipulation. In *Behavioural ecology: An evolutionary approach,* ed. J. R. Krebs and N. B. Davies. Oxford: Blackwell Scientific.

Denham, W. W. 1987. *West Indian green monkeys: Problems in historical biogeography.* Basel: Karger.

Dennett, D. C. 1971. Intentional systems. *J. Philos.* 68:68–87.

———. 1978a. Beliefs about beliefs. *Behav. Brain Sci.* 1:568–70.

———. 1978b. *Brainstorms.* Cambridge, Mass.: MIT/Bradford Books.

———. 1983. Intentional systems in cognitive ethology: The "Panglossian paradigm" defended. *Behav. Brain Sci.* 6:343–55.

———. 1987. *The intentional stance.* Cambridge, Mass.: MIT/Bradford Books.

———. 1988. The intentional stance in theory and practice. In *Machiavellian intelligence: Social expertise and the evolution of intellect in monkeys, apes, and humans,* ed. R. W. Byrne and A. Whiten. Oxford: Oxford University Press.

Dewsbury, D. A. 1982. Dominance rank, copulatory behavior and differential reproduction. *Q. Rev. Biol.* 57:135–59.

———. 1984. *Comparative psychology in the 20th century.* Stroudsburg, Pa.: Hutchinson Ross.

Dickinson, A. 1980. *Contemporary animal learning theory.* Cambridge: Cambridge University Press.

Dohl, J. 1968. Uber die fahigkeit einer schimpansin, umwege mit selbstandigen zwischenzielen zu uberblicken. *Z. Tierpsychol.* 25:89–103.

Donaldson, M. 1978. *Children's minds.* New York: W. W. Norton.

Drickamer, L. C. 1974. A ten-year summary of reproductive data for free-ranging *Macaca mulatta. Folia primatol.* 21:61–80.

Dufty, A. M. 1986. Singing and the establishment and maintenance of dominance hierarchies in captive brown-headed cowbirds. *Behav. Ecol. Sociobiol.* 19:49–55.

Dunbar, R. I. M. 1976. Some aspects of research design and their implications in the observational study of behaviour. *Behaviour* 58:79–98.

———. 1983a. Structure of gelada baboon reproductive units. II. Social relationships between reproductive females. *Anim. Behav.* 31:556–64.

———. 1983b. Structure of gelada baboon reproductive units. III. The male's relationship with his females. *Anim. Behav.* 31:565–75.

———. 1988. *Primate social systems.* Ithaca, N.Y.: Comstock Publishing.

———. In press. Functional significance of social grooming in primates. *Folia primatol.*

Dunford, C. 1977. Kin selection for ground squirrel alarm calls. *Am. Nat.* 111:782–5.

Duvall, S. W., I. S. Bernstein, and T. P. Gordon. 1976. Paternity and status in a rhesus monkey group. *J. Reprod. Fert.* 47:25–31.

Eglash, A. R., and C. T. Snowdon. 1983. Mirror-image responses in pygmy marmosets. *Am. J. Primatol.* 5:211–19.

Ehart, C. L., and I. S. Bernstein. 1986. Matrilineal overthrows in rhesus monkey groups. *Int. J. Primatol.* 7:157–81.

Ehart-Seward, C., and C. A. Bramblett. 1980. The structure of social space among a captive group of vervet monkeys. *Folia primatol.* 34:214–38.

Eimas, P. D., P. Siqueland, P. Jusczyk, and J. Vigorito. 1971. Speech perception in infants. *Science* 171:303–6.

Elgar, M. A. 1986. House sparrows establish foraging flocks by giving chirrup calls if the resources are divisible. *Anim. Behav.* 34:169–74.

Emlen, S. T. 1971. The role of song in individual recognition in the indigo bunting. *Z. Tierpsychol.* 28:241–6.

———. 1981. Altruism, kinship, and reciprocity in the white-fronted bee-eater. In *Natural selection and social behavior: Recent research and new theory,* ed. R. D. Alexander and D. Tinkle. New York: Chiron Press.

———. 1984. Cooperative breeding in birds and mammals. In *Behavioral ecology: An evolutionary approach,* ed. J. R. Krebs and N. B. Davies. Oxford: Blackwell Scientific.

Emlen, S. T., and P. H. Wrege. 1988. The role of kinship in helping decisions among white-fronted bee-eaters. *Behav. Ecol. Sociobiol.* 23:305–15.

Essock-Vitale, S. M. 1978. Comparison of ape and monkey modes of problem solution. *J. Comp. Physiol. Psychol.* 92:942–57.

Essock-Vitale, S., and R. M. Seyfarth. 1987. Intelligence and social cognition. In *Primate societies,* ed. B. B. Smuts, D. L. Cheney, R. M. Seyfarth, R. W. Wrangham, and T. T. Struhsaker. Chicago: University of Chicago Press.

Estep, D. Q., M. E. Johnson, and T. P. Gordon. 1981. The effectiveness of sampling methods in detecting copulatory behaviour in *Macaca arctoides. Am. J. Primatol.* 1:453–5.

Evans, E. P. 1906/1987. *The criminal prosecution and capital punishment of animals.* London: Faber and Faber.

Fady, J. C. 1969. Les jeux sociaux: le compagnon de jeux chez les jeunes. Observations chez *Macaca iris. Folia primatol.* 11:134–43.

Fagen, R. 1981. *Animal play behavior.* Oxford: Oxford University Press.

Fairbanks, L. A. 1980. Relationships among adult females in captive vervet monkeys: Testing a model of rank-related attractiveness. *Anim. Behav.* 28:853–9.

———. 1988. Vervet monkey grandmothers: Effects on mother-infant relationships. *Behaviour* 104:176–88.

Fairbanks, L. A., and M. T. McGuire. 1984. Determinants of fecundity and reproductive success in captive vervet monkeys. *Am. J. Primatol.* 7:27–38.

———. 1985. Relationships of vervet monkey mothers with sons and daughters from one through three years of age. *Anim. Behav.* 33:40–50.

———. 1988. Long-term effects of early mothering behavior on responsiveness to the environment in vervet monkeys. *Develop. Psychobiol.* 21:711–24.

Fancher, R. E. 1979. *Pioneers of psychology.* New York: W. W. Norton.

Fedigan. L. 1982. *Primate paradigms: Sex roles and social bonds.* Montreal: Eden Press.

———. 1983. Dominance and reproductive success in primates. *Yrbk. Phys. Anthrop.* 26:91–129.

Fein, D. A. 1972. Judgments of causality to physical and social picture sequences. *Develop. Psych.* 8:147.

Fein, D. A., B. Pennington, P. Markowitz, M. Braverman, and L. Waterhouse. 1986. Toward a neuropsychological model of infantile autism: Are the social deficits primary? *J. Am. Acad. Child Psychiatr.* 25:198–212.

Feldman, H., S. Goldin-Meadow, and L. R. Gleitman. 1978. Beyond Herodotus: The creation of a language by linguistically deprived deaf children. In *Action, gesture, and symbol: The emergence of language*, ed. A. Lock. New York: Academic Press.

Fiske, A. P. 1990. *The four elementary forms of sociality: Communal sharing, authority ranking, equality matching, and market pricing*. New York: Free Press.

Flavell, J. H. 1985. *Cognitive development*, 2d ed. Englewood Cliffs, N.J.: Prentice Hall.

———. 1988. The development of children's knowledge about the mind: From cognitive connections to mental representations. In *Developing theories of mind*, ed. J. W. Astington, P. L. Harris, and D. R. Olson. Cambridge: Cambridge University Press.

Flavell, J. H., S. G. Shipstead, and K. Croft. 1978. Young children's knowledge about visual perception: Hiding objects from others. *Child Develop.* 49:1208–11.

Fleagle, J. G. 1988. *Primate adaptation and evolution*. New York: Academic Press.

Fletemeyer, J. R. 1978. Communication about potentially harmful foods in free-ranging chacma baboons, *Papio ursinus*. *Primates* 19:223–6.

Fodor, J. A. 1975. *The language of thought*. Cambridge, Mass.: Harvard University Press.

———. 1983. *The modularity of mind*. Cambridge, Mass.: MIT/Bradford Books.

———. 1985. Precis of *The modularity of mind*. *Behav. Brain Sci.* 8:1–5.

Fodor, J. A. and Z. W. Pylyshyn. 1981. How direct is visual perception?: Some reflections on Gibson's "ecological approach". *Cognition* 9:139–96.

Fossey, D. 1983. *Gorillas in the mist*. Boston: Houghton Mifflin.

Fouts, R. S., D. H. Fouts, and D. J. Schoenfeld. 1984. Sign language conversational interactions between chimpanzees. *Sign Lang. Stud.* 34:1–12.

Fouts, R. S., A. D. Hirsch, and D. H. Fouts. 1982. Cultural transmission of a human language in a chimpanzee mother-infant relationship. In *Psychobiological perspectives*, ed. H. E. Fitzgerald, J. A. Mullins, and P. Gage. New York: Plenum Press.

Frame, L H., J. R. Malcolm, G. W. Frame, and H. van Lawick. 1979. Social organization of African wild dogs (*Lycaon pictus*) on the Serengeti plains, Tanzania, 1967–1978. *Z. Tierpsychol.* 50:225–49.

Frank, L. 1986. Social organization of the spotted hyaena (*Crocuta crocuta*). II. Dominance and reproduction. *Anim. Behav.* 34:1510–27.

Frank, R. H. 1988. *Passions within reason: The strategic role of the emotions*. New York: W. W. Norton.

Frederickson, W. T., and G. P. Sackett. 1984. Kin preferences in primates (*Macaca nemestrina*): Relatedness or familiarity? *J. Comp. Psychol.* 98:29–34.

Freeland, W. J. 1976. Pathogens and the evolution of primate sociality. *Biotropica* 8:12–24.

French, J. A. 1981. Individual differences in play in *Macaca fuscata*: The role of maternal status and proximity. *Int. J. Primatol.* 2:237–46.

Fretwell, S. D., and H. L. Lucas. 1970. On territorial behavior and other factors influencing habitat distribution in birds. *Acta Biotheoretica* 19:16–36.

von Frisch, K. 1967. *The dance language and orientation of bees*. Cambridge, Mass.: Harvard University Press.

Fujita, K. 1987. Species recognition by five macaque monkeys. *Primates* 28:353–66.

Gabow, S. L. 1972. Dominance order reversal between two groups of free-ranging rhesus monkeys. *Primates* 14:215–23.

Galef, B. G. 1988. Imitation in animals: History, definition, and interpretation of data from the psychological laboratory. In *Social learning: Biological and psychological perspectives*, ed. T. R. Zentall and B. G. Galef. Hillsdale, N.J.: Lawrence Erlbaum Associates.

Gallistel, C. R. 1989a. Animal cognition: The representation of space, time, and number. *Ann. Rev. Psychol.* 40:155–89.

———. 1989b. *The organization of learning*. Cambridge, Mass.: Bradford Books/ MIT Press.

Gallistel, C. R., and K. Cheng. 1985. A modular sense of place? *Behav. Brain Sci.* 8:11–12.

Gallup, G. G. 1982. Self-awareness and the emergence of mind in primates. *Am. J. Primatol.* 2:237–48.

Garcia, J., and R. A. Koelling. 1966. The relation of cue to consequence in avoidance learning. *Psychonom. Sci.* 4:123–4.

Gardner, H. 1985. The centrality of modules. *Behav. Brain Sci.* 8:22–23.

———. 1987. *The mind's new science: A history of the cognitive revolution*, 2d ed. New York: Basic Books.

Gardner, R. A., and B. T. Gardner. 1969. Teaching sign language to a chimpanzee. *Science* 165:664–72.

———. 1975. Early signs of language in child and chimpanzee. *Science* 187:752–3.

Gaulin, S. J., and R. W. Fitzgerald. 1989. Sexual selection for spatial-learning ability. *Anim. Behav.* 37:322–31.

Gelman, R. 1987. Cognitive development: Principles guide learning and contribute to conceptual coherence. Invited address to the American Psychological Association, August 1987.

Gelman, R., and E. Spelke. 1981. The development of thoughts about animate and inanimate objects. In *Social cognitive development*, ed. J. H. Flavell and L. Ross. Cambridge: Cambridge University Press.

Gill, F. B., and L. L. Wolf. 1977. Nonrandom foraging by sunbirds in a patchy environment. *Ecology* 58:1284–96.

Gillan, D. J. 1981. Reasoning in the chimpanzee. II. Transitive inference. *J. Exp. Psychol. Anim. Behav. Proc.* 7:150–64.

Gillan, D. J., D. Premack, and G. Woodruff. 1981. Reasoning in the chimpanzee. I. Analogical reasoning. *J. Exp. Psychol.: Anim. Behav. Proc.* 7:1–17.

Gleitman, H. 1986. *Psychology*, 2d ed. New York: W. W. Norton.

Gleitman, L. R., and E. Wanner. 1982. Language acquisition: The state of the art. In *Language acquisition: The state of the art,* ed. L. R. Gleitman and E. Wanner. Cambridge: Cambridge University Press.

Godin, J.-G. J., and M. H. A. Keenleyside. 1984. Foraging on patchily distributed prey by a cichlid (*Teleosti, Cichlidae*): A test of the ideal free distribution theory. *Anim. Behav.* 32:120–31.

Goodall, J. van Lawick 1968. The behaviour of free-living chimpanzees in the Gombe Stream Reserve. *Anim. Behav. Monogr.* 1:165–311.

———. 1970. Tool-using in primates and other vertebrates. In *Advances in the study of behavior, vol. 3,* ed. D. S. Lehrman, R. A. Hinde, and E. Shaw. New York: Academic Press.

———. 1971. *In the shadow of man.* London: Collins.

Goodall, J. 1973. Cultural elements in the chimpanzee community. In *Precultural primate behaviour,* ed. E. W. Menzel. Basel: S. Karger.

———. 1979. Life and death at Gombe. *Nat. Geogr.* 155:592–621.

———. 1983. Population dynamics during a 15 year period in one community of free-living chimpanzees in the Gombe National Park, Tanzania. *Z. Tierpsychol.* 61:1–60.

———. 1986. *The chimpanzees of Gombe: Patterns of behavior.* Cambridge, Mass.: Harvard University Press.

Goodall, J., A. Bandora, E. Bergmann, C. Busse, H. Matama, E. Mpongo, A. Pierce, and D. Riss. 1979. Intercommunity interactions in the chimpanzee population of the Gombe National Park. In *The great apes,* ed. D. Hamburg and E. R. McCown. Menlo Park, Calif.: Benjamin/Cummings.

Goodall, J. van Lawick, H. van Lawick, and C. Packer. 1973. Tool-use in free-living baboons in the Gombe National Park, Tanzania. *Nature* 241:212–13.

Gould, J. L. 1982. *Ethology: The mechanisms and evolution of behavior.* New York: W. W. Norton.

Gould, J. L., and C. G. Gould. 1982. The insect mind: Physics or metaphysics. In *Animal mind–human mind,* ed. D. R. Griffin. Berlin: Springer-Verlag.

———. 1988. *The honey bee.* New York: W. H. Freeman.

Gouzoules, H. 1975. Maternal rank and early social interactions of infant stumptail macaques, *Macaca arctoides. Primates* 16:405–18.

———. 1980. A description of genealogical rank changes in a troop of Japanese monkeys (*Macaca fuscata*). *Primates* 21:262–7.

Gouzoules, H., and S. Gouzoules. 1989. Design features and developmental modification of pigtail macaque, *Macaca nemestrina,* agonistic screams. *Anim. Behav.* 37:383–401.

Gouzoules, H., S. Gouzoules, and P. Marler. 1986. Vocal communication: A vehicle for the study of social relationships. In *The Cayo Santiago macaques: History, behavior, and biology,* ed. R. G. Rawlins and M. J. Kessler. Albany: State University of New York Press.

Gouzoules, S. 1984. Primate mating systems, kin associations, and cooperative behavior: Evidence for kin recognition? *Yrbk. Phys. Anthropol.* 27:99–134.
Gouzoules, S., and H. Gouzoules. 1987. Kinship. In *Primate societies,* ed. B. B. Smuts, D. L. Cheney, R. M. Seyfarth. R. W. Wrangham, and T. T. Struhsaker. Chicago: University of Chicago Press.
Gouzoules, S., H. Gouzoules, and P. Marler. 1984. Rhesus monkey (*Macaca mulatta*) screams: Representational signalling in the recruitment of agonistic aid. *Anim. Behav.* 32:182–93.
Green, S. 1975. Communication by a graded vocal system in Japanese monkeys. In *Primate behavior, vol. 4,* ed. L. A. Rosenblum. New York: Academic Press.
Grice, H. P. 1957. Meaning. *Phil. Rev.* 66:377–88.
———. 1969. Utterer's meanings and intentions. *Phil. Rev.* 78:147–77.
Griffin, D. R. 1976. *The question of animal awareness: Evolutionary continuity of mental experience.* New York: Rockefeller University Press.
———, ed. 1982. *Animal mind–human mind.* Berlin: Springer-Verlag.
———. 1984. *Animal thinking.* Cambridge, Mass.: Harvard University Press.
Guardo, C. J., and J. B. Bohan. 1971. Development of a sense of self-identity in children. *Child Develop.* 42:1909–21.
Gyger, M., S. J. Karakashian, and P. Marler. 1986. Avian alarm-calling: Is there an audience effect? *Anim. Behav.* 34:1570–72.
Gyger, M., P. Marler, and R. Pickert. 1987. Semantics of an avian alarm call system: The male domestic fowl, *Gallus domesticus. Behaviour* 102:15–40.
Hall, K. R. L., and I. DeVore. 1965. Baboon social behavior. In *Primate behavior,* ed. I. DeVore. New York: Holt, Rinehart and Winston.
Hall, K. R. L., and J. S. Gartlan. 1965. Ecology and behaviour of the vervet monkey, *Cercopithecus aethiops,* Lolui Island, Lake Victoria. *Proc. Zool. Soc. Lond.* 145: 37–56.
Halliday, T. R. 1976. The libidinous newt, *Triturus vulgaris. Anim. Behav.* 24: 398–414.
———. 1983. The study of mate choice. In *Mate choice,* ed. P. P. G. Bateson. Cambridge: Cambridge University Press.
Hamilton, C. R. 1977. An assessment of hemispheric specialization in monkeys. *Ann. N.Y. Acad. Sci.* 299:222–32.
Hamilton, W. D. 1964. The genetical evolution of social behavior. *J. Theor. Biol.* 7:1–51.
Hamilton, W. J., R. E. Buskirk, and W. H. Buskirk. 1975. Defensive stoning by baboons. *Nature* 256:488–9.
Hamilton, W. J., C. Busse, and K. S. Smith. 1982. Adoption of infant orphan chacma baboons. *Anim. Behav.* 30:29–34.
Hansen, E. W. 1976. Selective responding by recently separated juvenile rhesus monkeys to the calls of their mothers. *Dev. Psychobiol.* 9:83–88.
Harcourt, A. H. 1988. Alliances in contests and social intelligence. In *Machiavellian*

intelligence: Social expertise and the evolution of intellect in monkeys, apes, and humans, ed. R. W. Byrne and A. Whiten. Oxford: Oxford University Press.

Harlow, H. F. 1949. The formation of learning sets. *Psych. Rev.* 56:51–65.

Harlow, H. F., and M. K. Harlow. 1965. The affectional systems. In *The behavior of nonhuman primates, vol. 2,* ed. A. M. Schrier, H. F. Harlow, and F. Stollnitz. New York: Academic Press.

Harper, D. G. C. 1982. Competitive foraging in mallards: 'Ideal free' ducks. *Anim. Behav.* 30:575–84.

Hauser, M. D. 1986. Male responsiveness to infant distress calls in free-ranging vervet monkeys. *Behav. Ecol. Sociobiol.* 19:65–71.

———. 1988a. How infant vervet monkeys learn to recognize starling alarm calls: The role of experience. *Behaviour* 105:187–201.

———. 1988b. Invention and social transmission: New data from wild vervet monkeys. In *Machiavellian intelligence: Social expertise and the evolution of intellect in monkeys, apes, and humans,* ed. R. W. Byrne and A. Whiten. Oxford: Oxford University Press.

Hauser, M. D., D. L. Cheney, and R. M. Seyfarth. 1986. Group extinction and fusion in free-ranging vervet monkeys. *Am. J. Primatol.* 11:63–77.

Hauser, M. D., and L. A. Fairbanks. 1988. Mother-offspring conflict in vervet monkeys: Variation in response to ecological conditions. *Anim. Behav.* 36:802–13.

Hauser, M. D., and R. W. Wrangham. 1987. Manipulation of food calls in captive chimpanzees. *Folia primatol.* 48:207–10.

Hausfater, G. 1972. Intergroup behavior of free-ranging rhesus monkeys (*Macaca mulatta*). *Folia primatol.* 18:78–107.

———. 1975. Dominance and reproduction in baboons (*Papio cynocephalus*). *Contributions to primatology, vol.* 7. Basel: S. Karger.

Hayes, K. J., and C. Hayes. 1951. The intellectual development of a home-raised chimpanzee. *Proc. Am. Phil. Soc.* 95:105–9.

Hayes, K. J., and C. H. Nissen. 1971. Higher mental functions of a home-raised chimpanzee. In *Behavior of non-human primates,* ed. A. M. Schrier and F. Stollnitz. New York: Academic Press.

Hefner, H. E., and R. S. Hefner. 1984. Temporal lobe lesions and perception of species-specific vocalizations by macaques. *Science* 226:75–76.

Hegner, R. E. 1982. Central place foraging in the white-fronted bee-eater. *Anim. Behav.* 30:953–63.

Hegner, R. E., and S. T. Emlen. 1987. Territorial organization of the white-fronted bee-eater in Kenya. *Ethology* 76:189–222.

Hegner, R. E., S. T. Emlen, and N. J. Demong. 1982. Spatial organization of the white-fronted bee-eater. *Nature* 298:264–6.

Henzi, S. P., and J. W. Lucas. 1980. Observations on the inter-troop movement of adult vervet monkeys. (*Cercopithecus aethiops*). *Folia primatol.* 33:220–35.

Herrnstein, R. J. 1970. On the law of effect. *J. Exp. Anal. Beh.* 13:243–66.

———. 1979. Acquisition, generalization, and discrimination reversal of a natural concept. *J. Exp. Psychol.: Anim. Behav. Proc.* 5:116–29.

———. 1985. Riddles of natural categorisation. In *Animal intelligence,* ed. L. Weiskrantz. Oxford: Clarendon Press.

———. 1990. Levels of stimulus control: A functional approach. *Cognition.*

Herrnstein, R. J., and D. Loveland. 1964. Complex visual concepts in the pigeon. *Science* 146:549–51.

Herrnstein, R. J., and W. Vaughan, Jr. 1980. Melioration and behavioral allocation. In *Limits to action,* ed. J. E. R. Staddon. New York: Academic Press.

Herzog, M., and S. Hopf. 1984. Behavioral responses to species-specific warning calls in infant squirrel monkeys reared in social isolation. *Am. J. Primatol.* 7:99–106.

Hill, A. 1976. Non-aggressive tactile interactions of *Hippopotamus amphibius* Linn. with *Syncerus caffer* (Sparrman). *Mammalia* 40:161–72.

Hill, W. C. O. 1966. *Primates: Comparative anatomy and taxonomy, vol. VI.* Edinburgh: The University Press.

Hinde, R. A. 1973. On the design of check-sheets. *Primates* 14:393–406.

———. 1974. *Biological bases of human social behaviour.* New York: McGraw-Hill.

———. 1976a. Interactions, relationships and social structure. *Man* 11:1–17.

———. 1976b. On describing relationships. *J. Child Psychol. Psychiatr.* 17:1–19.

———. 1981. Animal signals: Ethological and games-theory approaches are not incompatible. *Anim. Behav.* 29:535–42.

———. 1982. *Ethology: Its nature and relations with other disciplines.* Oxford: Oxford University Press.

———. 1983a. A conceptual framework. In *Primate social relationships: An integrated approach,* ed. R. A. Hinde. Oxford: Blackwell Scientific.

———. 1983b. General issues in describing social behavior. In *Primate social relationships: An integrated approach,* ed. R. A. Hinde. Oxford: Blackwell Scientific.

———. 1987. *Individuals, relationships, and cultures.* Cambridge: Cambridge University Press.

Hinde, R. A., and J. Stevenson-Hinde, ed. 1973. *Constraints on learning: Limitations and predispositions.* New York: Academic Press.

Hockett, C. F. 1960. Logical considerations in the study of animal communication. In *Animal sounds and communication,* ed. W. E. Lanyon and W. N. Tavolga. Washington: American Institute of Biological Sciences.

Hogrefe, G. J., H. Wimmer, and J. Perner. 1986. Ignorance versus false belief: A developmental lag in attribution of epistemic states. *Child Develop.* 57:567–82.

Holmes, W. G., and P. W. Sherman. 1983. Kin recognition in animals. *Am. Sci.* 71:46–55.

Hood, L., and L. Bloom. 1979. What, when, and how about why. *Monogr. Res. Child Dev.* 44:1–41.

Hoogland, J. L. 1983. Nepotism and alarm-calling in the black-tailed prairie dog (*Cynomys ludovicianus*). *Anim. Behav.* 31:472–9.

Hopp, S. L. 1985. Differential sensitivity of Japanese macaques (*Macaca fuscata*) to variations along a synthetic vocal continuum. Ph. D. diss., Indiana University.

Horrocks, J., and W. Hunte. 1983. Maternal rank and offspring rank in vervet monkeys: An appraisal of the mechanisms of rank acquisition. *Anim. Behav.* 31:772–82.

Howard, R. D. 1974. The influence of sexual selection and interspecific competition on mockingbird song (*Mimus polyglottos*). *Evolution* 28:428–38.

Humphrey. N. K. 1974. Species and individuals in the perceptual world of monkeys. *Perception* 3:105–14.

———. 1976. The social function of intellect. In *Growing points in ethology*, ed. P. P. G. Bateson and R. A. Hinde. Cambridge: Cambridge University Press.

———. 1983. *Consciousness regained*. Oxford: Oxford University Press.

———. 1986. *The inner eye*. London: Faber and Faber.

Hutchins, M., and D. P. Barash. 1976. Grooming in primates: Implications for its utilitarian function. *Primates* 17:145–50.

Imakawa, S. 1988. Development of co-feeding relationships in immature free-ranging Japanese monkeys. *Primates* 29:493–504.

Isbell, L. A. 1990. Influences of predation and resource competition on the social system of vervet monkeys (*Cercopithecus aethiops*). Ph. D. diss., University of California, Davis.

Itakura, S. 1987a. Mirror guided behavior in Japanese monkeys (*Macaca fuscata fuscata*). *Primates* 28:149–62.

———. 1987b. Use of a mirror to direct their responses in Japanese monkeys (*Macaca fuscata fuscata*). *Primates* 28:343–52.

Jackendoff, R. S. 1987. *Consciousness and the computational mind*. Cambridge, Mass.: MIT/Bradford Books.

Jacobson, E., and D. Premack. 1970. Choice and habituation as measures of response similarity. *J. Exp. Psychol.* 85:30–35.

Jerison, H. 1985. Animal intelligence as encephalisation. In *Animal intelligence*, ed. L. Weiskrantz. Oxford: Clarendon Press.

Johnson-Laird, P. N. 1987. The mental representation of the meaning of words. *Cognition* 25:189–211.

———. 1988. *The computer and the mind: An introduction to cognitive science*. Cambridge, Mass.: Harvard University Press.

Johnston, T. D. 1981. Contrasting approaches to a theory of learning. *Behav. Brain Sci.* 4:125–73.

Jolly, A. 1966. Lemur social behavior and primate intelligence. *Science* 153:501–6.

———. 1985. *The evolution of primate behavior*, 2d ed. New York: Macmillan.

———. 1988. The evolution of purpose. In *Machiavellian intelligence: Social expertise and the evolution of intellect in monkeys, apes, and humans*, ed. R. W. Byrne and A. Whiten. Oxford: Oxford University Press.

———. 1990. Conscious chimpanzees? A review of recent literature. In *Cognitive ethology: The minds of other animals (essays in honor of D. R. Griffin)*, ed. C. A. Ristau. Hillsdale, N.J.: Lawrence Erlbaum Associates.

Judge, P. 1982. Redirection of aggression based on kinship in a captive group of pigtail macaques. *Int. J. Primatol.* 3:301.

———. 1983. Reconciliation based on kinship in a captive group of pigtail macaques. *Am. J. Primatol.* 4:346.

Jurgens, U., M. Maurus, D. Ploog, and P. Winter. 1967. Vocalization in the squirrel monkey (*Saimiri sciureus*) elicited by brain stimulation. *Exp. Brain Res.* 4: 114–17.

Jurgens, U., and D. Ploog. 1970. Cerebral representation of vocalization in the squirrel monkey. *Exp. Brain Res.* 10:532–54.

Kahneman, D., and A. Tversky. 1982. The psychology of preferences. *Sci. Am.* 160–74.

Kamil, A. C. 1987. A synthetic approach to the study of animal intelligence. *Nebr. Symp. on Motivation* 7:257–308.

Kamil, A. C., and R. Balda. 1985. Cache recovery and spatial memory in Clark's nutcrackers (*Nucifraga columbiana*). *J. Exp. Psych. Anim. Behav. Proc.* 11: 95–111.

Kamil, A. C., J. R. Krebs, and H. R. Pulliam. 1987. *Foraging behavior.* New York: Plenum Press.

Kamil, A. C. and H. L. Roitblat. 1985. The ecology of foraging behavior: Implications for animal learning and memory. *Ann. Rev. Psych.* 36:141–69.

Kamil, A. C., S. Yoerg, and K. Clements. 1988. Rules to leave by: Patch departure in foraging blue jays. *Anim. Behav.* 36:843–53.

Kaplan, J. N., A. Winship-Ball, and L. Sim. 1978. Maternal discrimination of infant vocalizations in the squirrel monkey. *Primates* 19:187–93.

Karakashian, S. J., M. Gyger, and P. Marler. 1988. Audience effects on alarm-calling in chickens. *J. Comp. Psychol.* 102:129–35.

Kaufmann, J. H. 1965. A three-year study of mating behavior in a free-ranging band of rhesus monkeys. *Ecology* 46:500–512.

Kavanaugh, M. 1980. Invasion of the forest by an African savannah monkey: Behavioural adaptations. *Behaviour* 73:238–60.

Kawai, M. 1958. On the system of social ranks in a natural group of Japanese monkeys. *Primates* 1:11–48.

Kawamura, S. 1959. The process of sub-culture propagation among Japanese macaques. *Primates* 2:43–60.

Kawanaka, K. 1973. Intertroop relations among Japanese monkeys. *Primates* 14: 113–59.

Keddy, A. C. 1986. Female mate choice in vervet monkeys (*Cercopithecus aethiops sabaeus*). *Am. J. Primatol.* 10:125–34.

Keddy Hector, A. 1989. The effects of sexual selection on the mating behavior of vervet monkeys (*Cercopithecus aethiops*) and fruit flies. Ph. D. diss., University of California, Los Angeles.

Keddy Hector, A., R. M. Seyfarth, and M. J. Raleigh. 1989. Male parental care, female choice, and the effect of an audience in vervet monkeys. *Anim. Behav.* 37:262–71.

Killeen, P. R. 1985. The modularity of behavior. *Behav. Brain Sci.* 8:22–23.

Kohler, W. 1925/1959. *The mentality of apes,* 2d ed. New York: Viking.

Koyama, N. 1967. On dominance rank and kinship of a wild Japanese monkey troop in Arashiyama. *Primates* 8:189–216.

———. 1970. Changes in dominance rank and division of a wild Japanese monkey troop in Arashiyama. *Primates* 11:335–90.

Krebs, J. R., and R. Dawkins. 1984. Animal signals: Mind reading and manipulation. In *Behavioural ecology: An integrated approach,* ed. J. R. Krebs and N. B. Davies. Oxford: Blackwell Scientific.

Krebs, J. R., A. Kacelnik, and P. Taylor. 1978. Test of optimal sampling by great tits. *Nature* 275:27–31.

Krebs, J. R., and R. H. McCleery. 1984. Optimization in behavioural ecology. In *Behavioural ecology: An evolutionary approach,* ed. J. R. Krebs and N. B. Davies. Oxford: Blackwell Scientific.

Krebs, J. R., D. F. Sherry, S. D. Healy, V. H. Perry, and A. L. Vaccarino. 1989. Hippocampal specialization of food-storing birds. *Proc. Natl. Acad. Sci.* 86:1388–92.

Kroodsma, D. E. 1976. The effect of large song repertoires on neighbor 'recognition' in male song sparrows. *Condor* 78:97–99.

Kroodsma, D. E., and R. A. Canady. 1985. Differences in repertoire size, singing behavior, and associated neuroanatomy among marsh wren populations have a genetic basis. *Auk* 102:439–46.

Kroodsma, D. E., and J. Verner. 1987. Use of song repertoires among marsh wren populations. *Auk* 104:63–72.

Kummer, H. 1968. *Social organization of hamadryas baboons.* Basel: S. Karger.

———. 1971. *Primate societies.* Chicago: Aldine.

———. 1975. Rules of dyad and group formation among captive gelada baboons (*Theropithecus gelada*). Symp. Vth Congr. Intl. Primatol. Soc. Tokyo: Japan Science Press.

———. 1982. Social knowledge in free-ranging primates. In *Animal mind-human mind,* ed. D. R. Griffin. Berlin: Springer-Verlag.

Kummer, H., W. Goetz, and W. Angst. 1970. Cross-species modification of social behavior in baboons. In *Old World monkeys: Evolution, systematics and behavior,* ed. J. R. Napier and P. H. Napier. New York: Academic Press.

———. 1974. Triadic differentation: An inhibitory process protecting pair bonds in baboons. *Behaviour* 49:62–87.

Ladefoged, P. 1975. *A course in phonetics.* New York: Harcourt, Brace Jovanovich.

Lancaster, J. B. 1975. *Primate behavior and the emergence of human culture.* New York: Holt, Rinehart and Winston.

Lashley, K. 1956. Cerebral organization and behavior. In *The brain and human be-*

havior, ed. H. Solomon, S. Cobb, and W. Penfield. Baltimore: Williams and Wilkins.

Latimer, W. 1977. A comparative study of the songs and alarm calls of some *Parus* species. *Z. Tierpsychol.* 45:414–33.

Lea, S. E. G. 1984. In what sense do pigeons learn concepts? In *Animal cognition,* ed. H. L. Roitblat, T. G. Bever, and H. S. Terrace. Hillsdale, N.J.: Lawrence Erlbaum Associates.

Lee, P. C. 1983a. Caretaking of infants and mother-infant relationships. In *Primate social relationships: An integrated approach,* ed. R. A. Hinde. Oxford: Blackwell Scientific.

———. 1983b. Context-specific unpredictability in dominance interactions. In *Primate social relationships: An integrated approach,* ed. R. A. Hinde. Oxford: Blackwell Scientific.

———. 1983c. Effects of the loss of the mother on social development. In *Primate social relationships: An integrated approach,* ed. R. A. Hinde. Oxford: Blackwell Scientific.

———. 1983d. Play as a means for developing relationships. In *Primate social relationships: An integrated approach,* ed. R. A. Hinde. Oxford: Blackwell Scientific.

Leger, D. W., and D. H. Owings. 1978. Responses to alarm calls by California ground squirrels. *Behav. Ecol. Sociobiol.* 3:177–86.

Leger, D. W., D. H. Owings, and D. L. Gelfand. 1980. Single-note vocalizations of California ground squirrels: Graded signals and situation-specificity of predator and socially evoked calls. *Z. Tierpsychol.* 52:227–46.

Leslie, A. 1984. Infant perception of a manual pick-up event. *Br. J. Develop. Psych.* 2:19–32.

———. 1987. Pretense and representation in infancy: The origins of 'theory of mind'. *Psychol. Rev.* 94:412–26.

———. 1988. Some implications of pretense for mechanisms underlying the child's theory of mind. In *Developing theories of mind,* ed. J. W. Astington, P. L. Harris, and D. R. Olson. Cambridge: Cambridge University Press.

Liberman, A. M. 1982. On finding that speech is special. *Am. Psychol.* 37:148–67.

Liberman, A. M., F. S. Cooper, D. P. Shankweiler, and M. Studdert-Kennedy. 1967. Perception of the speech code. *Psychol Rev.* 74:431–61.

Lisker, L., and A. Abramson. 1964. A cross-language study of voicing in initial stops: Acoustical measurements. *Word* 20:384–422.

Locke, J. 1690/1964. *An essay concerning human understanding.* Cleveland: World Publishing Co.

Lord, C. 1984. The development of peer relations in children with autism. In *Advances in applied developmental psychology,* ed. F. J. Morrison, C. Lord, and D. P. Keating. New York: Academic Press.

Lorenz, K. 1975. *Evolution and modification of behavior.* Chicago: University of Chicago Press.

Lyons, J. 1972. Human language. In *Non-verbal communication,* ed. R. A. Hinde. Cambridge: Cambridge University Press.
McGinnis, P. R. 1979. Sexual behaviour in free-living chimpanzees: Consort relationships. In *The great apes: Perspectives on human evolution.*, ed. D. A. Hamburg and E. R. McCowan. Menlo Park, Calif.: Benjamin/Cummings.
McGonigle, B. O., and M. Chalmers. 1977. Are monkeys logical? *Nature* 267:694–6.
McGrew, W. C., and C. E. G. Tutin. 1973. Chimpanzee tool use in dental grooming. *Nature* 241:477–8.
McKenna, J. J. 1978. Biosocial function of grooming behavior among the common langur monkey (*Presbytis entellus*). *Am. J. Phys. Anthrop.* 48:503–10.
MacKinnon, J. R. 1974. The ecology and behaviour of wild orangutans (*Pongo pygmaeus*). *Anim. Behav.* 22:3–74.
MacNamara, J. 1982. *Names for things.* Cambridge, Mass.: MIT Press.
Macphail, E. M. 1985. Vertebrate intelligence: The null hypothesis. In *Animal intelligence,* ed. L. Weiskrantz. Oxford: Clarendon Press.
Markl, H. 1985. Manipulation, modulation, information, cognition: Some of the riddles of communication. In *Experimental behavioral ecology and sociobiology,* ed. B. Holldobler and M. Lindauer. Sunderland, Mass.: Sinauer Associates.
Marler, P. 1955. Studies of fighting in chaffinches. 1. Behaviour in relation to the social hierarchy. *Br. J. Anim. Behav.* 3:111–17.
———. 1956a. Studies of fighting in chaffinches. 3. Proximity as a cause of aggression. *Br. J. Anim. Behav.* 4:23–30.
———. 1956b. The voice of the chaffinch and its function as a language. *Ibis* 98:231–61.
———. 1961. The logical analysis of animal communication. *J. Theor. Biol.* 1:295–317.
———. 1965. Communication in monkeys and apes. In *Primate behavior,* ed. I. DeVore. New York: Holt, Rinehart and Winston.
———. 1976a. An ethological theory of the origin of vocal learning. *Ann. N.Y. Acad. Sci.* 280:708–17.
———. 1976b. On animal aggression. *Am. Psychol.* 31:239–46.
———. 1977a. Primate vocalizations: Affective or symbolic? In *Progress in ape research,* ed. G. H. Bourne. New York: Academic Press.
———. 1977b. The structure of animal communication sounds. In *Recognition of complex acoustic signals,* ed. T. H. Bullock. Berlin: Springer-Verlag.
———. 1978. Affective and symbolic meaning: Some zoosemiotic speculations. In *Sight, sound and sense,* ed. T. A. Sebeok. Bloomington, Ind.: Indiana University Press.
———. 1981. Birdsong: The acquisition of a learned motor skill. *Trends Neurosci.* 4:88–94.
Marler, P., S. Karakashian, and M. Gyger. 1990. Do animals have the option of withholding signals when communication is inappropriate? In *Cognitive ethology:*

The minds of other animals (essays in honor of Donald R. Griffin), ed. C. A. Ristau. Hillsdale, N.J.: Lawrence Erlbaum Associates.

Marler, P., and S. Peters. 1977. Selective vocal learning in a sparrow. *Science* 198:519–21.

———. 1981. Birdsong and speech: Evidence for special processing. In *Perspectives on the study of speech*, ed. P. D. Eimas and J. L. Miller. Hillsdale, N.J.: Lawrence Erlbaum Associates.

Marshall, J. C. 1970. The biology of communication in man and animals. In *New horizons in linguistics*, ed. J. Lyons. Harmondsworth, England: Penguin.

Martin, R. D. 1983. Human brain evolution in an ecological context. *Fifty-Second James Arthur Lecture on the Evolution of the Human Brain*. American Museum of Natural History.

Mason, W. A. 1976. Review of *The question of animal awareness*, by D. R. Griffin. *Science* 194:930–31.

———. 1978a. Environmental models and mental modes: Representational processes in the great apes. In *The great apes: Perspectives on human evolution*, ed. D. A. Hamburg and E. R. McCown. Menlo Park, Calif.: Benjamin/Cummings.

———. 1978b. Ontogeny of social systems. In *Recent advances in primatology, vol. 1*, ed. D. J. Chivers and J. Herbert. New York: Academic Press.

———. 1986. Behavior implies cognition. In *Integrating scientific disciplines*, ed. W. Bechtel. Dordrecht, Netherlands: Martinus Nijhoff.

Mason, W. A., and J. H. Hollis. 1962. Communication between young rhesus monkeys. *Anim. Behav.* 10:211–21.

Massey, C. R., and R. Gelman. 1988. Preschoolers' ability to decide whether photographed unfamiliar objects can move themselves. *Develop. Psych.* 24:307–17.

Matsuzawa, T. 1985. Use of numbers by a chimpanzee. *Nature* 315:57–59.

———. 1988. Spontaneous pattern construction in a chimpanzee. In *Understanding chimpanzees*, ed. J. Goodall. Chicago: Chicago Academy of Sciences.

May, B., D. B. Moody, and W. C. Stebbins. 1989. Categorical perception of conspecific communicative sounds by Japanese monkeys, *Macaca fuscata*. *J. Acoust. Soc. Am.* 85:837–47.

Maynard Smith, J. 1974. The theory of games and the evolution of animal conflict. *J. Theor. Biol.* 47:209–21.

———. 1979. Game theory and the evolution of behaviour. *Proc. Roy. Phil. Soc. Lond., B.* 205:474–88.

———. 1982. *Evolution and the theory of games*. Cambridge: Cambridge University Press.

———. 1984. Game theory and the evolution of behavior. *Behav. Brain Sci.* 7: 95–125.

———. 1986. Ownership and honesty in competitive interactions. *Behav. Brain Sci.* 9: 742–44.

Maynard Smith, J., and G. R. Price. 1973. The logic of animal conflicts. *Nature* 246:15–18.

Medin, D. L., and E. E. Smith. 1984. Concepts and concept formation. *Ann. Rev. Psychol.* 35 : 113–38.

Mehlman, P. T., and B. Chapais. 1988. Differential effects of kinship, dominance, and the mating season on female allogrooming in a captive group of *Macaca fuscata. Primates* 29 : 195–217.

Melchior, H. R. 1971. Characteristics of Arctic ground squirrel alarm calls. *Oecologia* 7 : 184–90.

Menzel, E. W. 1966. Responsiveness to objects in free-ranging Japanese monkeys. *Behaviour* 26 : 130–49.

———. 1971. Communication about the environment in a group of young chimpanzees. *Folia primatol.* 15 : 220–32.

———. 1972. Spontaneous invention of ladders in a group of young chimpanzees. *Folia primatol.* 17 : 87–106.

———. 1973. Further observations on the use of ladders in a group of young chimpanzees. *Folia primatol.* 19 : 450–57.

Menzel, E. W., E. S. Savage-Rumbaugh, and J. Lawson. 1985. Chimpanzee (*Pan troglodytes*) spatial problem solving with the use of mirrors and televised equivalents of mirrors. *J. Comp. Psychol.* 99 : 211–17.

Michotte, A. 1963. *The perception of causality.* New York: Basic Books.

Miles, L. 1983. Apes and language: The search for communicative competence. In *Language in primates: Perspectives and implications,* ed. J. de Luce and T. H. Wilder. New York: Springer Verlag.

Miller, A. 1949. The nature of tragedy. *N.Y. Herald Tribune,* March 27, 1949.

Miller, R. E. 1967. Experimental approaches to the physiological and behavioral concomitants of affective communication in rhesus monkeys. In *Social communication among primates,* ed. S. A. Altmann. Chicago: University of Chicago Press.

———. 1971. Experimental studies of communication in the monkey. In *Primate behavior, vol. 2,* ed. L. A. Rosenblum. New York: Academic Press.

Milton, K. 1981. Distribution patterns of tropical plant food as an evolutionary stimulus to primate mental development. *Am. Anthropol.* 83 : 534–48.

———. 1988. Foraging behaviour and the evolution of primate intelligence. In *Machiavellian intelligence: Social expertise and the evolution of intellect in monkeys, apes, and humans,* ed. R. W. Byrne and A. Whiten. Oxford: Oxford University Press.

Mineka, S., and M. Cook. 1988. Social learning and the acquisition of snake fear in monkeys. In *Social learning: Biological and psychological perspectives,* ed. T. R. Zentall and B. G. Galef. Hillsdale, N.J.: Lawrence Erlbaum Associates.

Mitani, J. 1987. Species discrimination of male song in gibbons. *Am. J. Primatol.* 13 : 413–23.

Mitani, J., and P. Marler. 1989. A phonological analysis of male gibbon singing behavior. *Behaviour* 109 : 20–45.

Moller, A. P. 1988. False alarm calls as a means of resource usurpation in the great tit *Parus major*. *Ethology* 79:25–30.

Morton, E. S. 1977. On the occurrence and significance of motivation-structural rules in some bird and mammal sounds. *Am. Nat.* 111:855–69.

Moss, C. J. 1988. *Elephant memories*. Boston: Houghton Mifflin.

Moss, C. J., and J. H. Poole. 1983. Relationships and social structure of African elephants. In *Primate social relationships: An integrated approach*, ed. R. A. Hinde. Oxford: Blackwell Scientific.

Moynihan, M. 1970. Some behavioral patterns of platyrrhine monkeys. II. *Saguinus geoffroyi* and some other tamarins. *Smithson. Contrib. Zool.* 28:1–27.

Munn, C. A. 1986a. Birds that cry 'wolf'. *Nature* 319:143–5.

———. 1986b. The deceptive use of alarm calls by sentinel species in mixed species flocks of neotropical birds. In *Deception: Perspectives on human and nonhuman deceit*, ed. R. W. Mitchell and N. S. Thompson. Albany: State University of New York Press.

Myers, R. E. 1976. Comparative neurology of vocalization and speech: Proof of a dichotomy. *Ann. N.Y. Acad. Sci.* 280:745–57.

Myers, S. A., A. Horel, and H. S. Pennypacker. 1965. Operant control of vocal behavior in the monkey *Cebus albifrons*. *Psychonomic. Sci.* 3:389–90.

Nagel, T. 1974. What is it like to be a bat? *Philos. Rev.* 83:435–50.

Nelson, K. 1973. Some evidence for the cognitive primacy of categorization and its functional basis. *Merril-Palmer Q.* 19:21–40.

Nice, M. N. 1943. Studies in the life history of the song sparrow. II. The behavior of the song sparrow and other passerines. *Trans. Linn. Soc. N.Y.* 6:1–284.

Nicolson, N. 1987. Infants, mothers, and other females. In *Primate societies*, ed. B. B. Smuts, D. L. Cheney, R. M. Seyfarth, R. W. Wrangham, and T. T. Struhsaker. Chicago: University of Chicago Press.

Nishida, T. 1983. Alpha status and agonistic alliance in wild chimpanzees (*Pan troglodytes schweinfurthii*). *Primates* 24:318–36.

———. 1987. Local traditions and cultural transmission. In *Primate societies*, ed. B. B. Smuts, D. L. Cheney, R. M. Seyfarth, R. W. Wrangham, and T. T. Struhsaker. Chicago: University of Chicago Press.

Noe, R. 1986. Lasting alliances among adult male savannah baboons. In *Primate ontogeny, cognition, and social behavior*, ed. J. Else and P. C. Lee. Cambridge: Cambridge University Press.

van Noordwijk, M., and C. van Schaik. 1987. Competition among female longtailed macaques, *Macaca fascicularis*. *Anim. Behav.* 35:577–89.

Nottebohm, F. 1979. Origins and mechanisms in the establishment of cerebral dominance. In *Handbook of behavioral neurobiology, vol. 2*, ed. M. S. Gazzaniga. New York: Plenum Press.

Oden, D. L., R. K. R. Thompson, and D. Premack. 1988. Spontaneous transfer of matching by infant chimpanzees (*Pan troglodytes*). *J. Exp. Psych. Anim. Behav. Proc.* 14:140–45.

Oki, J., and Y. Maeda. 1973. Grooming as a regulator of behavior in Japanese macaques. In *Behavioral regulators of behavior in primates,* ed. C. R. Carpenter. Lewisburg PA: Bucknell University Press.

Olivier, R. C. D., and A. Laurie. 1974. Habitat utilization by hippopotamus in the Mara River. *E. Afr. Wildl. J.* 12:249–71.

Olton, D. S. 1985. The temporal context of spatial memory. In *Animal intelligence,* ed. L. Weiskrantz. Oxford: Clarendon Press.

Olton, D. S., and R. J. Samuelson. 1976. Remembrance of places passed: Spatial memory in rats. *J. Exp. Psychol. Anim. Behav. Proc.* 2:97–116.

Owings, D., and D. Hennessy. 1984. The importance of variation in sciurid visual and vocal communication. In *Biology of ground dwelling squirrels: Annual cycles, behavioral ecology and sociality,* ed. J. O. Murie and G. R. Michener. Lincoln: University of Nebraska Press.

Owings, D. H., and D. W. Leger. 1980. Chatter vocalizations of California ground squirrels: Predator- and social-role specificity. *Z. Tierpsychol.* 54:163–84.

Owings, D. H., and R. A. Virginia. 1978. Alarm calls of California ground squirrels. *Z. Tierpsychol.* 46:58–70.

Owren, M. J. 1990a. Acoustic classification of alarm calls by vervet monkeys (*Cercopithecus aethiops*) and humans. I. Natural calls. *J. Comp. Psychol.* 104:20–28.

———. 1990b. Acoustic classification of alarm calls by vervet monkeys (*Cercopithecus aethiops*) and humans. II. Synthetic calls. *J. Comp. Psychol.* 104:29–40.

Owren, M. J., and R. H. Bernacki. 1988. The acoustic features of vervet monkey alarm calls. *J. Acoust. Soc. Am.* 83:1927–35.

Owren, M. J., S. L. Hopp, J. M. Sinnott, and M. R. Petersen. 1988. Absolute auditory thresholds in three Old World monkey species (*Cercopithecus aethiops, C. neglectus, Macaca fuscata*) and humans (*Homo sapiens*). *J. Comp. Psychol.* 102:99–107.

Owren, M. J., S. L. Hopp, and R. M. Seyfarth. 1990. Categorical vocal signaling in nonhuman primates. In *Nonverbal vocal communication: Comparative and developmental approaches,* ed. H. Papousek, U. Jurgens, and M. Papousek. New York: Cambridge University Press.

Packer, C. 1977. Reciprocal altruism in olive baboons. *Nature* 265:441–3.

———. 1979. Inter-troop transfer and inbreeding avoidance in *Papio anubis. Anim. Behav.* 27:1–36.

Packer, C., and L. Ruttan. 1988. The evolution of cooperative hunting. *Am. Nat.* 132:159–98.

Parker, G. A. 1974. Assessment strategy and the evolution of animal conflicts. *J. Theor. Biol.* 47:223–43.

Parker, S. T., and K. R. Gibson. 1977. Object manipulation, tool use, and sensorimotor intelligence as feeding adaptations in cebus monkeys and great apes. *J. Hum. Evol.* 6:623–41.

Passingham, R. E. 1982. *The human primate.* San Francisco: W. H. Freeman.

Paton, D. 1986. Communication by agonistic displays. II. Perceived information and the definition of agonistic displays. *Behaviour* 99:157–75.

Patterson, F. G. 1978. The gestures of a gorilla: Language acquisition in another pongid. *Brain Lang.* 5:72–97.

Patterson, F. G., and E. Linden. 1981. *The education of Koko.* New York: Holt, Rinehart and Winston.

Patterson, T. L., L Petrinovich, and D. K. James. 1980. Reproductive value and appropriateness of response to predators by white-crowned sparrows. *Behav. Ecol. Sociobiol.* 7:227–31.

Penfield, W., and L. Roberts. 1966. *Speech and brain mechanisms.* New York: Atheneum.

Pepperberg, I. M. 1983. Cognition in the African grey parrot: Preliminary evidence for auditory/vocal comprehension of the class concept. *Anim. Learn. Behav.* 11:179–85.

———. 1987. Acquisition of the same/different concept by an African grey parrot (*Psittacus erithacus*): Learning with respect to categories of color, shape, and material. *Anim. Learn. Behav.* 15:423–32.

Pereira, M. 1988. Effects of age and sex on intra-group spacing behavior in juvenile savannah baboons. *Anim. Behav.* 36:184–204.

Perner, J. 1988. Developing semantics for theories of mind, from propositional attitudes to mental representations. In *Developing theories of mind,* ed. J. W. Astington, P. L. Harris, and D. R. Olson. Cambridge: Cambridge University Press.

Perner, J., S. R. Leekam, and H. Wimmer. 1987. Three-year-olds' difficulty with false belief: The case for a conceptual deficit. *Br. J. Develop. Psych.* 5:125–37.

Petersen, M. R., M. D. Beecher, S. R. Zoloth, D. B. Moody, and W. C. Stebbins. 1978. Neural lateralization of species-specific vocalizations by Japanese macaques (*Macaca fuscata*). *Science* 202:324–7.

Petersen, M. R., M. D. Beecher, S. R. Zoloth, S. Green, P. Marler, D. B. Moody, and W. C. Stebbins. 1984. Neural lateralization of vocalizations by Japanese macaques: Communicative significance is more important than acoustic structure. *Behav. Neurosci.* 98:779–90.

Petrinovich, L. 1974. Individual recognition of pup vocalizations by northern elephant seal mothers. *Z. Tierpsychol.* 34:308–12.

Platt, M. M., and R. L. Thompson. 1985. Mirror response in a Japanese macaque troop (Arashiyama West). *Primates* 26:300–314.

Ploog, D. 1981. Neurobiology of primate audio-visual behavior. *Brain Res. Rev.* 3:35–61.

Pola, Y. V., and C. T. Snowdon. 1975. The vocalizations of pygmy marmosets (*Cebuella pygmaea*). *Anim. Behav.* 23:826–42.

Popp, J. W. 1987. Choice of opponents during competition for food among American goldfinches. *Ethology* 75:31–36.

Post, D. G., G. Hausfater, and S. A. McCuskey. 1980. Feeding behavior of yellow baboons (*Papio cynocephalus*): Relationship to age, gender and dominance rank. *Folia primatol.* 34:170–95.

Povinelli, D. J., K. E. Nelson, and S. T. Boysen. 1990. Inferences about guessing and knowing by chimpanzees. (*Pan troglodytes*). *J. Comp. Psychol.* 104:203–10.

Premack, D. 1975. On the origins of language. In *Handbook of psychobiology,* ed. M. S. Gazzaniga and C. B. Blakemore. New York: Academic Press.

———. 1976. *Intelligence in ape and man.* Hillsdale, N.J.: Lawrence Erlbaum Associates.

———. 1983a. Social cognition. *Ann. Rev. Psychol.* 34:351–62.

———. 1983b. The codes of man and beast. *Behav. Brain Sci.* 6:125–67.

———. 1986. *Gavagai.* Cambridge, Mass.: MIT/Bradford Books.

———. 1988. 'Does the chimpanzee have a theory of mind?' revisited. In *Machiavellian intelligence: Social expertise and the evolution of intellect in monkeys, apes, and humans,* ed. R. W. Byrne and A. Whiten. Oxford: Oxford University Press.

Premack, D. and A. Premack. 1982. *The Mind of an ape.* New York: W. W. Norton.

Premack, D., and G. Woodruff. 1978. Does the chimpanzee have a theory of mind? *Behav. Brain Sci.* 1:515–26.

Pusey, A. E. 1983. Mother-offspring relationships in chimpanzees after weaning. *Anim. Behav.* 31:363–77.

Pusey, A. E., and C. Packer. 1987. Dispersal and philopatry. In *Primate societies,* ed. B. B. Smuts, D. L. Cheney, R. M. Seyfarth, R. W. Wrangham, and T. T. Struhsaker. Chicago: University of Chicago Press.

Quine, W. V. O. 1960. *Word and object.* Cambridge, Mass.: MIT Press.

Rakowitz, S. 1990. The development of deception in children. Ph. D. diss., University of Pennsylvania.

Raleigh, M., and M. T. McGuire. 1989. Female influences on male dominance acquisition in captive vervet monkeys (*Cercopithecus aethiops sabaeus*). *Anim. Behav.* 37:59–67.

Rasmussen, K. L. R. 1980. Consort behavior and mate selection in yellow baboons (*Papio cynocephalus*). Ph. D. diss., University of Cambridge.

Rensch, B., and J. Dohl. 1967. Spontanes ofnen verschedener Kristenverschlusse durch einen Schimpansen. *Z. Tierpsychol.* 24:476–89.

Rescorla, R. A. 1985. Associationism in animal learning. In *Perspectives on learning and memory,* ed. L. Nilsson and T. Archer. Hillsdale, N.J.: Lawrence Erlbaum Associates.

———. 1988. Pavlovian conditioning: It's not what you think it is. *Am. Psychol.* 43:151–60.

Reynolds, P. C. 1981. *On the evolution of human behavior.* Berkeley: University of California Press.

van Rhijn, J. G., and R. Vodegal. 1980. Being honest about one's intentions: An evolutionarily stable strategy for animal conflicts. *J. Theor. Biol.* 85:623–41.

Ristau, C. A. 1990. Injury feigning and other anti-predator behaviors by plovers:

Intentional behavior? In *Cognitive ethology: The minds of other animals (essays in honor of Donald R. Griffin)*, ed. C. A. Ristau. Hillsdale, N.J.: Lawrence Erlbaum Associates.

Roberts, W. A., and D. S. Mazmanian. 1988. Concept learning at different levels of abstraction by pigeons, monkeys, and people. *J. Exp. Psychol. Anim. Behav. Proc.* 14:247–60.

Robinson, B. W. 1967. Vocalization evoked from forebrain in *Macaca mulatta. Physiol. Behav.* 2:345–54.

Robinson, J. G. 1981. Vocal regulation of inter- and intragroup spacing during boundary encounters in the titi monkey, *Callicebus moloch. Primates* 22:161–72.

———. 1982. Intrasexual competition and mate choice in primates. *Am. J. Primatol.* 1 (suppl.): 131–44.

———. 1984. Syntactic structures in the vocalizations of wedge-capped capuchin monkeys *Cebus nigrivittatus. Behaviour* 90:46–79.

———. 1988. Group size in wedge-capped capuchin monkeys (*Cebus olivaceus*) and the reproductive success of males and females. *Behav. Ecol. Sociobiol.* 23:187–97.

Robinson, S. R. 1981. Alarm communication in Belding's ground squirrels. *Z. Tierpsychol.* 56:150–68.

Rodman, P. S. 1977. Feeding behaviour of orangutans of the Kutai Nature Reserve, East Kalimantan. In *Primate ecology: Studies of feeding and ranging behaviour in lemurs, monkeys, and apes,* ed. T. H. Clutton-Brock. London: Academic Press.

Roitblat, H. 1987. *Introduction to comparative cognition.* New York: W. H. Freeman.

Romanes, G. 1882. *Animal intelligence.* London: Kegan, Paul, Trench.

Rosch, E. H. 1973. On the internal structure of perceptual and semantic categories. In *Cognitive development and the acquisition of language,* ed. T. E. Moore. New York: Academic Press.

———. 1977. Classification of real-world objects: Origins and representations in cognition. In *Thinking: Readings in cognitive science,* ed. P. N. Johnson-Laird and P. C. Wason. Cambridge: Cambridge University Press.

Rotenberg, K. J. 1982. Development of character constancy of self and others. *Child Develop.* 53:505–15.

Rowell, T. E. 1962. Agonistic noises of the rhesus monkey (*Macaca mulatta*). *Symp. Zool. Soc. Lond.* 8:91–96.

———. 1966. Hierarchy in the organization of a captive baboon group. *Anim. Behav.* 14:430–43.

———. 1968. Grooming by adult baboons in relation to reproductive cycles. *Anim. Behav.* 16:585–8.

———. 1972. *Social behaviour of monkeys.* Baltimore: Penguin Books.

Rowell, T. E. and R. A. Hinde. 1962. Vocal communication by the rhesus monkey (*Macaca mulatta*). *Proc. Zool. Soc. Lond.* 138:279–94.

Rowher, S. 1982. The evolution of reliable and unreliable badges of fighting ability. *Am. Zool.* 22:531–46.

Rozin, P. 1976. The evolution of intelligence and access to the cognitive unconscious. In *Progress in psychology, vol. 6,* ed. J. N. Sprague and A. N. Epstein. New York: Academic Press.

Rozin, P., and J. Schull. 1988. The adaptive-evolutionary point of view in experimental psychology. In *Handbook of experimental psychology: Motivation,* ed. R. Atkinson, R. J. Herrnstein, G. Lindsey, and R. D. Luce. New York: John Wiley.

Rumbaugh, D., ed. 1977. *Language learning by a chimpanzee: The Lana project.* New York: Academic Press.

Rutter, M. 1983. Cognitive deficits in the pathogenesis of autism. In *Autism: A reappraisal of concepts and treatment,* ed. M. Rutter and E. Schopler. New York: Plenum Press.

Ryden, O. 1978. Differential responsiveness of great tit nestlings, *Parus major,* to natural auditory stimuli. *Z. Tierpsychol.* 47:236–53.

Ryle, G. 1949. *The concept of mind.* London: Hutchinson.

Sade, D. S. 1965. Some aspects of parent-offspring and sibling relations in a group of rhesus monkeys, with a discussion of grooming. *Am. J. Phys. Anthrop.* 23:1–18.

———. 1967. Determinants of dominance in a group of free-ranging rhesus monkeys. In *Social communication among primates,* ed. S. Altmann. Chicago: University of Chicago Press.

———. 1972a. A longitudinal study of social behavior of rhesus monkeys. In *The functional and evolutionary biology of primates,* ed. R. H. Tuttle. Chicago: Aldine.

———. 1972b. Sociometrics of *Macaca mulatta* I. Linkages and cliques in grooming matrices. *Folia primatol.* 18:196–223.

Sade, D. S., K. Cushing, P. Cushing, J. Dunaif, A. Figueroa, J. R. Kaplan, C. Laurer, D. Rhodes, and J. Schneider. 1976. Population dynamics in relation to social structure on Cayo Santiago. *Yrbk. Phys. Anthrop.* 20:253–62.

Samuels, A., J. B. Silk, and J. Altmann. 1987. Continuity and change in dominance relations among female baboons. *Anim. Behav.* 35:785–93.

Sands, S. F., C. E. Lincoln, and A. A. Wright. 1982. Pictorial similarity judgments and the organization of visual memory in the rhesus monkey. *J. Exp. Psych: Gen.* 111, 369–89.

Savage-Rumbaugh, E. S. 1986. *Ape language: From conditioned response to symbol.* New York: Columbia University Press.

Savage-Rumbaugh, E. S., and K. McDonald. 1988. Deception and social manipulation in symbol-using apes. In *Machiavellian intelligence: Social expertise and the evolution of intellect in monkeys, apes, and humans,* ed. R. W. Byrne and A. Whiten. Oxford: Oxford University Press.

Savage-Rumbaugh, E. S., J. L. Pate, J. Lawson, S. T. Smith, and S. Rosenbaum. 1983. Can a chimpanzee make a statement? *J. Exp. Psychol.: Gen.* 112:457–92.

Savage-Rumbaugh, E. S., D. M. Rumbaugh, and S. Boysen. 1978. Linguistically mediated tool use and exchange by chimpanzees (*Pan troglodytes*). *Behav. Brain Sci.* 4:539–54.

Savage-Rumbaugh, E. S., D. M. Rumbaugh, S. T. Smith, and J. Lawson. 1980. Reference: The linguistic essential. *Science* 210:922–5.

Schrier, A. M. 1984. Learning how to learn: The significance and current status of learning set formation. *Primates* 25:95–102.

Schrier, A. M., R. Angarella, and M. L. Povar. 1984. Studies of concept formation by stumptail monkeys: Concepts, humans, monkeys, and the letter A. *J. Exp. Psychol. Anim. Behav. Proc.* 10:564–84.

Schrier, A. M., and P. M Brady. 1987. Categorization of natural stimuli by monkeys (*Macaca mulatta*): Effects of stimulus set size and modification of exemplars. *J. Exp. Psych.: Anim. Behav. Proc.* 13, 136–43.

Schusterman, R. J. 1988. Animal language research: Marine mammals re-enter the controversy. In *Evolutionary biology and intelligence*, ed. H. J. Jerison and I. Jerison. Heidelberg: Springer-Verlag.

Scott, L. M. 1984. Reproductive behavior of adolescent female baboons (*Papio anubis*) in Kenya. In *Female primates*, ed. M. Small. New York: Alan Liss.

Searle, J. 1983. *Intentionality: An essay in the philosophy of mind*. Cambridge: Cambridge University Press.

Seligman, M. E. P. 1970. On the generality of the laws of learning. *Psych. Rev.* 77: 406–18.

Seligman, M. E. P., and J. L. Hager. 1972. *Biological boundaries of learning*. New York: Appleton-Century-Crofts.

Seyfarth, R. M. 1976. Social relationships among adult female baboons. *Anim. Behav.* 24:917–38.

———. 1977. A model of social grooming among adult female monkeys. *J. Theor. Biol.* 65:671–98.

———. 1978a. Social relationships among adult male and female baboons. I. Behaviour during sexual consortship. *Behaviour* 64:204–26.

———. 1978b. Social relationships among adult male and female baboons. II. Behaviour throughout the female reproductive cycle. *Behaviour* 64:227–47.

———. 1980. The distribution of grooming and related behaviours among adult female vervet monkeys. *Anim. Behav.* 28:798–813.

———. 1981. Do monkeys rank each other? *Behav. Brain Sci.* 4:447–8.

———. 1983. Grooming and social competition in primates. In *Primate social relationships: An integrated approach*, ed. R. A. Hinde. Oxford: Blackwell Scientific.

Seyfarth, R. M., and D. L. Cheney. 1980. The ontogeny of vervet monkey alarm-calling behavior: A preliminary report. *Z. Tierpsychol.* 54:37–56.

———. 1984a. Grooming, alliances and reciprocal altruism in vervet monkeys. *Nature* 308:541–3.

———. 1984b. The acoustic features of vervet monkey grunts. *J. Acoust. Soc. Am.* 75:1623–28.

———. 1986. Vocal development in vervet monkeys. *Anim. Behav.* 34:1640–58.

———. 1988a. Do monkeys understand their relations? In *Machiavellian intelli-*

gence: Social expertise and the evolution of intellect in monkeys, apes, and humans, ed. R. W. Byrne and A. Whiten. Oxford: Oxford University Press.

———. 1988b. Empirical tests of reciprocity theory: Problems in assessment. *Ethol. Sociobiol.* 9:181–8.

———. 1990. The assessment by vervet monkeys of their own and another species' alarm calls. *Anim. Behav.* 40:754–64.

Seyfarth, R. M., D. L. Cheney, and R. A. Hinde. 1978. Some principles relating social interaction and social structure among primates. In *Recent advances in primatology, vol. 1,* ed. D. J. Chivers and J. Herbert. New York: Academic Press.

Seyfarth, R. M., D. L. Cheney, and P. Marler. 1980a. Monkey responses to three different alarm calls: Evidence for predator classification and semantic communication. *Science* 210:801–3.

———. 1980b. Vervet monkey alarm calls: Semantic communication in a free-ranging primate. *Anim. Behav.* 28:1070–94.

Shatz, M., H. Wellman, and S. Silber. 1983. The acquisition of mental verbs: A systematic investigation of the first reference to mental state. *Cognition* 14:301–21.

Sherman, P. W. 1977. Nepotism and the evolution of alarm calls. *Science* 197: 1246–53.

———. 1985. Alarm calls of Belding's ground squirrels to aerial predators: Nepotism or self-preservation? *Behav. Ecol. Sociobiol.* 17:313–23.

Sherman, P. W., and W. G. Holmes. 1985. Kin recognition: Issues and evidence. In *Experimental behavioral ecology and sociobiology* ed. B. Holldobler and M. Lindauer. Stuttgart: G. Fischer Verlag.

Sherrod, L. R. 1981. Issues in cognitive-perceptual development: The special case of social stimuli. In *Infant social cognition,* ed. M. E. Lamb and L. R. Sherrod. Hillsdale, N.J.: Lawrence Erlbaum Associates.

Sherry, D. F. 1985. Food storage by birds and mammals. In *Advances in the study of behavior, vol. 15,* ed. J. S. Rosenblatt, C. Beer, M. C. Busnell, and P. J. B. Slater. New York: Academic Press.

Sherry, D. F., and D. L. Schacter. 1987. The evolution of multiple memory systems. *Psych. Rev.* 94:439–54.

Shettleworth, S. J. 1984. Learning and behavioral ecology. In *Behavioural ecology: An evolutionary approach,* ed. J. R. Krebs and N. B. Davies. Sunderland, Mass.: Sinauer Associates.

Shettleworth, S. J., and J. R. Krebs. 1982. How marsh tits find their hoards: The role of site preference and spatial memory. *J. Exp. Psych. Anim. Behav. Proc.* 8:354–75.

———. 1986. Stored and encountered seeds: A comparison of two spatial memory tasks in marsh tits and chickadees. *J. Exp. Psych. Anim. Behav. Proc.* 12:248–57.

Shettleworth, S. J., J. R. Krebs, D. W. Stephens, and J. Gibson. 1988. Tracking a fluctuating environment: A study of sampling. *Anim. Behav.* 36:87–105.

Shipley, E. F., C. S. Smith, and L. R. Gleitman. 1969. A study in the acquisition of language: Free responses to commands. *Language* 45:322–42.

Shultz, T. R., and K. Cloghesy. 1981. Development of recursive awareness of intention. *Develop. Psych.* 17:465–71.

Sigg, H. 1980. Differentiation of female positions in hamadryas one-male-units. *Z. Tierpsychol.* 53:265–302.

———. 1986. Ranging patterns in hamadryas baboons: Evidence for a mental map. In *Primate ontogeny, cognition, and social behavior,* ed. J. Else and P. C. Lee. Cambridge: Cambridge University Press.

Sigg, H., and A. Stolba. 1981. Home range and daily march in a hamadryas baboon troop. *Folia primatol.* 36:40–75.

Sigg, H., A. Stolba, J. J. Abegglen, and V. Dasser. 1982. Life history of hamadryas baboons: Physical development, infant mortality, reproductive parameters, and family relationships. *Primates* 23:473–87.

Silk, J. B. 1982. Altruism among female *Macaca radiata:* Explanations and analysis of patterns of grooming and coalition formation. *Behaviour* 79:162–88.

———. 1987. Social behavior in evolutionary perspective. In *Primate societies,* ed. B. B. Smuts, D. L. Cheney, R. M. Seyfarth, R. W. Wrangham, and T. T. Struhsaker. Chicago: University of Chicago Press.

Silk, J. B., A. Samuels, and P. Rodman. 1981. The influence of kinship, rank, and sex on affiliation and aggression between adult female and immature bonnet macaques (*Macaca radiata*). *Behaviour* 78:111–77.

Sinnott, J. M., W. C. Stebbins, and D. B. Moody. 1975. Regulation of voice amplitude by the monkey. *J. Acoust. Soc. Am.* 58:412–14.

Skinner, B. F. 1957. *Verbal behavior.* New York: Appleton-Century-Crofts.

———. 1974. *About behaviorism.* New York: Knopf.

Smith, E. E., and D. L. Medin. 1981. *Categories and concepts.* Cambridge, Mass.: Harvard University Press.

Smith, W. J. 1965. Meaning, message, and context in ethology. *Am. Nat.* 99:405–9.

———. 1969. Messages of vertebrate communication. *Science* 165:145–50.

———. 1977. *The behavior of communicating: An ethological approach.* Cambridge, Mass.: Harvard University Press.

———. 1986. Signaling behavior: Contributions of different repertoires. In *Dolphin cognition and behavior: A comparative approach,* ed. R. J. Schusterman, J. A. Thomas, and F. G. Wood. Hillsdale, N.J.: Lawrence Erlbaum Associates.

———. 1990. Animal communication and the study of cognition. In *Cognitive ethology: The minds of other animals (essays in honor of Donald R. Griffin),* ed. C. A. Ristau. Hillsdale, N.J.: Lawrence Erlbaum Associates.

Smuts, B. B. 1980. Effects on social behavior of loss of high rank in wild adult female baboons (*Papio anubis*). Paper presented at the Annual Meeting of the Animal Behavior Society, Ft. Collins, Colo., June 9–13.

———. 1983. Dynamics of 'special relationships' between adult male and female olive baboons. In *Primate social relationships: An integrated approach,* ed. R. A. Hinde. Oxford: Blackwell Scientific.

———. 1985. *Sex and friendship in baboons.* New York: Aldine.

———. 1987a. Gender, aggression, and influence. In *Primate societies,* ed. B. B. Smuts, D. L. Cheney, R. M. Seyfarth, R. W. Wrangham, and T. T. Struhsaker. Chicago: University of Chicago Press.

———. 1987b. Sexual competition and mate choice. In *Primate societies,* ed. B. B. Smuts, D. L. Cheney, R. M. Seyfarth, R. W. Wrangham, and T. T. Struhsaker. Chicago: University of Chicago Press.

Snowdon, C. T. 1982. Linguistic and psycholinguistic approaches to primate communication. In *Primate communication,* ed. C. T. Snowdon, C. H. Brown, and M. R. Petersen. New York: Cambridge University Press.

Snowdon, C. T., and J. Cleveland. 1984. 'Conversations' among pygmy marmosets. *Am. J. Primatol.* 7:15–20.

Snowdon, C. T., J. A. French, and J. Cleveland. 1986. Ontogeny of primate vocalizations: Models from birdsong and human speech. In *Current perspectives in primate social behavior,* ed. D. Taub and F. E. King. New York: Van Nostrand Reinhold.

Snowdon, C. T., and A. Hodun. 1981. Acoustic adaptations in pygmy marmoset contact calls: Locational cues vary with distance between conspecifics. *Behav. Ecol. Sociobiol.* 9:295–300.

Snowdon, C. T., and Y. Pola. 1978. Interspecific and intraspecific responses to synthesized pygmy marmoset vocalizations. *Anim. Behav.* 26:192–206.

Sodian, B., and H. Wimmer. 1987. Children's understanding of inference as a source of knowledge. *Child Develop.* 58:424–33.

Sorce, J. F., R. N. Emde, and M. Frank. 1982. Maternal referencing in normal and Down's syndrome infants: A longitudinal study. In *The development of attachment and affiliative systems,* ed. R. N. Emde and R. Harmon. New York: Plenum Press.

Spetch, M. L., and W. K. Honig. 1988. Characteristics of pigeons' spatial working memory in an open-field task. *Anim. Learn. Behav.* 16:123–31.

Stacey, P. B. 1986. Group size and foraging efficiency in yellow baboons. *Behav. Ecol. Sociobiol.* 18:175–87.

Stammbach, E. 1978. On social differentiation in groups of captive female hamadryas baboons. *Behaviour* 67:322–38.

———. 1987. Desert, forest, and montane baboons: Multilevel societies. In *Primate societies,* ed. B. B. Smuts, D. L. Cheney, R. M. Seyfarth, R. W. Wrangham, and T. T. Struhsaker. Chicago: University of Chicago Press.

———. 1988a. An experimental study of social knowledge: Adaptation to the special manipulative skills of single individuals in a *Macaca fascicularis* group. In *Machiavellian intelligence: Social expertise and the evolution of intellect in monkeys, apes, and humans,* ed. R. W. Byrne and A. Whiten. Oxford: Oxford University Press.

———. 1988b. Group responses to specially skilled individuals in a *Macaca fascicularis* group. *Behaviour* 107:241–66.

Stammbach, E., and H. Kummer. 1982. Individual contributions to a dyadic interaction: An analysis of baboon grooming. *Anim. Behav.* 30:964–71.

Steklis, H. D., and M. J. Raleigh. 1979. Behavioral and neurobiological aspects of primate vocalization and facial expression. In *Neurobiology of social communication in primates,* ed. H. D. Steklis and M. J. Raleigh. New York: Academic Press.

Stenmark, G., T. Slagsvold, and J. T. Lifjeld. 1988. Polygyny in the pied flycatcher *Ficedula hypoleuca:* A test of the deception hypothesis. *Anim. Behav.* 36: 1646–57.

Stephens, D. W., and J. R. Krebs. 1986. *Foraging theory.* Princeton: Princeton University Press.

Stern, B. R., and D. G. Smith. 1984. Sexual behavior and paternity in three captive groups of rhesus monkeys (*Macaca mulatta*). *Anim. Behav.* 32:23–32.

Stewart, K. J., and A. H. Harcourt. 1987. Gorillas: Variation in female relationships. In *Primate societies,* ed. B. B. Smuts, D. L. Cheney, R. M. Seyfarth, R. W. Wrangham, and T. T. Struhsaker. Chicago: University of Chicago Press.

Stich, S. P. 1983. *From folk psychology to cognitive science.* Cambridge, Mass.: MIT Press.

Strong, P. N., and M. Hedges. 1966. Comparative studies in simple oddity learning. 1. Cats, raccoons, monkeys, and chimpanzees. *Psychonom. Sci.* 5:13–14.

Struhsaker, T. T. 1967a. Auditory communication among vervet monkeys (*Cercopithecus aethiops*). In *Social communication among primates,* ed. S. A. Altmann. Chicago: University of Chicago Press.

———. 1967b. Behavior of vervet monkeys (*Cercopithecus aethiops*). *Univ. Calif. Pub. Zool.* 82:1–74.

———. 1967c. Ecology of vervet monkeys (*Cercopithecus aethiops*) in the Masai-Amboseli Game Reserve, Kenya. *Ecology* 48:891–904.

———. 1967d. Social structure among vervet monkeys. *Behaviour* 29:83–121.

———. 1971. Social behavior of mother and infant vervet monkeys (*Cercopithecus aethiops*). *Anim. Behav.* 19:233–50.

———. 1973. A recensus of vervet monkeys in the Masai-Amboseli Game Reserve, Kenya. *Ecology* 54:930–32.

———. 1975. *The red colobus monkey.* Chicago: University of Chicago Press.

———. 1976. A further decline in numbers of Amboseli vervet monkeys. *Biotropica* 8:211–14.

Strum, S. C. 1981. Processes and products of change: Baboon predatory behavior at Gilgil, Kenya. In *Omnivorous primates,* ed. R. S. O. Harding and G. Teleki. New York: Columbia University Press.

———. 1982. Agonistic dominance in male baboons: An alternative view. *Int. J. Primatol.* 3:175–202.

———. 1984. Why males use infants. In *Primate paternalism,* ed. D. M. Taub. New York: Van Nostrand Reinhold.

Sullivan, K. 1985. Selective alarm-calling by downy woodpeckers in mixed-species flocks. *Auk* 102:184–7.

Suomi, S. J., and C. Ripp. 1983. A history of motherless mother monkey mothering at the University of Wisconsin Primate Laboratory. In *Child abuse: The nonhuman primate data,* ed. M. Reite and N. G. Caine. New York: Alan Liss.

Sutton, D., C. Larson, E. M. Taylor, and R. C. Lindeman. 1973. Vocalizations in rhesus monkeys: Conditionability. *Brain Res.* 52:225–31.

Sutton, D., H. H. Samson, and C. R. Larson. 1978. Brain mechanisms in learned phonotation of *Macaca mulatta.* In *Recent advances in primatology, vol. 1,* ed. D. J. Chivers and J. Herbert. New York: Academic Press.

Takahata, Y. 1982. Social relations between adult males and females of Japanese monkeys in the Arashiyama B troop. *Primates* 23:1–23.

Talmadge-Riggs, G., P. Winter, D. Ploog, and W. Mayer. 1972. Effect of deafening on the vocal behavior of the squirrel monkey (*Saimiri sciureus*). *Folia primatol.* 17:404–20.

Taub, D. M. 1980. Female choice and mating strategies among wild Barbary macaques (*Macaca sylvanus L.*). In *The macaques: Studies in ecology, behavior, and evolution,* ed. D. G. Lindburg. New York: Van Nostrand Rinehold.

Taylor, M. 1988. The development of children's understanding of the seeing-knowing distinction. In *Developing theories of mind,* ed. J. W. Astington, P. L. Harris, and D. R. Olson. Cambridge: Cambridge University Press.

Teleki, G. 1981. The omnivorous diet and eclectic feeding habits of chimpanzees in Gombe National Park, Tanzania. In *Omnivorous primates,* ed. R. S. O. Harding and G. Teleki. New York: Columbia University Press.

Terrace, H. S. 1979. *Nim.* New York: Knopf.

Terrace, H. S., L. A. Pettito, R. J. Sanders, and T. G. Bever. 1979. Can an ape create a sentence? *Science* 206:891–902.

Testa, T. J. 1974. Causal relationships and the acquisition of avoidance responses. *Psychol. Rev.* 81:491–505.

Thielke, G. A. 1976. *Bird sounds.* Ann Arbor: University of Michigan Press.

Thompson, N. S. 1986. Deception and the concept of natural design. In *Deception: Perspectives on human and nonhuman deceit,* ed. R. W. Mitchell and N. S. Thompson. Albany: State University of New York Press.

———. 1988. Deception and descriptive mentalism. *Behav. Brain Sci.* 11:266.

Thornhill, R. 1979. Adaptive female-mimicking behavior in a scorpion fly. *Science* 205:412–14.

Tiles, J. E. 1987. Meaning. In *The Oxford companion to the mind,* ed. R. L. Gregory. Oxford: Oxford University Press.

Tinbergen, N. 1951. *The study of instinct.* New York: Oxford University Press.

———. 1953. *The herring gull's world.* London: Collins.

Tomasello, M., M. Davis-Dasilva, L. Camak, and K. Bard. 1987. Observational learning of tool-use by young chimpanzees. *Hum. Evol.* 2:175–83.

Trivers, R. L. 1971. The evolution of reciprocal altruism. *Q. Rev. Biol.* 46:35–57.

———. 1985. *Social evolution.* Menlo Park, Calif.: Benjamin/Cummings.

Turner, L. W. 1973. Vocal and escape responses of *Spermophilus beldingi* to predators. *J. Mammal.* 54:990–93.

Uexkull, J. von. 1934/1957. A stroll through the worlds of animals and men. In *Instinctive behavior,* ed. C. H. Schiller. New York: International Universities Press.

Vaitl, E. 1978. Nature and implications of the complexly organized social system in non-human primates. In *Recent advances in primatology, vol. 1,* ed. D. J. Chivers and J. Herbert. New York: Academic Press.

Vaughan, W., Jr., and S. L. Greene. 1984. Pigeon visual memory capacity. *J. Exp. Psych.: Anim. Behav. Proc.* 10:256–71.

Vessey, S. H. 1971. Free-ranging rhesus monkeys: Behavioral effects of removal, separation, and reintroduction of group members. *Behaviour* 40:216–27.

de Villiers, J. G., and P. A. de Villiers. 1978. *Language acquisition.* Cambridge, Mass.: Harvard University Press.

Visalberghi, E., and D. M. Fragaszy. 1990. Do monkeys ape? In *Comparative developmental psychology of language and intelligence in primates,* ed. S. Parker and K. Gibson. Cambridge: Cambridge University Press.

Visalberghi, E., and L. Trinca. 1989. Tool use in capuchin monkeys: Distinguishing between performing and understanding. *Primates* 30:511–21.

de Waal, F. B. M. 1982. *Chimpanzee politics.* New York: Harper and Row.

———. 1986a. Conflict resolution in monkeys and apes. In *Primates, the road to self-sustaining populations,* ed. K. Benirschke. Berlin: Springer-Verlag.

———. 1986b. Deception in the natural communication of chimpanzees. In *Deception: Perspectives on human and non-human deceit,* ed. R. W. Mitchell and N. S. Thompson. Albany: State University of New York Press.

———. 1986c. Imaginative bonobo games. *Zoonooz* 59:6–10.

———. 1987. Dynamics of social relationships. In *Primate societies,* ed. B. B. Smuts, D. L. Cheney, R. M. Seyfarth, R. W. Wrangham, and T. T. Struhsaker. Chicago: University of Chicago Press.

———. 1989. *Peacemaking among primates.* Cambridge, Mass.: Harvard University Press.

de Waal, F. B. M., and L. M. Luttrell. 1986. The similarity principle underlying social bonding among female rhesus monkeys. *Folia primatol.* 46:215–34.

———. 1988. Mechanisms of social reciprocity in three primate species: Symmetrical relationship characteristics or cognition? *Ethol. Sociobiol.* 9:101–18.

de Waal, F. B. M., and R.-M. Ren. 1988. Comparison of the reconciliation behavior of stumptail and rhesus macaques. *Ethology* 78:129–42.

de Waal, F. B. M., and A. van Roosmalen. 1979. Reconciliation and consolation among chimpanzees. *Behav. Ecol. Sociobiol.* 5:55–66.

de Waal, F. B. M., and D. Yoshihara. 1983. Reconciliation and redirected affection in rhesus monkeys. *Behaviour* 85:224–41.

Waldman, B., P. Frumhoff, and P. Sherman. 1988. Problems of kin recognition. *Trends Ecol. Evol.* 3:8–13.

Walters, J. R. 1981. Inferring kinship from behaviour: Maternity determinations in yellow baboons. *Anim. Behav.* 29:126–36.

———. 1986. Kin recognition in nonhuman primates. In *Kin recognition in animals,* ed. D. J. Fletcher and C. D. Michener. New York: John Wiley and Sons.

Walters, J. R. 1987. Transition to adulthood. In *Primate societies,* ed. B. B. Smuts, D. L. Cheney, R. M. Seyfarth, R. W. Wrangham, and T. T. Struhsaker. Chicago: University of Chicago Press.

———. 1990. Anti-predator behavior of lapwings: Field evidence of discriminative abilities. *Wilson Bull.* 102:49–70.

Walters, J. R. and R. M. Seyfarth. 1987. Conflict and cooperation. In *Primate societies,* ed. B. B. Smuts, D. L. Cheney, R. M. Seyfarth. R. W. Wrangham, and T. T. Struhsaker. Chicago: University of Chicago Press.

Warren, J. M. 1973. Learning in vertebrates. In *Comparative psychology: A modern survey,* ed. D. A. Dewesbury and D. A. Rethlingshafer. New York: McGraw-Hill.

———. 1977. Handedness and cerebral dominance in monkeys. In *Lateralization in the nervous system,* ed. S. Harnad, R. W. Doty, J. Jaynes, L. Goldstein, and G. Krauthamer. New York: Academic Press.

Waser, P. 1976. *Cercocebus albigena:* Site attachment, avoidance, and intergroup spacing. *Am. Nat.* 110:911–35.

———. 1977. Individual recognition, intragroup cohesion, and intergroup spacing: Evidence from sound playback to forest monkeys. *Behaviour* 60:28–74.

Washburn, M. F. 1908. *The animal mind.* New York: Macmillan.

Washburn, S. L. 1982. Language and the fossil record. *Anthrop. UCLA* 7:231–8.

Wason, P. C. 1983. Realism and rationality in the selection task. In *Thinking and reasoning: Psychological Approaches,* ed. J. St. B. T. Evans. London: Routledge and Kegan Paul.

Wason, P. C., and P. N. Johnson-Laird. 1972. *Psychology of reasoning.* London: B. T. Batsford.

Weisbard, C., and R. Goy. 1976. Effect of parturition and group composition on competitive drinking order in stumptail macaques. *Folia primatol.* 25:95–121.

Weiskrantz, L. 1985. Categorisation, cleverness and consciousness. In *Animal intelligence,* ed. L. Weiskrantz. Oxford: Clarendon Press.

Weiss, M. L. 1987. Nucleic acid evidence bearing on hominoid relationships. *Yrbk. Phys. Anthrop.* 30:41–74.

Wellman, H. M. 1988. First steps in the child's theorizing about the mind. In *Developing theories of mind,* ed. J. W. Astington, P. L. Harris, and D. R. Olson. Cambridge: Cambridge University Press.

Wellman, H. M., and K. Bartsch. 1988. Young children's reasoning about beliefs. *Cognition* 30:239–77.

West-Eberhard, M. J. 1975. The evolution of social behavior by kin selection. *Q. Rev. Biol.* 50:1–33.

Westergaard, G. C., and D. M. Fragaszy. 1987. The manufacture and use of tools by capuchin monkeys (*Cebus apella*). *J. Comp. Psychol.* 102: 152–9.

Western, D. 1983. *A wildlife guide and a natural history of Amboseli.* Nairobi: General Printers.

Western, D., and C. van Praet. 1973. Cyclical changes in the habitat and climate of an East African ecosystem. *Nature* 241: 104–6.

Whiten, A., and R. W. Byrne. 1988a. Taking (Machiavellian) intelligence apart: Editorial. In *Machiavellian intelligence: Social expertise and the evolution of intellect in monkeys, apes, and humans,* ed. R. W. Byrne and A. Whiten. Oxford: Oxford University Press.

———. 1988b. The Machiavellian intelligence hypothesis: Editorial. In *Machiavellian intelligence: Social expertise and the evolution of intellect in monkeys, apes, and humans,* ed. R. W. Byrne and A. Whiten. Oxford: Oxford University Press.

———. 1988c. The manipulation of attention in primate tactical deception. In *Machiavellian intelligence: Social expertise and the evolution of intellect in monkeys, apes, and humans,* ed. R. W. Byrne and A. Whiten. Oxford: Oxford University Press.

Whitten, P. L. 1982. Female reproductive strategies among vervet monkeys. Ph. D. diss., Harvard University.

———. 1983. Diet and dominance among female vervet monkeys (*Cercopithecus aethiops*). *Am. J. Primatol.* 5: 139–59.

Whitten, P. L. 1987. Infants and adult males. In *Primate societies,* ed. B. B. Smuts, D. L. Cheney, R. M. Seyfarth, R. W. Wrangham, and T. T. Struhsaker. Chicago: University of Chicago Press.

Wilkinson, G. R. 1984. Reciprocal food sharing in vampire bats. *Nature* 308: 181–4.

Wilson, E. O. 1971. *The insect societies.* Cambridge, Mass.: Harvard University Press.

———. 1975. *Sociobiology: The new synthesis.* Cambridge, Mass.: Harvard University Press.

Wilson, W. A. 1975. Discriminative conditioning of vocalizations in *Lemur catta. Anim. Behav.* 23: 432–6.

Wimmer, H., G. Hogrefe, and B. Sodian. 1988. A second stage in children's conception of mental life: Understanding informational access as origins of knowledge and belief. In *Developing theories of mind,* ed. J. W. Astington, P. L. Harris, and D. R. Olson. Cambridge: Cambridge University Press.

Wimmer, H., and J. Perner. 1983. Beliefs about beliefs: Representation and constraining function of wrong beliefs in young children's understanding of deception. *Cognition* 13: 103–28.

Wing, L. and J. Gould. 1979. Severe impairments of social interaction and associated abnormalities in children: Epidemiology and classification. *J. Aut. Develop. Disord.* 9: 11–29.

Winter, P., P. Handley, D. Ploog, and D. Schott. 1973. Ontogeny of squirrel monkey calls under normal conditions and under acoustic isolation. *Behaviour* 47:230–39.

Wittgenstein, L. 1958. *Philosophical investigations*. Translated by G. E. M. Anscombe. Oxford: Oxford University Press.

Woodruff, G., and D. Premack. 1979. Intentional communication in the chimpanzee: The development of deception. *Cognition* 7:333–62.

Wrangham, R. W. 1975. The behavioural ecology of chimpanzees in the Gombe National Park, Tanzania. Ph. D. diss., Cambridge University.

———. 1977. Feeding behaviour of chimpanzees in Gombe National Park, Tanzania. In *Primate ecology: Studies of feeding and ranging behaviour in lemurs, monkeys, and apes,* ed. T. H. Clutton-Brock. New York: Academic Press.

Wrangham, R. W. 1980. An ecological model of female-bonded primate groups. *Behaviour* 75:262–300.

———. 1981. Drinking competition in vervet monkeys. *Anim. Behav.* 29:904–10.

Wright, A. A., H. C. Santiago, P. J. Urcuioli, and S. F. Sands. 1983. Monkey and pigeon acquisition of same/different concept using pictorial stimuli. In *Quantitative analyses of behavior: Discriminating processes, vol. IV,* ed. M. L. Commons, R. J. Herrnstein, and A. R. Wagner. Cambridge, Mass.: Ballinger.

Wright, R. V. S. 1972. Imitative learning of a flaked-tool technology—the case of an orangutan. *Mankind* 8:296–306.

Yamada, M. 1963. A study of blood-relationship in the natural society of the Japanese macaque: An analysis of co-feeding, grooming, and playmate relationships in Minoo-B. troop. *Primates* 4:43–65.

Yamaguchi, S., and R. Myers. 1972. Failure of discriminative vocal conditioning in rhesus monkeys. *Brain Res.* 37:109–14.

Yasukawa, K. 1979. A fair advantage in animal confrontations. *New Sci.* 84:366–8.

Yates, J., and N. Tule. 1979. Perceiving surprising words in an unattended auditory channel. *Q. J. Exp. Psychol.* 31:281–6.

Yerkes, R. M. 1917. *The mental life of monkeys and apes.* Behavior Monographs, 3. New York: Holt.

———. 1925. *Almost human.* New York: Macmillan.

York, A. D., and T. E. Rowell. 1988. Reconciliation following aggression in patas monkeys, *Erythrocebus patas. Anim. Behav.* 36:502–9.

Zoloth, S. R., M. R. Petersen. M. D. Beecher, S. Green, P. Marler, D. B. Moody, and W. Stebbins. 1979. Species-specific perceptual processing of vocal sounds by monkeys. *Science* 204:870–72.

INDEX